Urban
Construction
Project
Management

Other Books in the McGraw-Hill Construction Series

Building Anatomy: An Illustrated Guide to How Structures Work by Iver Wahl

Building Information Modeling: Planning and Managing Construction Projects with 4D CAD and Simulations by Willem Kymmell

Construction Safety Engineering Principles: Designing and Managing Safer Job Sites by David V. MacCollum

Defect-Free Buildings: A Construction Manual for Quality Control and Conflict Resolution by Robert S. Mann

Emerald Architecture: Case Studies in Green Building by *GreenSource: The Magazine of Sustainable Design* (a McGraw-Hill Construction publication)

The Engineering Guide to LEED–New Construction: Sustainable Construction for Engineers by Liv Haselbach

McGraw-Hill Construction Locator: Building Codes, Construction Standards, and Government Regulations by Joseph A. MacDonald

Solar Power in Building Design: The Engineer's Complete Design Resource by Peter Gevorkian

About McGraw-Hill Construction

McGraw-Hill Construction, part of The McGraw-Hill Companies (NYSE: MHP), connects people, projects, and products across the design and construction industry. Backed by the power of Dodge, Sweets, *Engineering News-Record* (*ENR*), *Architectural Record, GreenSource, Constructor,* and regional publications, the company provides information, intelligence, tools, applications, and resources to help customers grow their businesses. McGraw-Hill Construction serves more than 1,000,000 customers within the $4.6 trillion global construction community. For more information, visit www.construction.com.

Urban
Construction
Project
Management

Richard Lambeck, P.E.
Clinical Assistant Professor
New York University

John Eschemuller, P.E.
Clinical Associate Professor
New York University

McGraw Hill

New York   Chicago   San Francisco   Lisbon   London   Madrid   Mexico City
Milan   New Delhi   San Juan   Seoul   Singapore   Sydney   Toronto

Library of Congress Cataloging-in-Publication Data

Lambeck, Richard.
    Urban construction project management / Richard Lambeck, John Eschemuller.
        p.   cm.
    ISBN 978-0-07-154468-9 (alk. paper)
    1. Construction industry—Management.   2. Project management.
    3. Construction industry—United States—Management.   4. Project
management—United States.   I. Eschemuller, John.   II. Title.
    HD9715.8.A2 L36   2009
    690.068′4—dc22   2008037504

Urban Construction Project Management

McGraw-Hill books are available at special quantity discounts to use as premiums and sales
promotions, or for use in corporate training programs. To contact a special sales representative, please visit the Contact Us page at www.mhprofessional.com.

1 2 3 4 5 6 7 8 9 0 DOC/DOC 0 1 4 3 2 1 0 9 8

ISBN 978-0-07-154468-9
MHID 0-07-154468-2

This book is printed on acid-free paper.

Sponsoring Editor
  Joy Bramble Oehlkers

Editorial Supervisor
  Stephen M. Smith

Production Supervisor
  Richard C. Ruzycka

Developmental Editor
  Rebecca Behrens

Project Manager
  Somya Rustagi, International
  Typesetting and Composition

Copy Editor
  Susan Fox-Greenberg

Proofreader
  Nigel Peter O'Brien, International
  Typesetting and Composition

Indexer
  Broccoli Information Management

Art Director, Cover
  Jeff Weeks

Composition
  International Typesetting and Composition

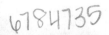

ABOUT THE AUTHORS

Richard Lambeck, P.E., is the principal of RL Project Management, Inc., a project management consulting firm, and is also a Clinical Assistant Professor of Construction Management at New York University Schack Institute of Real Estate. Mr. Lambeck has over 40 years of experience in the construction industry and is a Professional Engineer in New York.

John Eschemuller, P.E., is President of John Eschemuller Consulting Services Limited, LLC, and is also a Clinical Associate Professor of Construction Management at New York University Schack Institute of Real Estate. Mr. Eschemuller has over 38 years of experience in the construction industry and is a Professional Engineer in New York.

DEDICATION

This is book is dedicated to the past, present, and future construction personnel who have created the great buildings and structures for humanity.

Contents

19 Design-Build 371

20 Requisitions 381

21 Project Punch List and Close-out 399

22 Technology 411

23 Green Construction 423

Acknowledgments

A book as complex as *Urban Construction Project Management* could not be written without guidance and assistance from our colleagues, friends, and family. Therefore, we would like to acknowledge the following people: the family of Richard Lambeck; the family of John Eschemuller; Cesar DeCamps; Dan Buffey; Robert Morganstern, NYU Schack Institute of Real Estate; Kenneth Patton, NYU Schack Institute of Real Estate; Melissa Gasparotto, NYU Schack Institute of Real Estate; Andrea Harpole, NYU Schack Institute of Real Estate; Scott St. Martin, NYU Schack Institute of Real Estate; NYU Schack Institute of Real Estate faculty, staff, and students; Lou Milo, Milrose Consulting; Gus Podias, Marson Constructing Company; Anthony Iandoli, Lehr Construction Company; David Slegowski, United Bank of Switzerland; Gary Morrissey, The Treiber Group LLC; Marvin Levine, Levine Construction; Troy Stedman, Mortenson Construction; Gregg Knutson, Mortenson Construction; George Feddish, Cushman & Wakefield; Larry Shapiro, Howard I. Shapiro and Associates, P.C.; Neal Eisman, Goetz Fitzpatrick LLP; Larry Blinn, Jr., Regional Scaffolding; Ken Colyer, Structuretone; Norbet Young, McGraw-Hill Construction; Ravi Bhatia, Project Strategix; Ted Hammer, HLW International; Joseph Mizzi, Frank Sciame Construction; Robert Burkavage, F.J. Sciame; Steven Levy, AIG; Melanie Fordin; Candice Fordin; Sheila Parsons; Raymond E. Levitt, Sanford University; Charles Rizzio, Rizzio Group; Marie Willett, Construction Specification Institute; Jeff Robinson, PAS, Inc.; Jay Jason, Jason Swift, LLP; Rebecca Behrens, McGraw-Hill Professional; Joy Bramble Oehlkers, McGraw-Hill Professional; Cary Sullivan, McGraw-Hill Professional; Sam Battaglia, Turner Construction; Mickey Reiger, Westfield Insurance; and Jennifer Calderella, HRH Insurance.

Introduction

OVERVIEW

Urban Construction Project Management is designed to help construction management companies, general contractors, owners and their representatives, architects, engineers, developers, large corporations, and students studying construction management in universities throughout the United States and the world better understand and manage construction of complex projects built in large urban environments. The book is organized into 23 chapters. The chapters include exhibits, checklists, and photographs to which the user can refer when dealing with the subject matter on a real life project in an urban environment. The chapters address such issues as project organization; problem solving and risk analysis; testing; permits and building codes; safety; logistics; contracts; insurance and bonds; meetings and communications; budgets; schedules; requisitions; and green (LEED) construction.

The book is designed to explain the major aspects of a construction project within an urban environment, and includes several checklists of items to consider while planning and constructing a project in the urban environment. Copies of these checklists are also available online.

Most of the construction contracts in the urban environment are as a construction manager (CM) providing preconstruction, construction, and project close-out professional services, or as a general contractor (GC) with a lump sum bid. If the owner retains a CM, then the CM has the opportunity to provide preconstruction planning services for the project. The proper planning of a construction project is the key to its success. Building in an urban environment presents many unique challenges, especially in cities such as New York, Chicago, Boston, Dallas, San Francisco, and Washington, D.C., as well as large urban cities around the world. In these large metropolitan areas, there is a large concentration of people, high-rise buildings, narrow sidewalks and streets, mass transportation systems, underground utilities, local building ordinances and restrictions, adjoining public and residential areas, and union issues, which provide for unique opportunities and challenges. These are addressed in this book to assist CM/GCs with building in the urban environment. Some of the large urban projects that have been built recently and which have had to deal with these unique challenges are New York Times Corporate Headquarters, New York; Hearst Tower, New York; AOL Time Warner Complex, New York; Bloomberg Corporate Headquarters, New York; Bank of America Corporate Headquarters, New York; Solaire, New York; Hilton San Diego Convention Center Hotel, San Diego, California; Burg Dubai, Dubai; Taipei 101, Taipei; Moscow Triumph Palace, Moscow, Russia; Palm Jumeirah, Dubai; and Merch Serono: Ecological Antidote, Geneva, Switzerland.

HISTORY OF CONSTRUCTION

As long as men and women have been on earth, they have had an ongoing quest to construct buildings of various types to be able to provide housing, shelter, meeting places, houses of prayer, and workplaces, as well as areas that meet their special needs and monuments to their empires and cultures. People first found shelter by living in caves. However, this became a problem with the invention of fire and animals wanting to share their abode. As humanity started to explore other venues, the need for temporary shelters became apparent. At that time, only basic raw materials and methods were available, so the shelter was built with tree limbs and leaves. The Great Pyramid at Giza, which was built by the Egyptians over 4500 years ago, is one of the great wonders of the world. With the use of large stone granite and limestone blocks (some weighing over 2.5 tons), the Egyptians created a lasting monument to the Pharaoh. This was accomplished without the use of the wheel, steel tools, or cranes. The Pyramid at Giza was constructed using very simple surveying tools, bronze tools, and a huge amount of free labor. Most of the labor was made available when the Nile overflowed, in that the workers could no longer cultivate the land for agriculture. When the floods receded and during harvest time these workers went back to tilling the fertile land. Approximately 5000 people were full-time workers and were used to cut the granite and limestone slabs. This number also included a primitive construction management team. It is estimated that approximately another 21,000 were part-time workers. Construction has always been an industry that is very labor intensive, and uses raw materials that have been readily available. However, the Egyptians had to go to Aswan (500 miles away) to obtain the granite for the structurally critical elements of the pyramid. They used boats and human and animal power to slide large blocks of stone across the desert sand and built slopes of sand to access elevations and grades. The pyramids were sophisticated structures with precise axes for their layout (a slope of 51°). They were structurally intricate and the workers utilized a lot of innovative tools and techniques to accomplish the construction.

The Egyptians understood geometry, astronomy, and the load-bearing capacity of materials and the management of people. Then from Roman times to the Dark Ages, humans learned how to more effectively build structures. They were capable of using pulleys, cranes, and the wheel to construct massive government buildings and fabulous religious buildings (Notre Dome in Paris as one example). From there humans discovered the ability to smelt cast iron, allowing buildings to be built to a height of several stories. The Bessemer steel process allowed for the beginning of steel structures reaching new heights and the beginning of the modern skyscraper. A skyscraper would not have been possible without the invention of the elevator and the safety brake by Mr. Otis to carry people above six stories. Electrical distribution and lighting systems developed by Thomas Edison gave people the ability to work in high-rise buildings during all hours and seasons. The ability to supply water via reservoirs such as the Croton Reservoir in upstate New York, water from rivers and lakes, such as the Great Lakes supplying cities like Chicago with water, along with pumps to deliver water with sufficient hydrostatic pressure above six building stories and sanitary systems to remove waste products made the skyscraper a viable working building. With harnessing the power of steam to power derricks for lifting heavy objects, and the fireproofing of steel to protect the steel structures, the skyscraper became a very competitive building type. This was especially true when land prices started to escalate and six-story structures were no longer financially viable. The use of high-strength concrete, metal decking, electrical raceways, curtain walls for the building envelope, new life-saving systems, and sophisticated elevator systems have all assisted with the evolution of a safe and quickly constructed skyscraper in the urban environment. All of these inventions have allowed humans to go from occupying a primitive cave to the pyramids to the Eiffel Tower, the first high-rise buildings in New York and Chicago, such as the Woolworth Building and the MetLife Building, to the modern high-rise skyscraper. The first high-rise building in Chicago started the evolution of commercial office buildings as we know them today. Buildings that were tall and slender; had core support space for elevator shafts; and had mechanical, electrical, and telecommunications systems, windows, staircases, and bathrooms become feasible and the size and sky were the limit. Since the beginning of time, humans have never been satisfied with the status quo, so we are seeking higher and more technological complex structures to satisfy our demand for quality space to live and work in the changing world.

CONSTRUCTION TODAY

Many buildings and construction techniques that we use today have not changed much since Egyptian times. The construction process still uses raw materials of varying sorts that are now available from around the world with the ease of transportation and shipping, but it still requires an extensive amount of labor to manufacture and erect the materials and deliver a finished building. We have introduced cast iron and then steel into the structural framework of buildings, elevators and escalators for vertical transportation, piles and excavation to bedrock to allow for construction of taller structures, non-load-bearing curtain walls made of stone, concrete, metal, glass, and bricks to enclose a building with an attractive façade, pumps to deliver water to higher elevations in a building, and life safety systems to make the buildings safer for the occupants.

The construction industry has evolved from small- to medium-sized GCs who perform all aspects of construction on a project including demolition, excavation, foundations, structures, electrical, plumbing, and finishes to a very specialized construction industry of today, where the CMs and GCs have become overall managers of the construction process and broker out the work to specialized subcontractors, who perform their unique construction work on the project. Most CMs and GCs today only perform the overall project management, project oversight, protection, clean up, and general conditions (see Chapter 1) type of support work for the overall project, while each trade performs its specialized work. The CM and GC of today has become the "broker of the construction process."

AVAILABILITY OF LAND

Another major influence on the types and sizes of buildings being built today in a large urban environment is the availability of land on which to build, many stakeholders (see list of project stakeholders in Chapter 1) to consider and satisfy, and increased land values. Given the technological advancements in structural systems, stronger materials, elevators, plumbing, heating, ventilating, air conditioning systems, life safety systems, etc., buildings can be built taller and bigger, often with mixed-use occupancies. An example is the AOL Time Warner complex in New York City, which replaced the Coliseum at Columbus Circle. The AOL Time Warner complex replaced a single-use exhibition facility with a mixed-use complex that houses retail space, restaurants, a hotel, commercial office space, residential apartments, and pedestrian space.

CONSTRUCTION IN THE URBAN ENVIRONMENT

The majority of the experience of the authors is in the New York City metropolitan area, and therefore many examples, forms, processes, checklists, and references are of the construction process in New York City. The authors have performed work in other parts of the United States and the world, and each urban environment used processes, procedures, and regulations similar to those in New York City. New York always seems to be in the forefront of what is happening in the construction industry. Examples include the following:

1. New York City has promoted safety from the building department's Building Enforcement Safety Team (BEST) squad.
2. New York City has enacted many local laws to prompt life safety in buildings for occupants; façade inspections; and scaffold, hoist, crane, and derrick inspections.
3. Construction occurs in an environment where 10 million people transverse the city every day, along with another 1 million people who go to and from work each day.
4. There are 700 miles of subway running beneath the city, along with surface bus transportation, five major commuter railroads, and ferries.
5. Underground central utilities provide steam, electrical power, telecommunications, cable television, water, storm sewer, and sanitary sewer services.
6. New York City is the ultimate urban environment, where new building space is extremely limited.
7. There are climate changes from 0 to 100°F within four seasons.
8. Streets are congested with traffic, impeding the flow of construction trucks carrying materials to the site.
9. Streets and sidewalks are difficult to close, which makes for unique challenges for access to the site, construction materials, personnel, and equipment.
10. Many construction sites have access from only one side.

11. Influence of many levels of government at the city, state, and federal levels, and special interest groups, such as local community boards, landmarks preservation, local neighborhood business associations, Metropolitan Transportation Authority, Metro North, Long Island Railroad, Port Authority, etc.

12. Millions of tourists visit the city each year, and the city wants to keep traffic flowing and the city looking pristine, especially during the holiday period from Thanksgiving to New Year's.

As it is often said, "If you as the CM/GC can figure out how to build in New York City, you can build anywhere." The techniques used to construct a building in New York City are applicable to any major city in the United States and the world.

Urban Construction Project Management

1 Project Organization
(How does that huge project get done?)

HOME OFFICE AND PROJECT-SPECIFIC PERSONNEL ASSIGNED TO THE PROJECT

The organization for a large construction project usually consists of personnel assigned to support the project from the home office and personnel assigned directly to the project and often located at the construction site. It is essential that the proper resources be made available to the project leader in a timely manner to ensure the success of the project.

Home office personnel include staff that is assigned in the home office exclusively, and may be working on many different projects concurrently. Exhibit 1-1 contains a list of home office personnel in this category.

Project personnel assigned directly to the project are usually on the project site working on a part- and/or full-time basis. Exhibit 1-2 contains a list of project-specific personnel in this category.

PRECONSTRUCTION ORGANIZATION

On large construction projects in the urban environment, the owner often prefers to retain the construction manager/general contractor (CM/GC) at the inception of the project, to work with the owner, architect, engineers, and other consultants. The CM/GC can add value by assisting with all of the fundamental construction processes, which would include services such as:

1. Estimating and budgets
2. Scheduling
3. Logistics
4. Long lead items

Exhibit 1-1

List of home office personnel assigned to a construction project.

Title	Included
1. Scheduling Department	
2. Estimating Department	
3. Purchasing Department	
4. Administrative Department	
5. Operations Department	
6. Legal Department	
7. Accounting Department	
8. Human Resources Department	
9. Building Department Expediting	
10. MEPS Department (mechanical, electrical, plumbing, sprinkler, and other technical disciplines)	
11. Purchasing Department	
12. Executives of the Corporation	
13. Project Executive	

Exhibit 1-2

List of project-specific personnel assigned to a project.

Title	Included (Yes/No)	Name
1. Project Manager		
2. Account Executive		
3. Project Estimators		
4. Project General Superintendent		
5. Superintendents		
6. Assistant Superintendents		
7. Plan Clerks		
8. Labor Foreman		
9. Laborers		
10. Timekeeper		
11. Project Scheduler		
12. Secretaries		
13. Administrative Assistants		
14. Project Accountants		
15. Site Safety Manager		
16. Scheduler		

5. Insurance

6. Bid and award strategy

7. Filing and permits

8. Project phasing

9. Planning for temporary and permanent utility services

10. Identifying the personnel and organizations associated with the project (stakeholders) and their respective issues

11. Planning for project administration and control procedures

12. Developing the project manual and policies

13. Risk identification and management

14. Quality control program

15. Safety program

16. Procurement and sourcing direction

17. Value engineering

Staff will usually be assigned for a specific period of time to be involved with their particular areas of expertise. The resources assigned during this phase of the project will usually operate out of the home office, since the project site is often not available at this time. The people assigned to perform preconstruction services for a project usually are the same personnel who will be involved with the construction phase of the project. Therefore, it is wise for upper management to carefully consider the assignment of staff, not only short term but long term as well because construction projects will often last two to three years, including the preconstruction, construction, and project close-out phases. A sample organization chart for the preconstruction phase of a project is illustrated in Exhibit 1-3.

Exhibit 1-4 contains a list of preconstruction services.

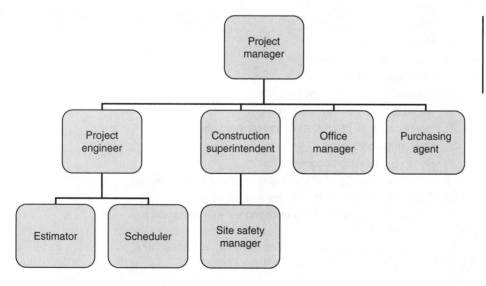

Exhibit 1-3

Organizational chart–preconstruction phase.

Exhibit 1-4

List of precon-
struction services.

Item	Included
1. Review of project scope with the owner and design team	
2. Review of the program for the project with the owner and consultants	
3. Conceptual development, design development, and 80% bid budgets	
4. Conceptual development, design development, and 80% bid schedules	
5. Risk analysis and management	
6. Identification and procurement of long lead items	
7. Value engineering	
8. Prequalification of subcontractors	
9. Project safety program	
10. Project quality control/quality assurance programs	
11. Development of site logistics	
12. Identification of project phasing and early deliverables	
13. Identification of all required filings and permits	
14. Fire safety plan during construction	
15. Review of existing conditions at the site	
16. Review of as-built drawings	
17. Evaluation of constructability and means and methods	
18. Assist the owner with the coordination of their work	
19. Assist with the development of the insurance program for the project	
20. Assist with any bonding required for the project	
21. Definition of any special administrative programs associated with the project	
22. Development of periodic reports and documentation processes for the project	
23. Cash flow management	
24. Requisition processes	
25. Shop drawing processes	
26. Document control	
27. Management information systems	
28. Negotiation of contract with owner	
29. Identification of project stakeholders and their concerns	
30. Field investigations	
31. Review of existing documentation	
32. Assist the owner with identification of hazardous materials	
33. Assist the owner with the abatement process	
34. Identification of items to be coordinated with the owners programs	
35. Coordination of owner's furniture, fixtures, and equipment	

Item	Included
36. Determination of appropriate contingency for the project budget	
37. Determination if a guaranteed maximum price and schedule are required	
38. Identification of cost for preconstruction and construction services	

Exhibit 1-4
(Continued)

COST OF PRECONSTRUCTION SERVICES

The cost of the construction manager's personnel time during this phase is usually billed in one of the following ways: at cost, plus an agreed to mark up; as a percentage of the budget for the project; a monthly fee; or a lump sum amount. The cost of preconstruction services is often merged into the overall CM/GC's fee and/or general conditions for the overall project. The percentage or lump sum amount of money paid to the CM/GCs to cover their overhead and profit is referred to as the fee. The costs associated with providing the CM/GC management and field personnel assigned directly to the construction project, along with administrative related equipment and expenses are referred to as the general conditions. It is important to have an understanding of how the preconstruction services will be paid for. Often the contract between the owner and construction manager will not be finalized at this time, and a separate, smaller, initial agreement for preconstruction services only should be considered. Preconstruction services usually range between 0.5 and 1.0% of the overall project cost. Some CM/GC companies offer to perform the preconstruction services for a project at no cost, with the expectation that they will get the construction phase of a project, which is not always the case. The CM/GCs should put their best foot forward and assign qualified and experienced personnel, who will be assigned to the project for its entire duration to provide professional preconstruction services.

One of the tasks to be performed during preconstruction services is risk assessment for the project. Many people think about risk in terms of financial issues such as insurance and bonding. Risks are not limited to finances, and can include any matter that adversely affects the project such as design deficiencies, construction deficiencies, material deficiencies, labor disputes, contract provisions (see Chapter 9), strikes, acts of sabotage or terrorism, material shortages, personnel shortages, geotechnical problems, and governmental agency problems (see Chapter 2 for further detail of the risk management process).

BID AND AWARD PHASE

During the bid and award phase, the various subcontractors are screened and prequalified, a bidders list is developed, the bid package is assembled, and bids are received and analyzed to recommend award of the particular subcontract. Exhibit 1-5 contains a list of bid and award services.

Exhibit 1-5

List of bid and award services.

Item	Included
1. Develop the prequalification questionnaire for qualification of subcontractors.	
2. Develop the final list of bidders.	
3. Develop a bid strategy to ensure adequate coverage of subcontractors.	
4. Ensure compliance with special administrative programs.	
5. Develop the bid packages.	
6. Bid and award long lead items.	
7. Assign long lead items to subcontractors when awarded.	
8. Develop alternate bid prices to be obtained.	
9. Develop unit prices to be obtained.	
10. Receive, analyze, and level all bids.	
11. Determine the responsive bidder.	
12. Make recommendations for award.	
13. Develop subcontractors' contract form.	
14. Analyze bids and contract commitments.	
15. Recommend any corrective actions if over budget.	
16. Ensure compliance with MWBE programs.	
17. Obtain unit prices where applicable.	
18. Obtain organization chart of key personnel from each subcontractor.	
19. Obtain coordinated schedule from each subcontractor.	
20. Obtain schedule of values against the bid for requisitioning.	

CONSTRUCTION PHASE ORGANIZATION

During the construction phase of the project, the staff will usually be a combination of home office and project-specific personnel. The organizational structure will often be a matrix management structure, with personnel being assigned to the project from various departments within the organization. The employees in this instance will have two superiors, the departmental boss to whom they report and the project executive or project manager to whom they are assigned for the project. Exhibit 1-6 illustrates the matrix management reporting relationship of an employee to both the project team and the traditional organizational superior. Matrix organizations optimizes resource management and ensures that specialized professionals remain under the umbrella of a functional department head who is responsible for training, support, and staffing based on the needs of the overall organization and not just a specific project. A significant downside is that the project manager is often tasked with responsibility without significant control over key resources for a project.

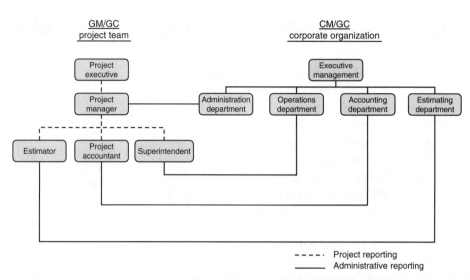

GM/GC
project team

CM/GC
corporate organization

----- Project reporting
_____ Administrative reporting

Exhibit 1-6
Matrix manage-
ment reporting
relationship.

This can lead to challenging situations for all personnel involved. The employee has two managers, and there is often a conflict as to which one they will serve first, who will perform their annual appraisal, and who determines their annual raise, bonus and consideration for a promotion. Often employees who are assigned to an off-site construction project for a period of time find themselves "out of site and out of mind." Employees who find themselves in this situation are smart to spend some time each week back at the home office to remind their superiors of their presence. A sample organization chart for construction services for a project is detailed in Exhibit 1-7.

Exhibit 1-8 contains a list of construction phase services.

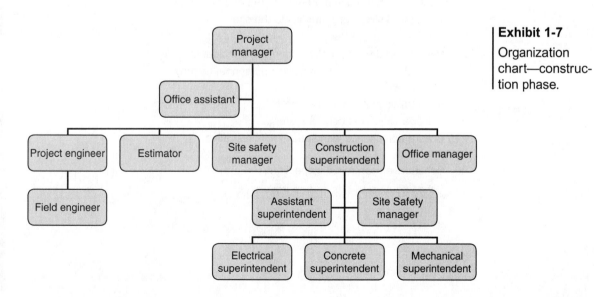

Exhibit 1-7
Organization
chart—construc-
tion phase.

Exhibit 1-8

List of construction services.

Item	Included
1. Mobilize the project.	
2. Hold a kick-off meeting with all subcontractors.	
3. Set up field office, shanties, and project logistics.	
4. Implement safety program.	
5. Implement security program.	
6. Implement quality control/quality assurance programs.	
7. Chair the project meeting.	
8. Conduct all required project-related meetings.	
9. Monitor the schedule of the project.	
10. Prepare one-week look-ahead schedules for the project.	
11. Provide project management services.	
12. Prepare list of submissions for each subcontractor.	
13. Prepare monthly requisitions and manage the process.	
14. Provide cash flow projections.	
15. Review and process all change orders.	
16. Manage the submission of all shop drawings, cuts, and product literature for review and approval.	
17. Process all requests for information (RFI) and their resolution.	
18. Manage the work of all of the subcontractors working on the site.	
19. Manage all documents.	
20. Prepare all monthly progress reports.	
21. Manage the logistics of the project.	
22. Track all construction materials stored on and off site.	
23. Track the delivery of materials to the site.	
24. Take progress photos.	
25. Report and document all accidents and incidents.	
26. Solicit the support of home office personnel as may be required.	
27. Maintain all logs and field reports.	
28 Manage all project control systems.	
29. Ensure compliance with special administrative programs.	
30 Schedule all controlled inspections.	
31. Schedule all sign-offs and approvals.	

DETERMINING THE COST OF PERSONNEL ASSIGNED TO THE PROJECT—GENERAL CONDITIONS

The cost of the construction manager's personnel time during the construction phase of the project is usually billed as part of the general conditions for the project. The general conditions for a project can be billed in a variety of ways, similar to the preconstruction costs described previously. The contractual relationship between the owner and construction manager will define how the costs for the project staff will be billed, as well as which costs are considered part of the overhead and fee, and which ones are part of the general conditions. Make sure that you have a firm understanding of the terms of the contract to determine which employees in the home office and their associated expenses can be billed to the project. Many contracts read that only those employees assigned directly to the project on a full time basis at the project site can be reimbursed as part of the general conditions costs. If an employee, such as the project accountant, is working out of the home office because he or she needs access to the project accounting and administrative systems within the company to properly perform his or her work, then make sure that is provided for in the contract. The CM/GC does not want to be arguing over such matters during the construction phase, when the project executive and project manager need to focus their time and energy on getting the project built, rather than dealing with auditors, lawyers, and contract compliance personnel.

PROJECT CLOSE-OUT PHASE

The close-out phase of a project is a very important phase for the successful completion and turn over of the project to the owner. It is often not given the attention that it requires by upper management of a CM/GC. Because of the shortage of qualified professional personnel available, the CM/GC is often in a hurry to reassign these personnel to another project, without properly performing the close-out. The same key staff of personnel, who were involved with the construction phase of the project, should be involved with the close-out of the project. These personnel may be involved on a part time basis, as may be required. The CM/GC may have provided excellent services during the preconstruction and construction phases of the project, and then not performed the project close-out phase in a very competent and professional manner, especially if the overall general conditions costs for the project have been used up or exceeded. The owner tends to remember the last items of work performed, and often the CM/GC is remembered not for the 99 things that the CM/GC did right, but for the 1 thing that the owner perceives the CM/GC did wrong. The CM/GC's reputation is at stake, and at the end of the day, the CM/GC wants a satisfied client who will give the CM/GC a good recommendation and retain the CM/GC for the next project. That is the best advertising and marketing approach that the CM/GC can have.

A sample organization chart for providing project close-out services is illustrated in Exhibit 1-9.

Exhibit 1-10 contains a detailed list of project close-out services for a project.

Exhibit 1-9

Organization
chart—project
close phase.

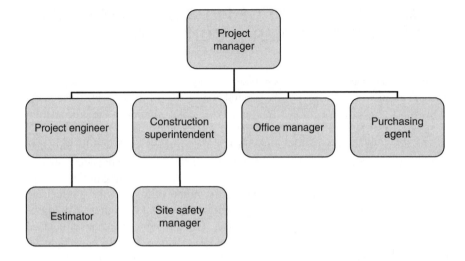

Exhibit 1-10

List of project
close-out services.

Item	Included
1. Develop a preliminary punch list.	
2. Pre-punch the project as it is being built.	
3. Coordinate the preparation of the final punch list.	
4. Advise the owner when the project is available for beneficial occupancy.	
5. Coordinate all controlled inspections and documentation.	
6. Manage the development of the as-built documentation.	
7. Submit all operating and maintenance manuals.	
8. Arrange for all training.	
9. Arrange for the turnover of all attic stock.	
10. Coordinate with the expeditors the obtaining of the temporary and/or final certificates of occupancy.	
11. Provide for all guarantees and warranties.	
12. Provide for final clean up.	
13. Close out all contracts.	
14. Resolve any outstanding change orders and claims.	
15. Obtain final waiver of liens.	
16. Apply for reduction and payment of retainer.	
17. Obtain final payment.	
18. Post close-out activities as may be required.	
19. Archive all project files.	

DUTIES AND RESPONSIBILITIES OF THE CM/GC'S PERSONNEL ASSIGNED TO THE PROJECT

The CM/GC personnel assigned to the project must function as a well-coordinated and integrated team. The actions of each team member affect the rest of the team and the overall project. An item of work budgeted by the estimator must also be included in the overall project schedule by the scheduler. The appropriate subcontractor must be identified to bid and award the work to. The item must be constructed under the management and supervision of the project manager, superintendent, and safety manager. Finally, the work must be invoiced by the project accountant. The following is a list of duties and responsibilities for the various construction managers' personnel who are involved with a construction project during the different phases as just described:

Project Executive (PE)

It is the responsibility of the project executive (PE) to act as the chief liaison between the CM/GC and the owner, architect, engineer, specialty consultants, and other members of the project team. It is the PE's responsibility to ensure that the team is working harmoniously and meeting the overall project's requirements, such as completing the project on schedule, within budget, and according to quality specifications. It is also the PE's responsibility to acquire the proper resources to perform the work, often from department heads, or arrange for the hiring of personnel with the Human Resources Department when the resources are not available from within the organization. The PE needs to keep upper management informed about the overall project status and bring major issues to upper management's attention. The PE should also cultivate and maintain a positive relationship with all members of the project team (account executive, owner, lenders, brokers, consultants, and trade associations). The PE is ultimately responsible for the satisfaction of the client. The PE may be involved with more than one project at a time. The PE is usually assigned to the home office.

Project Manager (PM)

The project manager (PM) is responsible for the daily management, supervision, coordination, communication, and successful completion of the construction project to meet the schedule, budget, administrative requirements, safety, estimating, quality control, bid and award process, and contract administration. This is where the rubber meets the road. The PM must own the schedule and budget for the project, along with the projects logistics and administration. The PM will be involved with all aspects of the project, including preconstruction services, construction phase, project close-out phase, and see the project through from its inception to its completion. The PM will also be involved with the safety, quality control, logistics, temporary facilities, utilities, insurance, bonding, Equal Employment Opportunities (EEO), community relations, unions and special interest groups. The PM must establish and maintain working relationships with the owner, architect, engineer, specialty consultants, subcontractors, suppliers, vendors, etc. to facilitate the construction process. The PM must possess both technical and managerial skills to perform the job properly and professionally. It is often said that a good PM needs to walk on water to do the job properly. This is usually a seasoned construction person, with a college degree in the construction industry or related field. All of the members of the team report to the PM, who is the team leader. The PM is usually assigned to the home office during preconstruction services, and then to the project site for the construction and project close-out phases.

Account Executive (AE)

The account executive (AE) title is unique to some CM/GCs, and is not found in all organizations or regions of the United States and the other countries. Many CM/GCs in a large urban environment have found over time that it is difficult to find a PM who has both excellent and professional construction, technical, and managerial skills combined. Even when you do find this unique person, there are often so many demands placed on the PM in the course of the daily activities of a large construction project, that there is just not enough time to perform all of the construction, technical, and managerial duties simultaneously. Therefore, some firms divide the overall responsibilities of the PM as described previously, and leave the PM to the construction and technical aspects of the project, and assign an AE to deal with the administration of the project. The AE is responsible for managing client relationships on a day-to-day basis and to ensure that a consistent level of professional service is being provided.

There is a management philosophy that says, "Do what you do well, and leave to others what they do well." Seldom does one person have all the skills required to be a good PM, and when they do, they often do not have enough time to do them simultaneously, especially with the fast tracking and pace of many projects. Therefore, splitting up the responsibility for the administration and the construction of the project is sometimes done, allowing each party to concentrate on their area of expertise.

Estimator

The estimator for the project will be involved with putting together conceptual budgets, design development budgets, and progress construction document estimates, and preparing the project budgets including subcontractor trade costs, general conditions costs, contingencies, fees, insurances, and analysis of the subcontractors bids. The estimator must have a full understanding of all aspects of the construction process, the delineation of the work among the subcontractor trades and jurisdictions, cost of construction materials, cost of construction personnel, and appropriate mark ups in order to put together meaningful estimates and budgets for the project.

During the preconstruction phase, the estimator will play a key role in working with the project team to define the project scope, design intent, architectural finishes, engineering systems, specialty items, long lead items, appropriate inflation, and contingency factors to develop a meaningful estimate for the project. If the estimate is too high or too low, it can significantly affect the feasibility and scope of the project and cause a creditability problem for the CM/GC. The estimator must have a sixth sense and a magic scale to weigh and assess matters to come up with a meaningful estimate for the project, even when there is little documentation from the design team on which to base it. Estimates during the preconstruction phase of the project are often done on an order of magnitude basis or by cost per square foot, cost per Btu of cooling or heating, and cost per kilowatt of electrical load. Often, allowances may be made where sufficient definition of the exact scope of certain items, such as the lobby, plaza, roof garden, and signage is not available. In these cases, a target allowance is helpful for the entire project team to manage the design and engineering of the area or system. The estimator will usually work out of the home office during this phase of the project.

During the bid and award process, the estimator will be involved with developing the list of perspective bidders, screening and finalizing the list of bidders, developing bid packages, receiving bids, analyzing and leveling bids, recommending award of subcontracts, and reviewing the bid for completeness, qualification and exclusions, and unit prices (see Chapter 16). This is a very important and critical task; if the subcontractor is not scoped out properly and is awarded without a complete project purchase, the amount of change orders and claims thereafter can be significant and have a negative impact on the project. The estimator will usually work out of the on-site project office for all phases of the project.

During the construction phase, the estimator will be involved with pricing change orders for bulletins, sketches, field information memos (FIMs), requests for information (RFIs), and reviewing subcontractor's claims for extra work as it relates to their contracts. All allowances that may have been included in a subcontract need to be accounted for and reconciled as the final scope of work and project is defined.

During the project close-out phase, the estimator will be finalizing all subcontractor final contract amounts, including all base contract amounts, approved change orders, adjustment of allowances, back charges, etc. to determine the final contract price for each subcontractor.

Scheduler

The project scheduler is responsible for working with the project team to develop a meaningful schedule for the project. All items that are estimated must be scheduled, and all items of work that are scheduled must be estimated. This is a good check between the estimator and scheduler to ensure a complete scope of work, estimate, and appropriate schedule. During the preconstruction phase of the project, the scheduler will develop a conceptual schedule, which will identify the major activities and milestones for the project, along with all of the major trades who will be performing work on the project. As the design and construction documentation progress, the schedule will be refined to include more detail as the sequence of each trades work in each area of the project. A schedule will be developed during the bid and award process on which all subcontractors will be expected to base their bid and incorporate their activities and schedule and sequence of activities accordingly. Once the project is awarded and a construction base line schedule is established, the scheduler will monitor and report on the progress of the work, as it relates to the base line schedule. Major deviations to the schedule will be monitored and reported on to ensure that appropriate and timely action can be taken to correct a problem before it gets out of hand and there is insufficient time and resources to bring the project back on schedule. The scheduler will utilize project management software (see Chapter 15) to develop, monitor, and control the schedule. The project management software has the capability of cost loading and resource loading the schedule to further monitor and control the progress and cost of the work simultaneously. The scheduler, like the estimator, will work out of the home office during the preconstruction phase, and at the on-site project office for the balance of the phases.

Accountant

The project accountant is the keeper of the books and all costs, and is responsible for developing the periodic billing (requisition) for the project, tracking all payments, disbursing all payments once payments are received from the owner, requesting advances for deposits on order and contracts where appropriate, obtaining partial and final waivers of lien, tracking all allowances, tracking all monies charged to the general conditions for the project, tracking all monies charged against the project contingencies, and developing all cost control and reports that may be required by the owner for the project. These reports are usually developed on a monthly basis and submitted with the requisition for the project. Project management software, as previously described in the description of the scheduler's responsibilities, is also available to the accountants to assist in the performing of their duties and responsibilities. The accountant, like the scheduler and estimator, will work out of the home office during the preconstruction phase, and at the on-site project office for the balance of the phases.

General Superintendent (GS)

The general superintendent (GS) for a large construction project is responsible for all field construction activities. All superintendents, assistant superintendents, plan clerks, timekeepers, labor foremen, etc. report to the GS. The GS is responsible for the on-site field administration, supervision, and technical management of all construction operations, including direct supervision of all field construction personnel. The GS will be involved with the planning of the work, scheduling, logistics, coordination and execution of the work, safety, quality control, punch list, security of the site, and inspections approvals. The GS is often a person who has been involved in the construction industry for a while, and is an experienced superintendent in many disciplines. The GS is usually involved with the preconstruction services of a project, working from the home office, and with the construction and project close-out phases working from the construction site. Some CM/GCs utilize the GS on a more extensive basis to cover all superintendent assignments across all projects within the organization.

Superintendent

There are usually a number of superintendents assigned to a project. They all work under the direction of the GS in an assigned area of the project, which may be by areas to be constructed, or by particular areas of specialization. The superintendent is usually a person with a fair amount of experience in the construction industry, and who has worked as an assistant superintendent for some time. The superintendents may be involved in the preconstruction services of a project, working under a GS in the home office, and during the construction and project close-out phases from the construction site.

Assistant Superintendent (AS)

The assistant superintendent (AS) works under a superintendent assigned to a specific area or technical discipline of the project. The AS will assist the superintendent with performing their duties and responsibilities on a daily basis. The AS becomes another set of eyes and ears for the superintendent, who cannot be in more than one location at a time. An AS is usually a person with a limited amount of experience in the construction industry, and may have worked as a plan clerk, time keeper, and labor foreman previously on other projects. The AS is usually not involved in the preconstruction phase of the project, and is assigned to the construction site during the project's construction and close-out phases.

Mechanical, Electrical, Plumbing, Sprinkler (MEPS) Engineer

The mechanical, electrical, plumbing, sprinkler (MEPS) engineer is responsible for the technical support of the project team for the planning, coordination, construction, start up, and turn over of the mechanical, electrical, plumbing, sprinkler, life safety, elevators, security, and telecommunications of a project. They are usually graduate engineers and/or personnel who have had extensive experience in the design and installation of these technical systems. It is important for this individual to be able to interface with the design team to ensure that the design and engineering intent of the construction documents is being built and properly incorporated into the project. The MEPS engineer also has to have a sense of all aspects of the project and how the engineering is incorporated into the overall infrastructure of the project. The MEPS engineer is involved in the budget process, bid evaluations, value engineering, quality assurance, and system start up and commissioning of systems. Without the proper infrastructure, the systems will not operate properly and will not fulfill the needs of the project and program requirements.

Project Engineer

The project engineer title is used by some companies to describe an entry level administrative position that supports the PM in tracking submittals, schedules, shop drawings, RFIs, and other project documentation.

Labor Foreman

The labor foreman is a key person to the construction phase of the project. He is responsible for the assignment of all laborers working on the project to ensure that the proper protection is in place for safety, keeping the hoist log for all deliveries, ensuring that the construction areas are kept clean, and arranging for all trash removal. In addition, the labor foreman is a focal point of communications at the field construction level for all other foremen working for the subcontractors at the project construction site. The labor foreman helps the subcontractors communicate and resolve matters related to logistics, and the need for protection, storage, staging, clean up, power, and water.

Timekeeper

The timekeeper is responsible for the logging of all time for all trade personnel assigned to the project working on the CM/GC's staff that is providing logistical support to the project. These personnel may include the personnel hoist operator, the material hoist operator, the teamster, the master mechanic, laborers, and delivery truck drivers. assigned to work on the project. Many of these personnel can be members of various unions or non-union, depending on the project and its location, and are assigned to work on a given construction project. They are placed on the CM/GC's payroll for a specific project only, and then reassigned to other projects thereafter. Simply stated, without a proper time sheet or ticket, they will not be paid.

Plan Clerk

The plan clerk for a project is responsible for receiving and distributing all project documentation, including construction drawings and specifications to the subcontractors, maintaining document logs, request for information logs, sketch logs, shop drawing logs, field information memo logs, submittals of cuts and samples, and approval status. The plan clerk is the central depository of all construction documentation for a project. The plan clerk also arranges for all messenger and blueprint services for the project. Project management systems are of great assistance to the plan clerks in performing their duties and responsibilities, and tracking all appropriate documentation. In today's computer and technology age, much of the project documentation and files are available electronically and greatly saves the cost of paper (as well as some trees in the forest!).

Site Safety Manager

The site safety manager is responsible for the overall safe work practices and procedures of all personnel and firms working on the construction site. The CM/GC will develop a site-specific safety plan, and each subcontractor will be responsible for complying with the plan, holding safety meetings, training personnel, ensuring proper safety equipment is being utilized by all workers at the site, inspecting and maintaining equipment, investigating all incidents and accidents, and developing corrective action programs to minimize any injuries, fatalities, or accidents. Many CM/GC firms are implementing a safety training program for all personnel prior to allowing them to work on a specific site.

PROJECT STAKEHOLDERS

Every construction project is unique and has unique personnel who will be working directly, indirectly, or be affected by the project (project stakeholders). It is important to identify all of the project stakeholders, and understand the focus and concerns each has on the project. The stakeholders must be dealt with both individually and collectively. It is important to develop a "win-win" strategy for all of the stakeholders involved in the project and to have each of them be supportive and committed to the project. Exhibit 1-11 contains a sample list of project stakeholders.

Name	Name and Address of Stakeholder	Comments
1. Owner		
2. Building manager		
3. Developer		
4. Lending institution		
5. Architect		
6. Mechanical, electrical, plumbing, sprinkler (MEPS) engineer		
7. Structural engineer		
8. Life safety engineer		
9. Audio-visual consultant		
10. Security consultant		
11. Acoustical consultant		
12. Building department expeditor		
13. Fire department expeditor		
14. Local department of traffic		
15. Local department of buildings		
16. Local mass transit		
17. Local community board		
18. Insurance company		
19. Bonding company		
20. Construction manager/ general contractor		
21. Occupational safety and health administration (OSHA)		
22. Owner's representative		
23. Furniture company		
24. Moving company		
25. IT department		
26. Security department		
27. Auditing department		
28. Facilities department		
29. Accounting department		

Exhibit 1-11

List of stakeholders for a typical construction project.

ROLE OF THE OWNER'S REPRESENTATIVE

An owner's representative is an individual who looks out for the owner's interests in all matters relating to the construction of the project. The owner's representative is independent of the design team and the CM/GC and reports directly to the owner. Thus, when problems arise between the design team and the CM/GC, the owner's representative should take an unbiased point of view in resolving any potential dispute. However, like all stakeholders of a construction project, each party has a bias to protect their own turf. The owner's representative is no different and thus the CM/GC must be aware of this situation when dealing with the stakeholders.

The owner's representative usually has expertise in the areas of construction, architecture, or engineering. The owner's representative can be very helpful in obtaining information that may be required from the owner and in expediting items such as shop drawings, response to RFIs, payment of requisitions, and obtaining final payment.

Some of the responsibilities of the owner's representative include:

1. Review the drawings with the design team.
2. Attend all construction meetings.
3. Review and comment on schedules.
4. Review and comment on all costs submitted.
5. Review and approve all requisitions.
6. Review and approve all change orders.
7. Inspect the project during construction.
8. Make sure the CM/GC conforms to the construction documents.
9. Review and approve all value engineering items submitted by the design team and the CM/GC.
10. Review and try to expedite shop drawings, submittals, and RFI resolution.
11. Obtain all owner's furnishing and equipment for the project per the schedule.
12. Review the commissioning process.
13. Involve in the punch list process.
14. Assimilate all close-out documents.
15. Responsible for final payment.
16. Move into the project.

SUMMARY

- The project organization must be planned for each project to determine which personnel will be assigned to the project and work in the office, versus those who will be assigned directly to the project site.
- Qualified and professional personnel must be assigned to each phase of the project—preconstruction, construction, and project close-out—to ensure the success of the project.
- There needs to be continuity in the team members throughout each phase of the project.
- Personnel assigned to a project site will often have two superiors, the project executive for the project and the departmental manager in the home office. This creates a matrix management arrangement, which must be managed by all parties.

- The experience, education, and skill set of each person assigned to the project is unique, and it is extremely important that the proper person be chosen for each position, and that there be synergy in the entire project team.

- Often employees who are assigned to an off-site construction project for a period of time find themselves "out of site and out of mind."

- Ensure that there is a firm understanding of the terms of the contract to determine which employees in the home office and their associated expenses can be billed to the project.

- There is a management philosophy that says, "Do what you do well, and leave to others what they do well." Seldom does one person have all the skill sets required to be a good PM, and when they do, they often do not have enough time to do them simultaneously, especially with the fast tracking and pace of many projects.

2 Risk Assessment and Problem Solving
(The bane of every construction project.)

RISK MANAGEMENT

Risk management is a methodology that has gained formal acceptance in construction project management circles during the last several years. Risk management must be done throughout the construction project. A risk management plan will provide the guidelines for risk management for the construction project team. The risk management plan will also serve as a baseline for the construction project to track variances, validate assumptions based on objective information and the potential probability, and impact of the risk of the events. Risk management, monitoring, and reporting need to be integrated into project cost control and project management reporting systems for the project. On large and complex construction projects, such as those encountered in the urban environment, risk management must be performed in a structured manner, or matters will soon become unwieldy. In many respects, the construction of a large complex project is about identifying, mitigating, shifting, and managing project risks.

Risk can be defined as any event that is likely to adversely affect the project's ability to achieve its defined objectives, which is: completing the project on time, within budget, at high quality, and with a satisfied client who is willing to give the construction manager/general contractor (CM/GC) repeat business and a good reference. Risk is also the uncertainty of the outcome of an event, which can cause a potential problem to the project or project team. The aim of risk assessment and management is to ensure that risks are identified at project inception, their potential impacts allowed for and, where possible, the risks minimized or eliminated. Risk management planning describes how risk will be identified, structured, tracked, acted on in a timely and prudent manner, and resolved.

Risk management plans should be in place to deal quickly and effectively with risks as they arise as follows:

1. Identify and assess the risks in terms of impact and probability.
2. Identify project risks and categorize them.

3. Ensure that the CM/GC, architect, engineer, and owner have the opportunity to be involved and communicate an appropriate agreement of the type of risk and contingency plans to deal with that risk.

4. Identify project risks such as re-occurring risks, risks with small implications, risks with manageable implications, risks with very large implications, risks with catastrophic implications, and devise a risk management plan to deal with them.

5. Ensure control of risks by planning how risks are to be managed through the life cycle of the project.

6. Allocate responsibility for managing each risk with the party best able to do so.

7. Establish and maintain a risk management plan by the CM/GC.

8. Establish procedures for actively managing and monitoring risks throughout the project and during completion and occupancy.

9. Develop contingency plans to deal with the risks if they were to materialize.

10. Escalate the risk to upper management if it is of significant importance and potential adverse impact to the project.

11. Monitor risks throughout the project life cycle, as new risks may arise that have not been previously identified or anticipated.

12. Update risk information throughout the life of the project.

13. Shift risks down to the subcontractors, suppliers, and vendors.

Exhibit 2-1 is a summary of the risk management process.

Exhibit 2-2 is a risk management flowchart.

RISK ASSESSMENT

One of the tasks to be performed during preconstruction services is risk assessment for the project. Many people think about risk in terms of financial issues such as insurance and bonding. Risks are not limited to finances, and can include any matter that adversely affects the project such as design deficiencies, construction deficiencies, material deficiencies, labor disputes, contract provisions (see Chapter 9), strikes, acts

EXHIBIT 2-1
Summary of the risk management process.

The risk management process can be defined in the following manner:

1. Identify risks to the project.
2. Complete the risk assessment form.
3. Investigate the risk to the project.
4. If the risk is determined to be large, escalate the risk to upper management.
5. Continuously monitor the risk.
6. Assign an appropriate risk management mitigating action, or contingency plan.
7. Implement the appropriate risk management action at the proper time.
8. Bring closure to the risk item.

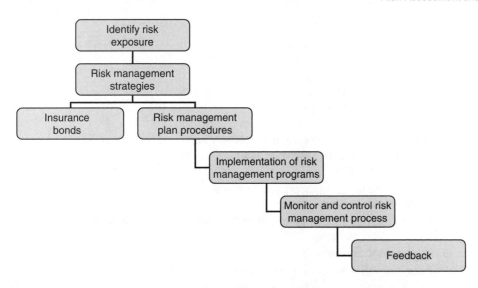

Exhibit 2-2
Risk management flowchart.

of sabotage or terrorism, material shortages, personnel shortages, geotechnical problems, and governmental agency problems.

In order to properly manage the potential risks to the project, a risk management plan needs to be developed to formally identify, quantify, assess, and mitigate the risks during the execution of the project. The risk management planning process entails completing a number of actions to reduce the likelihood of occurrence and the severity of the impact of each risk. This process will enable the CM/GC to identify, document, review, and mitigate these risks, and any others that may arise in the ever-changing world in which we live and conduct business. An outline of a risk management plan for a construction project, which must be customized for each specific project, is shown in Exhibit 2-3.

A model used in the construction industry to manage risk is the construction risk management system (CRMS). This model provides an effective systematic framework for quantitatively identifying, evaluating, and responding to risks in construction projects. Risk management should be seen as managing risk proactively, rather than responding to risk events after they happen. Hence, the theme of risk management approach is to act instead of react to project risks. Many contractors think of risk management as insurance management, where the main objective is to find the optimal economic insurance coverage for the insurable risks. Risk management, if done properly, goes way beyond that and looks to scientifically and systematically approach the management of risks faced by contractors, and deal with both insurable and uninsurable risks by choosing the appropriate techniques for dealing with the risks to the project and the contractor. Contractors must evaluate their own risk tolerance and the degree of exposure to risk with which they are comfortable and can be exposed to financially. Remember that there should be a balance with all risks and rewards, and if one does not take prudent risks, there may not be the potential for reward.

Exhibit 2-3

Sample outline
for a risk
management plan.

Item	Included
1. Identify risks to the project, especially during the preconstruction planning.	
2. Consider arranging for a risk workshop to assist in identifying project risks.	
3. Fill out a risk assessment form.	
4. Document the risk in the risk management log for proper tracking and administration.	
5. Investigate the risk.	
6. Quantify the risks, likelihood, and potential impact on the project.	
7. Review the risk to determine its overall priority (low, medium, or high).	
8. Take actions to resolve the low and medium risks.	
9. Escalate high risks to upper management.	
10. Describe the earliest indicator or trigger condition that might indicate that the risk is turning into a significant problem.	
11. Continuously identify and update monthly the top ten risks to the project.	
12. Review risks and approve the appropriate set of actions.	
13. Where possible avoid, transfer, or mitigate risks.	
14. The project manager will assign the actions to the members of the project team and oversee the implementation.	
15. Members of the project team will implement approved actions with diligence and timeliness.	
16. Advise the owner of any impact to the project's budget, schedule, and logistics.	
17. Obtain approvals as may be required for any contractual costs that may be incurred.	

Tools of Risk Management

The following is a list of tools used to manage risk for a construction project:

1. Contract language
2. Insurance
3. Administrative procedures
4. Operational procedures
5. Bonding and surety
6. Safety programs and loss prevention

7. Controlling claims and litigation
8. Risk management plan
9. Risk avoidance and mitigation
10. Risk transfer

Categories of Risk

Risks are sometimes broken down into three categories depending on their size and frequency as follows:

1. *Small, reoccurring, routine risk.* These risks taken together usually pose no significant financial impact and the costs associated with them are absorbed by the CM/GC or the project budget contingency and loss control procedures. An example of this might be stolen small tools from the project site.
2. *Accidental risk.* These are usually small in number, but large in cost, both individually and taken together. These are usually covered by insurance and loss control procedures. An example of this might be a pedestrian being struck by some construction material while walking by the site.
3. *Catastrophic risk.* These risks are usually small in number, but could have significant impact on the CM/GC and/or project, depending on the company's liquidity, retained earnings, profitability, and level and types of insurance coverage. An example of a catastrophic risk is a building catching fire during the construction process, or a tower crane collapsing during the construction process.

RISKS ENCOUNTERED DURING THE PROJECT LIFE CYCLE

During the life cycle of the project it is likely that new risks may be encountered that had not been previously identified. Exhibit 2-4 is a sample list of such risks.

After a new project risk has been identified, the member of the project team who identified the risk should raise the issue to the person on the project team responsible for risk management, usually the project manager (PM). The PM should then review the risk to determine its validity and level of priority. The PM should then take the appropriate action to resolve the issue, if that action is within their capabilities. These actions are:

1. Identify and evaluate the risks.
2. Complete risk assessment form and log them into the risk management log.
3. Investigate project risks and develop alternative approaches.
4. Assess impact on cost, schedule, and project logistics.
5. Assess if there is any governing code considerations.
6. Obtain approval of the owner.

Exhibit 2-4

Sample risks encountered during the life cycle of the project.

Item	Included
1. Change in program or customer requirements.	
2. A sudden shortage in skilled personnel.	
3. A sudden shortage in construction materials.	
4. Unexpected project expenses and inflation of materials.	
5. Poor performance on the part of a subcontractor, supplier, or vendor.	
6. Failure of project equipment.	
7. Significant accident at the project site.	
8. Act of sabotage or terrorism.	
9. Changing governmental authority requirements.	
10. Surveyor errs with the wrong elevation or benchmark, which presents coordination conflicts beyond the control of the construction manager/ general contractor.	
11. Coordination drawings are not completed prior to the work being installed, resulting in conflicts in the field.	
12. Conflicts exist in the installation of major piping, conduits, ductwork, etc. that result in a significant change in the design concept of the project and routing of services.	
13. Long lead items are not being produced in a timely manner by the equipment fabricator.	
14. Subcontractor purchased the wrong materials.	
15. Encountering poor subsoil or rock conditions, which are different from those anticipated.	
16. Encountering water not previously indicated.	
17. Encountering underpinning, bracing, and shoring problems with adjacent buildings due to unanticipated conditions.	
18. Jurisdictional disputes with unions workers.	
19. Poor quality control.	

Exhibit 2-5 is a sample of a risk assessment form for identifying and managing project risks.

Exhibit 2-6 is a sample of a risk management log to log in risks when they are identified and classified on the risk assessment form.

The following guidelines should be used to complete the risk forms for a construction project:

1. The risk form provides a complete description of the risk identified.
2. The likelihood of the risk is rated as its probability of occurrence.

Project Name:_____Project #_____

Risk ID: _____
        Number in the log, often year, followed by project number and item number, e.g., 2008-017-00

Exhibit 2-5

Risk assessment form.

Classification: _____ *Risk category: high, medium, low, recurring, nonrecurring*

Report Date:_____ Reported By:_____
        (The date the risk was reported)

Probability:_____
        What is the likelihood of this risk adversely affecting the contractor?

Impact:_____
        What is the damage if the risk becomes real?

Risk Exposure_____
        Multiply probability times loss to estimate the risk exposure and potential cost.

First Indicator:

> *Describe the earliest indicator or trigger condition that might indicate that the risk is turning into a problem.*

Mitigation approach:

> *State one or more approaches to control, avoid, minimize, or otherwise mitigate the risk. Mitigation approaches may reduce the probability or the impact.*

Date Started: _____*State the date the mitigation plan implementation was initiated.*

Date to Complete:_____*State the date by which the mitigation plan is to be implemented.*

Person Responsible: _____
        Assign each risk mitigation action to an individual for resolution.

Current Status:_____
        Describe the status and effectiveness of the risk mitigation actions.

Contingency Plan:

> *Describe the actions that will be taken to deal with the situation if the risk factor develops into a problem.*

Trigger for Contingency Plan:

> *State the conditions under which the contingency plan will begin to be implemented.*

Signed:_____

Exhibit 2-6

Risk assessment form.

CSI CODE	SUBDIVISION	HAZARD IDENTIFICATION	RISK PRESENCE YES/NO	RISK ASSESSMENT (HIGH, MEDIUM, OR LOW) FREQUENCY	RISK ASSESSMENT (HIGH, MEDIUM, OR LOW) SEVERITY	RISK CONTROL PLAN

3. The impact of the risk is rated as the Frequency of Exposure to Hazard × Severity of the Outcome × Probability of Occurrence = Impact to Project or Degree of Risk.

4. A list of actions needed to prevent the risk from occurring is recommended.

5. Contingency actions should be listed to minimize the impact on the project should the risk actually occur.

6. The completed risk form should be immediately sent to the PM for review and follow-up action with upper management if appropriate.

It is important that the PM be given not only the responsibility, but also the authority to go with it, to act on all items and matters that can adversely affect the project. Remember that time is money, but money does not always buy time. It is therefore of utmost importance to identify, react, and resolve matters in a timely manner. You cannot expect the PMs to be able to do their jobs if they do not have the control of the resources, that is, the money, personnel, equipment, and systems, to resolve it.

Health, Safety, and Environmental Risks

During the construction process, the CM/GC will often encounter various health, safety, and environmental issues on a project. Each of these issues presents potential risks to the CM/GC and the overall project. Exhibit 2-7 is a summary of health, safety, and environmental risks for a project.

Construction Defects as Potential Risks

Construction defects are sometimes encountered which present potential risks to the CM/GC and the overall project. These risks are summarized in Exhibit 2-8.

Item	Included
1. Workers may be exposed to hazardous materials such as:	

1. Workers may be exposed to hazardous materials such as:
   - Lead or lead paint
   - Silica or concrete dust
   - Lead paint
   - Hazardous waste, such as petrochemical products, medical waste, and industrial waste
   - Contaminated soil
   - Asbestos
   - Fumes from torches used for cutting and welding
   - Fumes from paving, asphalt and tar products

2. Workers may be exposed to the following types of activities:
   - Working in a confined space
   - Demolition with a lot of dust, debris, and sharp objects
   - Excavation and trenching
   - Heavy lifting of steel, concrete, curtain wall, or miscellaneous metals
   - Lockout of electrical and mechanical equipment
   - Working on mechanical and electrical equipment while it is operating
   - Working near heavy equipment such as cranes, derricks, scaffolding, and hoists
   - Working near drilling, chipping, and blasting of rock

3. Construction workers may work in the following environmental exposures:
   - Cold or hot extremes of temperatures
   - Laser equipment
   - Noisy equipment
   - Equipment with ionizing radiation
   - Ergonomically difficult or unnatural
   - Aerial lifts, ladders, personnel hoists, scaffolding, scissor lifts, and platforms suspended off the ground

4. Construction workers will utilize tools and equipments such as:
   - Compressed gas cylinders
   - Electric power tools
   - Forklifts
   - Ladders
   - Powder actuated tools
   - Rigging equipment
   - Temporary heating systems
   - Respiratory protection
   - Heavy equipment and motors

Exhibit 2-7

List of health, safety, and environmental risks.

Exhibit 2-8

Sample construction defects as potential risks.

Item	Included
1. Material deficiencies	
2. Poor quality or substandard work	
3. Subsurface or geotechnical problems	
4. Defects in wall systems such as shear failure, drag failure, cracks, settlement, water infiltration, inadequate framing, and break in the piping within a wall cavity	
5. Floor or ceiling defects such as inadequate support, excessive spans, overloading during construction, undersized framing members, rotting of the structure, cracking, water intrusion, and excessive sloping	
6. Defects in decks and balconies such as improper deck-to-wall transitions, threshold transfers, improper coatings and finishes, improper flashing, use of improper materials, and lack of protection during the installation process	
7. Roofs defects such as improper installation, improper flashing, improper slope, improperly installed gutters and drainage systems wet insulation, wet substrate below the roof, deteriorated substrate materials, and lack of protection during installation	
8. Window and doors defects such as improper installation, improper anchorage, improper flashing, water infiltration, corrosion of materials, use of dissimilar materials, trapped moisture during the installation process, and breaking of the seal on insulated glass units.	

WHY CM/GCs FAIL

CM/GCs sometimes fail because of numerous issues and risks that they encounter while running a construction business. The construction process is, by its very nature, a very risky environment. Exhibit 2-9 summarizes why CM/GCs sometimes fail.

CM/GC COMPANY'S WELLNESS

The overall wellness of the CM/GCs is critical to their ability to perform their duties and responsibilities in the construction process, satisfy their clients and employees, and build efficiently and effectively. Exhibit 2-10 is a checklist for the wellness of CM/GCs.

Item	Included
1. Not having a strategic, business, marketing, and annual planning process	
2. Growing too fast and over-expansion	
3. Insufficient working capital	
4. Insufficient human resources	
5. Working for problematic owners	
6. Obsession with business volume without proper project evaluation	
7. Selection of bad projects for the wrong reasons	
8. Insufficient diversification	
9. Phantom profits	
10. Management not having good business skills	
11. Not staying focused in your market niche	
12. Not having a risk management plan and procedures	
13. Not planning for the orderly succession of management	
14. Not training personnel	
15. Poor project performance	
16. Poor safety record	
17. Poor quality control program	
18. Poor management of the environment and stakeholders	
19. Onerous contract provisions	
20. Inability to pre-qualify for construction projects	
21. Criminal activities	

Exhibit 2-9
Summary of why CM/GCs fail.

Item	Included
1. Strategic, business, marketing, and annual plan for the firm	
2. Qualified managerial professional personnel to run the business	
3. Qualified technical professional personnel to market, procure, construct, administer, and close-out the construction projects	
4. Training and development program for personnel	
5. A corporate culture that sets a high standard and a supportive work environment	
6. Perform annual appraisals of personnel against their pre-established goals and objectives	

Exhibit 2-10
Checklist for wellness of the CM/GCs company.

(Continued)

Exhibit 2-10
(*Continued*)

Item	Included
7. Working SMART (strategic, measurable, achievable, realistic, and tactical)	
8. Performing a SWOT (strengths, weakness, opportunities, and threats) analysis for the overall organization	
9. Proper and fair compensation and benefits plan for all employees	
10. Minimize turnover by providing opportunities for training, advancement, development, and recognition of personnel	
11. Focus on market niches where the firm performs well	
12. Build on relationships with existing clients	
13. Adequate financing of the firm	
14. Customer satisfaction for referral for future work and references	
15. Safety program with excellent results	
16. Quality control program with excellent results	
17. Project management systems to monitor and control the entire construction process	
18. Administration of special administrative programs for the project	
19. Equal Employment Opportunity (EEO) and Minority Woman Business Enterprises (MWBE) goals and objectives	
20. Commitment to the community	
21. Proper accounting systems	
22. Adequate life insurance for key personnel	
23. Shareholder agreements for multiple stockholders, and an orderly succession	
24. Adequate succession planning for upper management	
25. Overall profitability of the company	
26. Keeping pace with technology changes	
27. Monitoring and adopting to a changing environment	

SUMMARY

- Plan a project properly, minimize surprises, develop appropriate contingency and action plans, be flexible, and expect that in spite of all best efforts things will happen!
- In the preconstruction phase of the project, it is important to identify risks to the project and develop a risk management plan to manage these risks.

- Enough adverse conditions will happen on a day-to-day basis that will keep the project team well occupied.
- Having appropriate contingency plans developed for possible events that propose a risk to a project will go a long way in managing the process.
- One must be flexible and adaptable to the construction process, as there are always surprises every step of the way.
- Risk assessment forms and register must be prepared early in the project so that risks can be identified and appropriate actions can be taken when unforeseen conditions arise.
- Appropriate contingency plans should be developed to manage risk.
- The CRMS model of risk management consists of the following processes: risk identification, risk analysis and evaluation, risk response management, and risk system administration.
- Allocate responsibility for managing each risk with the party best able to do so.
- During the life cycle of the project it is likely that new risks may be encountered that had not been previously identified.
- The more complex and numerous the project interfaces, the greater the uncertainty and the greater the probability of risk events occurring.

3 Testing and Quality Control
(Why do we have to test?)

WHAT IS TESTING?

Testing, in terms of construction projects, is an impartial and systematic way to prove that the construction components that have been installed meet the criteria established by the construction documents, the various agencies, user groups, quality control groups, and building codes (hereto referred to as organizations). All of these organizations publish detailed testing procedures to be followed during the construction process. A failure of any of the components used on a construction site may have catastrophic results. As you may know, the construction industry unfortunately has an extensive history of building failures, due to one or more components not performing to established standards. Hence, testing has become routine and a compulsory part of sound and responsible construction. A testing flowchart is indicated in Exhibit 3-1.

TESTING GROUPS

Testing procedures are grouped as follows:

1. Mixtures (i.e., concrete, soil)
2. Connections (i.e., welding, bolts)
3. Assemblies (i.e., curtain walls, roofs)
4. Material (i.e., steel, cement, aluminum, glass)
5. Load capacity (i.e., piles, concrete cylinders, steel)
6. Pressures (i.e., pipes, duct work)
7. Flow (i.e., air balancing, water balancing, sprinklers)
8. Systems (i.e., life safety, electrical)
9. Performance (i.e., elevators, mechanical equipment, electrical equipment)

The following are examples of testing for the aforementioned testing groups.

Exhibit 3-1

Testing flowchart.

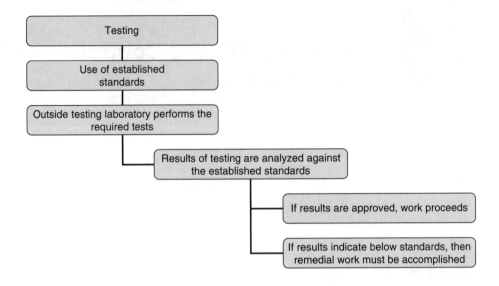

Mixtures

Since mixtures are made up of a blend of materials that are typically manufactured separately but combined to act as a composite material, testing may require additional steps and techniques. A good example is concrete, a composite of (1) cement, (2) sand, (3) aggregate, and (4) water mixed together to form a cohesive strong material. This material requires checks on temperature of the mixture, the ratio of water to cement, and the percentage of weight of all materials to confirm that the composite mix meets the requirements of the job.

Connections

In the case of connections, tests are performed to determine the capacity of a connector to join two pieces of material together to sustain a required load. For steel beam construction, a typical connection may be welded. These welds need to be tested for subsurface imperfections that may reduce the holding strength of the connection. Various methods for testing weld connections, especially for hidden problems, include x-ray, dye penetration, and ultrasonic or magnetic particle procedures. Additionally, a micrometer or a measuring ruler checks the thickness of the weld (as specified by the structural calculations and shop drawing).

Assemblies

An assemblage of various materials requires testing of the entire constructed assembly to ensure that all materials are acting in concert with one another without failure. For example, when fully assembled a typical curtain wall includes (1) glass, (2) frame, (3) gaskets, (4) caulking, (5) setting blocks, (6) weep holes, (7) structural supports, and (8) anchorage. Testing is performed on the entire assembly to check for water leakage, air infiltration, and glass and mullion failures, and to verify that it has met all design criteria.

Material

The basic material that is used for construction has to meet industry standards. Thus, structural steel has to conform to an A36 (or higher) standard. In other words, the manufacturer process must produce steel that will have a yield strength of at least 36,000 psi (pounds per square inch).

Load Capacity

Some materials and components are required to sustain direct loads without failing. An example is piles, which are used in foundations for buildings. Piles must take the dead load and live load of the building without failure or settlement. Thus, piles have to be load tested to make sure that the imposed load meets the building code criteria and load bearing capacity of the soil and friction of the soil.

Pressures

Some installed building material or components must withstand external and internal pressures. For instance, a pipe that has water that is being pressurized by a pump must be examined for proper pipe thickness and proper joint connections so that no leaks or failure will occur. Usually pipes are tested at 1.5 times the normal operating pressures specified.

Flow

Enclosures that have air or water flowing through them must be tested for proper flow capacity. In HVAC (heating, ventilation, and air conditioning) ductwork, the air flowing through the system must be of sufficient capacity to heat or cool the occupants in a space. This test is accomplished by measuring the cubic feet per minute (cfm) that exits at the room's diffuser (room air outlet). This cfm is then compared to the cfm of the consulting engineer's design drawings.

Systems

A system is an amalgamation of independent components performing together as a unit for a specific function (i.e., speakers, strobes, elevator drops to lobby area, etc.). A good example is a fire alarm system which is typically comprised of various electrical and mechanical parts and components that must act together when a fire is detected. This would include fire and smoke detectors giving off signals, strobes lighting up, elevators immediately going to the lobby, and turning off of mechanical systems.

Performance

Performance guidelines are established for various parts, components, and systems. For example, the design team may call for a certain elevator speed to be maintained to accommodate the building's tenants without delays.

Each one of these types of tests or a combination of them will advise the project manager (PM) on how well each component is performing. The PM must know how these components are to perform in order to determine their effectiveness. Thus, the PM must be familiar with the drawings and specifications to understand the working capability of all the components that constitute the project.

MOCK-UPS AND WIND MODEL TESTS

In addition to tests that are performed in the field, certain assemblies of buildings are mocked up at full scale and fully tested. This would be true for the assembles that are associated with curtain wall construction. These mock-ups are required because it is extremely difficult to test in situ (on site when constructed). Also, it is more cost effective to check the assembly in the laboratory rather than finding a failure when the assembly is installed. In addition, in the urban environment, wind loads on a building are impacted by other structures in the area. To account for this variable, a model of the building is created, along with models of the surrounding buildings. These models are then tested in a wind tunnel. Sensors are placed on the building being tested and readings are taken of the pounds per square foot (psf) that would be imposed on the total outside surface of the building. The wind tunnel loads that are used for testing are based on historical wind data of the area in question and requirements from the local code.

When a mock-up of the curtain wall is constructed (full scale) in a laboratory setting, it usually consists of a typical two-story curtain wall section of the building. Any unusual condition may have to be mocked-up for testing as well. The mock-up assembly is tested to determine:

1. Amount of air and water infiltration observed
2. Drainage of the system
3. Structural capability of the mullions and glass
4. Expansion and contraction of the assemblies
5. Deflection of materials

Corrections are made to the mock-up, if required, so that all the assemblies are performing according to specified criteria. The components are then fabricated based on the results of the test. Thus, the installation of the curtain wall should perform according to the standards established by the organizations. This eliminates the potential for failure that could have occurred and the consequence of replacement of the curtain wall if the mock-up was not tested in the laboratory.

FIELD FABRICATION OF STRUCTURAL COMPONENTS (MIXTURE AND COMPONENTS)

Structural components that are fabricated on site by trades people constitute the greatest risk for a catastrophic failure. This is due to the fact that control of putting parts together in the field is not done with the same diligence and controlled environment as a factory-made component. Thus, great care must be taken to ensure that proper testing is performed so that a failure will not occur. The erection of a concrete structure is an excellent example where the use of a mixed type material must have adequate testing. Concrete is a very viable construction material if placed according to the standards established by the organizations. However, due to the complexity of mixing the ingredients at the plant and transporting it to the site, placing the concrete at the site requires numerous controls to obtain an excellent final product. The testing of concrete should include:

1. A trial concrete mix approved by the owner's engineer
2. Proper mixing procedures at the concrete plant
3. Timing for the transportation of the concrete mix
4. Designed and properly installed form work and shoring so that they will not collapse or deflect
5. Temperature monitoring of the concrete at the site (to make sure that flash setting will not occur)
6. Ambient temperature monitoring (too hot for flash setting and too cold for freezing)
7. Slump test to confirm water/cement ratio of the concrete
8. Supervision for concrete vibration and dropping height for the actual placement of the concrete
9. Monitoring the thickness of a concrete slab
10. Assurance that all the concrete encapsulates the reinforcing bars, especially when pouring columns
11. Placement of a sample of the concrete into concrete cylinders to determine the compressive strength of the concrete at 7, 14, and 28 days (via testing in the laboratory). This will be accomplished for design strength conformance and to know when the forms can be stripped
12. Checking the number and location of the reinforcing bars required for the pour
13. Proper curing of the concrete
14. Assurance that reinforcing bars are properly lapped
15. Assurance that all exterior exposed concrete is covered by 3 inches of concrete (2 inches for interior concrete) over the reinforcing steel

Even though steel sections are fabricated in a controlled environment at a plant, the steel members must be connected in the field by iron workers with bolts and/or welding. Thus, stringent testing is also required for a steel structure. Some of the tests that would have to be considered when erecting steel are the following:

1. Proper bolts are being utilized.
2. Required tightening (torque) of the bolts needs to be accomplished by code standards.
3. Steel sections as indicated on the approved shop drawings are in fact being installed.
4. Welds have to be checked for proper thickness and continuity.

5. All welders have to be certified.

6. Shear stud connectors have to be attached to the steel with proper spacing and welds.

7. The steel has to be fireproofed with approved material that will have proper thickness, adhesion, and density.

8. All columns are perfectly aligned (plumbed).

9. Correct steel is being used (i.e., A36).

10. Proper steel camber has been placed on the steel as specified by the consultants.

11. Splice plates must be of the approved thickness.

12. Inspection at the fabricator's shop would be helpful for checking beam camber and obtaining coupons.

TESTING PROCEDURE ORGANIZATIONS

Numerous organizations that are involved with the construction industry have testing as part of their specifications. Exhibit 3-2 is a list of those special organizations (not inclusive).

These organizations have set up procedures for testing every component on a construction site. The architects and consultants that have developed the contract drawings and specification usually include in these documents the tests that are required for the project. These testing procedures are taken from the organizations listed in Exhibit 3-2. In the urban environment, the local, state, and municipal governments have established additional testing requirements that must be followed.

In a majority of cases, outside testing laboratories are used to determine the capability of the components. The testing laboratories used should be completely independent from the party requesting the tests and are usually retained by the owner. The names of the local testing laboratories can be obtained through the Internet, local yellow pages, local contractors, or municipal agencies.

Exhibit 3-2

Testing procedure organizations.

The following is a list of organizations (not inclusive) that have testing as part of their specifications:

1. ACI (American Concrete Institute)—http://www.concrete.org/general/home.asp
2. AISC (American Institute of Steel Construction)—http://www.aisc.org/
3. ASTM (American Society of Testing and Materials)—http://www.astm.org/
4. NEC (National Electrical Code)—http://www.nfpa.org/catalog/product.asp?pid=7008SB&src=nfpa&order_src=A292
5. ASHRAE (American Society of Heating Refrigeration Air Conditioning Engineers)—http://www.ashrae.org/
6. IBC (International Building Code)—http://www.iccsafe.org/
7. NFPA (National Fire Prevention Association)—http://www.nfpa.org/
8. UL (Underwriters Laboratories Inc.)—http://www.ul.com/

9. Local building codes—Will vary depending upon location

10. ANSI (American National Standards Institute)—http://www.ansi.org/

11. AWS (American Welding Society)—http://www.aws.org/w/a/

12. SMACNA (Sheet Metal and Air Conditioning Contractors National Association)—http://www.smacna.org/

13. ASME (American Society of Mechanical Engineers)—http://www.asme.org/

14. SIGMA (Sealed Insulating Glass Manufacturers Association)—401 N. Michigan Ave., Suite 2400, Chicago, IL 60611; (312) 644-6610

15. AAMA (American Architectural Manufacturers Association)—http://www.aamanet.org/

16. FM (Factory Mutual)—http://www.fmglobal.com/

17. NEMA (National Electrical Manufacturers Association)—http://www.nema.org/

18. NIST (National Institute of Standards and Technology)—http://www.nist.gov/

19. PCI (Precast/Prestressed Concrete Institute)—http://www.pci.org/intro.cfm

20. ASCE (American Society of Civil Engineers)—http://www.asce.org/asce.cfm

Exhibit 3-2
(*Continued*)

PROJECT TESTS

Exhibit 3-3 is a list of proposed tests (not comprehensive) that should be accomplished on a building construction site, depending on the components used.

QUALITY CONTROL

Quality control (QC) is a process of inspecting and confirming that the finished installation or work has indeed met the design specifications enumerated in the contract documents. Thus, a thorough understanding of all the design documents is required prior to establishing a QC program for all the major components for a project. In addition, the PM must be knowledgeable of the approved (as prepared by the industry standards applicable to this component) installation procedures. The purpose of a QC program is to eliminate all possible defects. Unfortunately, this objective is never achieved. Nonetheless, the PM should strive for the "zero defects" goal. See Exhibit 3-4 for the QC flowchart.

In order for a QC program to be effective, the following has to be accomplished:

1. The drawings and associated notes have to be reviewed and understood.

2. The specification section related to the component being installed has to be read and reviewed for installation procedures and testing.

3. The related industry standards have to be read (i.e., ACI for concrete and AISC for steel).

(*Text continues on p. 57.*)

1. Site

Type of Field Test Required	Local Building Code Section	Specification or Industry Standard Section	Date Inspected	Location	Inspector's Name	Pass/Fail	Comments
Environmental Phase I	Report prepared based on existing available information provided by the local EPA, local fire department, coast and geodetic survey indicating the locale of potential hazardous waste sites and/or reported spills.						
Environmental Phase II	Soil samples are taken from the ground and then analyzed for potential hazardous material as determined by the local EPA.						
Soil borings	Soil borings will determine the compressive strength of the soil and rock as well as soil composition and depth of rock. This is used to determine the bearing capacity of the base foundation material.						
Test pits	Test pits are dug prior to utility installation to determine if there are any obstructions prohibiting the path of utilities. Test pits will allow for safe digging to ensure that no other utilities will be compromised during the excavation process. It also gives an indication of the type of soil up to four feet below the surface.						

Water table	Water table must be determined, through borings, in order for the contractor to determine the amount of dewatering required and the impact on the design of the foundation.							
2. Soils								
Type	Soil types can vary in strength. Rock is the strongest. Weathered rock has less capacity, sand even less, and clay has the lowest compressive strength.							
Compressive strength	Compressive strength gives the soil a value as to how much weight the soil can take without deflection.							
Expansive clay	Expansive clay has the ability to expand when fully saturated. This can shift a structure sitting on it.							
Settlement	All soil types are expected to settle over time after the building is erected. The soil must be tested to understand how much settlement is to be expected.							
Settlement monitoring Shear	Check the condition of a building and monitor the settlement of the structure. Shear is the soil's ability to want to cave into an open end.							
Compaction	Soil is tested for proper density as required.							

Exhibit 3-3
Required testing.

Type of Field Test Required	Local Building Code Section	Specification or Industry Standard Section	Date Inspected	Location	Inspector's Name	Pass/Fail	Comments
Rock (strength and recovery)	Rock is rated as to its strength. Weathered rock typically has lower strength, while bedrock typically has high strength. Recovery of the rock from the sample tube will indicate its total cohesion.						
3. Foundation							
Pile loads	Test piles, which are sample building piles, are driven in various locations throughout the site. Weights are then brought in and applied to these piles. The piles are measured for any settlement or movement to ensure that they are capable of supporting the required loads.						
Piling driving criteria	Dropping of the pile hammer must conform to the standards established.						
Shoring and underpinning	When special construction is required to temporary support sections of an existing structure, then the existing sections being supported must be monitored for any stress or movement.						

Monitoring of adjacent buildings	Adjacent structures that may be impacted by excavation, noise, or vibration, must be monitored for stress or movement.
Piezometer	Locate wells with gages to determine the elevation of water at a site.
4. Concrete	
Compressive strength	Cylinders of concrete are taken from a typical concrete pour and then taken to the lab. After 7, 14, and 28 days, these cylinders are then compressed to failure and measured to determine the strength of the concrete in psi (pounds per square inch).
Slump	A sample of the concrete that is delivered to the site is taken and placed into a cone-shaped instrument. This case is removed and the settlement of the concrete is measured from the top of the cone. The inspector will measure how far the concrete falls to determine the slump of the concrete.
Temperature	Temperature of the concrete is measured to determine that the concrete is not cool or too hot. Each could affect the strength of the concrete.
Concrete design mix	A concrete design mix is submitted for approval prior to delivery. The design mix will determine the proposed strength of concrete required.

Exhibit 3-3
(*Continued*)

Type of Field Test Required	Local Building Code Section	Specification or Industry Standard Section	Date Inspected	Location	Inspector's Name	Pass/ Fail	Comments
Admixtures — A sample is taken of the concrete mix to determine if the added admixture has been placed according to the design mix approved by the consulting engineer.							
Sieve analysis — The aggregate is passed through a series of grated pans. The intent is to make sure the aggregate is of sufficient graded sizes so that a good concrete matrix is developed.							
Sand and water — These materials need to be analyzed for their salt content or any other minerals that may impact the concrete mix or reinforcing bars.							
Curing — Concrete must be cured to ensure that the concrete retains its maximum strength. This is achieved by spraying a curing compound, wetting the concrete, or covering the concrete with burlap, hay, or plastic.							
Shoring and formwork — Formwork must be examined to ensure that the forms are properly installed and will not fail from the load of the concrete. All columns must be banded around the outside of the forms to prevent collapse. Shoring must be in place until the maximum strength of the							

	concrete is achieved. Some municipalities require the shoring to remain in place until 7 floors have been poured. A professional engineer may have to sign off on the shoring and formwork.						
Shoring foundation	The base soil material supporting the shoring must be level and strong enough to accommodate the imposed load.						
Core samples	If there is a doubt as to the compressive strength of the concrete, core samples can be drilled through the slab and tested similar to cylinders.						
Pre-placement	The area must be inspected to determine if the proper amount of steel reinforcing or wire mesh has been used and if there are adequate amounts of chairs in place to allow for proper concrete coverage.						
Concrete truck transfer time	The mix being transported in the truck should not be greater than the established time specified.						
Full load test	If the strength of the concrete is questionable after core samples are taken, then the structure may have to be load tested to determine if the concrete meets the load requirement established by the consulting engineer.						
Mortar	Compressive strength of the mortar is taken to make sure it meets the approved standards.						

Exhibit 3-3
(Continued)

5. Steel

Type of Field Test Required	Local Building Code Section	Specification or Industry Standard Section	Date Inspected	Location	Inspector's Name	Pass/Fail	Comments
Welds	Welds must be tested to ensure that there is no cracking. These welds can be examined by x-ray, ultrasonic testing, or dye penetration. Welds must also be examined to ensure that they are of the proper thickness as required.						
Welders certification	All welders must be certified for the material they will be welding.						
Bolt torque and compressive washers	Torque gun devices must be calibrated properly. Washers must be measured to ensure that they are properly compressed.						
Camber	Camber must be surveyed before and after the steel member is installed to ensure that the piece was made with the proper camber and that the camber was "removed" after the concrete was poured.						
Depth of concrete	The concrete slab must be surveyed to ensure slab thickness specified.						
Spray-on fireproofing (depth, density, and adhesion)	A pull test must be done by hanging a weight from the fireproofing to ensure that there is proper adhesion. The thickness and density of the fireproofing is to be measured. Check that all of the						

Item	Description						
	required steel members have the required coverage. Check the composition of the material.						
Fireproof paint (intumescent coatings)	Thickness of the paint must conform to approved standards.						
Alignment (by survey)	Steel columns, beams, and girders must be surveyed to ensure that they are in the proper location and floor elevation and are plumb.						
Thickness of members	Steel sections must be measured to ensure they are the correct beams or columns as required by the structural engineer.						
Metal deck	Check welds of decking to make sure that the deck is adhering to the steel.						
Shear studs connectors	Test that the studs have adequately adhered to the structural steel.						
Coupon	A piece of steel can be cut off and sent to a lab to determine its strength and chemical composition.						
6. HVAC							
Air balancing	Air must be balanced to ensure that the design cfm (cubic feet per minute) is exiting through the various registers.						

Exhibit 3-3
(Continued)

Type of Field Test Required	Local Building Code Section	Specification or Industry Standard Section	Date Inspected	Location	Inspector's Name	Pass/Fail	Comments
Water balancing	Condenser and chilled water flowing through the system must be checked to make sure the systems are providing a sufficient amount of water to satisfy the requirements of the cooling process.						
Static pressure	Pressure in the ducts has to be checked to make sure the design intent of the consulting engineers is being met.						
Temperature	Air and water temperatures must be tested to ensure that the criteria established by the consulting engineers are correct.						
Amps of equipment	Major equipment must be tested to ensure that they are drawing the right amount of amps. Improper amperage may result in equipment malfunction or power failure.						
Water treatment	Water must be tested to ensure that pH levels are acceptable. Also, make sure that the water is kept clean to reduce clogging of the water filters.						
Fan cubic feet per minute	Measurement made across the fan area to determine the total cubic feet per minute (cfm) being provided as required by the consulting engineer.						

Smoke purge test	Check to see if smoke is removed from a space within a designated time as determined by the fire department.			
7. Electrical				
Balance of phases	Electrical phases must be balanced to ensure that each phase of the leg is equal.			
Grounding	Grounding must be tested to ensure that in the event of escaping "free" current, it is sent to the ground.			
Hot potential test	Measure any voltage leakage that may occur in wire and equipment insulation.			
Life safety systems	Fire alarm and backup power must be tested to ensure that they are functioning and can be used in the event that they become needed. A "pull the plug" test can be done to simulate a power failure.			
Security system	All egress doors must be tested so that in the event of a fire, all doors will open.			
Infrared test	Check to determine if excessive heat is being emitted in electrical panels, wires, fuses, or equipment.			
Smoke detectors	Check the operation of smoke detectors to see if they send a signal to the console and if addressable, show the location of the test smoke detector.			

Exhibit 3-3
(Continued)

8. Piping

Type of Field Test Required	Local Building Code Section	Specification or Industry Standard Section	Date Inspected	Location	Inspector's Name	Pass/Fail	Comments
Pressure	Check pressure to make sure pipes will not burst or leak. The pressure is usually maintained at 1.5 times the normal working pressure for a certain period of time as dictated by the consulting engineer.						
Flow rates	Flow rate must be determined to make sure that adequate capacity is being supplied to the system. If not, the system will not work properly.						
Welds	Welds must be tested to ensure that there are no leaks.						
Torque on bolts for victolic fittings	Bolts must be tight to make sure leaks do not occur.						
pH factor	The acidity of the water has to be analyzed to make sure it does not cause corrosion of the pipes.						
Potability	Water must be tested to ensure that there is nothing harmful remaining in the pipes prior to use by the public.						
PRV (pressure valves)	Backflow preventor valves and reducing pressure-reducing valves must be checked to determine that they are functioning properly.						

Back flow preventors	Check the valves to make sure that water cannot go back into the supply system.							
Coupon	Take samples of the pipe to determine its thickness, chemical composition, and general condition.							
Fire hydrant	Check the pressure to make sure it meets fire department standards.							
Manifold standpipe pump test	Start the fire pump and determine the pressure at the top of the standpipe. It needs to meet fire department standards.							
9. Elevators								
Load	Load test must be done to ensure that all safety features are functioning.							
Braking	Brakes must be able to stop the car under a full load.							
Acceleration	Elevators must be tested to make sure that they will automatically slow down once a certain speed is reached.							
Clearances	Proper clearances must be left below the car, in the pit, in order for there to be enough room for the elevator buffers, which stop the car in the event of a failure.							
Firefighter key switch	Firefighter-only key switch must be installed in all elevators.							
Elevator recall	Under a fire alarm condition, all elevator cars must go to the lobby.							

Exhibit 3-3 *(Continued)*

Type of Field Test Required	Local Building Code Section	Specification or Industry Standard Section	Date Inspected	Location	Inspector's Name	Pass/ Fail	Comments
10. Roof							
Water							
Water tests should be done once the roof is complete. Fill the roof with water and look for any potential leaks.							
Welds for single-ply roof							
Welds must be checked that they are continuous with no openings.							
Ballast							
Rock ballast must be checked to ensure that all stones are round and smooth.							
Infrared test							
Check the condition of the roof to find out where leaks and wet insulation are located.							
11. Curtain Wall							
Mock-up in laboratory							
A mock-up must be done to ensure that the curtain wall passes the air infiltration and water leakage tests established by the consultants' criteria.							
Model testing for wind and load criteria							
Test a model of the building to determine the wind load that will impact the curtain wall and structural system due to the designed wind criteria.							

12. Sprinkler							
Hydraulic calculations	Calculations must be checked to determine that the sprinkler system has the correct amount of pressure and associated pipe sizes.						
Pressure test	System will be tested at 1.5 times the normal operating pressure and held for a certain period of time.						
Sprinkler rig	Check to make sure tamper switch is operational. Check flow gage by noting the gallons per minute (GPM).						
Fire pump	Check pressure capability and GPM						
13. Flame and Smoke Density							
Flame spread	Material specifications must be checked to determine that the proper flame spread rating is given to each piece of material.						
Smoke density	Material specifications must be checked to determine that the smoke density of the material meets the requirements of the approving agency.						
Fire stopping	Approved fireproof material must be applied to all slab penetrations and openings.						

Exhibit 3-3
(*Continued*)

Type of Field Test Required	Local Building Code Section	Specification or Industry Standard Section	Date Inspected	Location	Inspector's Name	Pass/Fail	Comments
14. Hazardous Material							
Asbestos — Must be removed or encapsulated by approved EPA standards.							
Asbestos handling — An EPA approved handler must transport the asbestos material to an approved site.							
Asbestos manifests — All handling of asbestos and transporting to a site must be signed off on approved manifests.							
Air monitoring — If any hazardous materials are encountered that requires removal and abatement, then air monitoring must be performed by a hygienist, to ensure proper containment and no release of containments.							
Lead paint — Potential lead paint samples have to be tested and then, if found, have to be removed by approved methods.							
Oil and petrochemicals — Potential harmful chemicals have to be tested and, if found, then they will have to be removed by approved methods. Oil tanks can be filled with sand.							
Mold — Must be removed by standards established by the Public Health Department.							

Exhibit 3-3
(*Continued*)

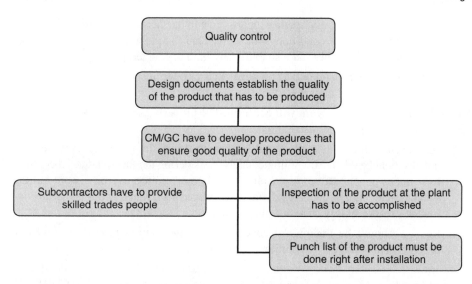

Exhibit 3-4
Quality control flowchart.

4. Shop drawings and other submittals must be reviewed by the PM and approved by the design team, making sure they conform to the design documents.

5. Only approved shop drawings from the design team will be utilized.

6. Installation procedures will be reviewed with the applicable subcontractor and approved by the PM.

7. All tests required for the component being installed will be enumerated on a log and approved by the PM. See Exhibit 3-3.

8. If any off-site inspection of a component is required, then qualified Construction Manager/General Contractor (CM/GC) personnel, design team, and/or testing laboratory will perform the requisite evaluations.

9. All the contractor's personnel who will be inspecting the components and their installation will be adequately trained in the proper installation procedures.

10. Based on past experiences, a list of potential problem areas and avoidance procedures should be established.

11. On complex component installation, a peer review and "brain storming" sessions are required to minimize potential problems. These sessions will require, as a minimum, the PM, the subcontractor, the superintendent, and the design team.

12. In certain cases, additional installation drawings and sequence of operation may have to be prepared in order to eliminate potential problems.

13. The area in which the component will be installed should be inspected for clearances, cleanliness, fit, safety, and logistical conformance.

14. All necessary applicable equipment, testing equipment, material, and shoring must be available at the location where the component will be installed.

15. All testing equipment must be calibrated prior to starting any work.

16. The subcontractor must have sufficient properly trained trades people to handle the installation work.

17. For all exterior work, the weather reports have to be checked on a constant basis to make sure no adverse conditions will occur. This can include not only high winds, tornados, and hurricanes but also extremely high or low temperatures.

18. Contingency solutions have to be developed prior to installation in case adverse conditions do occur unexpectedly.

19. The area in which the work will occur must be properly prepared in sufficient time so as not to impact the installation.

20. After installation, the work should be "punched out" immediately. Any defaults noted should be taken care of immediately, considering the fact that the subcontractor's trades people are readily available. This may require overtime work but the subcontractor should understand that defects have been found and it is the subcontractor's responsibility to rectify the problems.

21. Records should be maintained concerning the installation procedures and any problems encountered. Resolution procedures should also be enumerated.

The end product of a good QC program is to achieve a minor punch list. See Chapter 21 for information on punch lists. This transfers into reduced costs for the CM/GC and the subcontractors and a reduced project schedule. One of the primary objectives of the CM/GC is to make a good profit on the job. Secondary objectives are to complete the project on time (or sooner) and to have a happy client. These objectives can only be achieved with a first class QC program.

A copy of a sample steel QC report is shown in Exhibit 3-5.

SUMMARY

- Tests are performed to make sure that the components installed conform to industry standards, local codes, and/or the architect and engineer's standards.
- Testing procedures are grouped into nine different areas:
  - Mixtures
  - Connections
  - Assemblies
  - Material
  - Load capacity
  - Pressure
  - Flow
  - Systems
  - Performance
- A full-scale mock-up test in a laboratory setting of some critical assemblies may be required to see how they perform under adverse conditions.
- Tests of field-installed structural concrete and steel must be especially stringent.

Quality Control Checklist

Checklist For: _____ **Structural Steel**

Date of Inspection: _____

Location of Inspection: _____

Item	Description	Accepted	Rejected	Reason for Rejection	Date Corrected	Remarks
1.	Material approved					
2.	Samples available					
3.	Review layout submission					
4.	Field survey existing anchor bolts and leveling plates					
5.	Erection methods					
6.	Verify steel grade					
7.	Steel identified for erection					
8.	Shop painted as per specifications					
9.	Mill test reports					
10.	Verify bolt origin/manufacturer					
11.	Material properly stored?					
12.	Inspect for damage					
13.	Welders certification on file					
14.	Spoils removed from site					
15.	Testing lab required					
16.	Check embedded items					
17.	Verify calibration of tools					
18.	Method of aligning structure					
19.	Check camber					
20.	Verify temp. shore /brace					
21.	Check for touch-up paint					
22.	Record as-builts					
23.	Provide final test reports as required for obtaining of certificate of occupancy					
24.	Inspect metal floor and roof deck fastening welds					
25.	Inspect metal deck shear studs, quantity, and welds					

Authorized Subcontractor Signature _____

Exhibit 3-5

Quality control
checklist.

59

- Numerous organizations have established tests for making sure that every one of the nine testing areas are installed according to approved standards.

- Outside testing laboratories usually perform the required tests and then submit the results to the PM and the owner.

- QC is required so that the standards established by the design documents are adhered to.

- In order for QC to be effective, the CM/GC and the subcontractor must work as team so that work as a quality product can be produced.

- A QC checklist has to be prepared prior to the installation of any project component. The CM/GC's team must review and check off each item of the QC checklist during and after installation of the construction components.

- A good QC program will keep defects to a minimum.

4 Building Codes and Permits

(Why does the city want to know what we are doing? Why do they want us to follow regulations?)

BUILDING CODES

The first building code was part of the Code of Hammurabi (approximately 2000 BC), an ancient document that detailed the law of Babylon. In one section of the code it states, "If a builder builds a house for someone and the house collapses and kills the owner, then the builder shall be put to death." Today's codes are very similar in nature to the Code of Hammurabi, except the consequences for not meeting the code are not as severe. The codes that are used today were primarily established to protect the public from injury and potential failure of a structure. Codes are in a constant state of flux due to changes in technology but primarily to further reduce public risk usually as an aftereffect of a tragic event. Thus, after the great Chicago fire of 1871 and the San Francisco earthquake of 1906, the powers that be at the time saw the need for the implementation and improvement in the quality of codes in regard to fire protection and to construct buildings that would not fail due to earthquake loads. The September 11, 2001 tragedy in New York City started a debate on how best to make large buildings safer for the public. New York City created numerous committees staffed by professionals in the building industry, to study ways to enhance and strengthen the code. The committees evaluated the impact of the 9/11 tragedy on building construction and developed new methods for making buildings safer.

At one time, each local municipality had its own code usually based on some national building code standard. However, as of January 2008, 48 states and the District of Columbia have adopted the International Building Code (IBC) as their standard. Some municipalities (such as New York City) combined the IBC with sections of their own building code (New York City Building Code).

International Building Code (IBC)

In the 1970s, four different types of codes were being used in the United States. These included:

1. National Building Code (NBC)
2. United Building Code (UBC)
3. Southern Building Code (SBC)
4. Building Officials and Code Administrators (BOCA)

In the 1980s, the NBC and the BOCA merged to form the BOCA National Building Code (BNBC). In the 1990s, the three remaining codes formed a new organization to develop a single national code. The final code was the International Building Code (IBC).

The IBC included the requirements from critical organizational codes. These include:

1. National Fire Protection Association (NFPA)
2. National Electric Code (NEC)
3. International Plumbing Code
4. International Mechanical Code
5. American Society of Testing Materials (ASTM)
6. American National Standards Institute (ANSI)
7. American Society of Civil Engineers (ASCE)
8. American for Disabilities Act (ADA)—Federal Law

The standards developed by these organizational groups make up the bulk of the IBC. As changes occur in the construction industry and as new technological advances occur, the IBC will be updated every three years to accommodate these changes. Due to the lengthy legislative approval process for updating and modifying the local codes, certain municipalities issue executive orders to take care of these immediate changes.

The IBC consists of the following sections:

1. Building occupancy classification
2. Building heights and areas
3. Interior finishes
4. Foundation, wall, and roof construction
5. Fire protection systems
6. Material used in construction
7. Elevators and escalators
8. Already existing structures
9. Means of egress

Information on the IBC can be found on the International Code Council website (www.codecomply.com).

MEANS AND METHODS FOR THE CONSTRUCTION MANAGER/GENERAL CONTRACTOR (CM/GC)

The project manager (PM) must become very familiar with the local code. In some codes the "Means and Methods" (CM/GC's responsibility for constructing a project that is not part of the construction documents, i.e., erecting steel) may be detailed in very definitive terms. This is true when working in urban settings where public exposure is pronounced. Some items that may be covered in the code but are the responsibility of the CM/GC (versus the design team) are indicated in Exhibit 4-1.

PERMITS

Permits are the documents issued by the municipality giving the CM/GC the right to start construction. The requirement for issuing permits was based on the need to make sure the architects and engineers designing a project were in fact conforming to the local building code. The local officials also want some form of quality control to be established for the project. Of primary concern is that the safety of the public and the trades people are being maintained. See Exhibit 4-2 for the permit flowchart.

Permit Process

The requirement for a permit starts with the architect and engineers who will prepare the necessary construction documents for the project. This design team must have a very good understanding of the technical aspects of the code because when the architect and engineers

---

1. Location of protective public walkways and the requirement for the roof to sustain a 300 pound per square foot falling load
2. Fire standpipe riser to be placed contiguous to the hoist (for fire fighters' access in case of a fire)
3. Safety crane requirements
4. The location of safety nets
5. Scaffolding requirements
6. Maintenance of the site
7. Protection of adjoining property
8. Excavation operations
9. Erection operations
10. Demolition operation
11. Repair and alterations operations
12. Material handling and hoisting equipment
13. Explosive-powered and projectile tools
14. Explosives and blasting
15. Flammable and compressed gases, and other hazardous materials

Exhibit 4-1

Construction manager/general contractor's code compliance responsibilities.

Exhibit 4-2

Permit process flowchart.

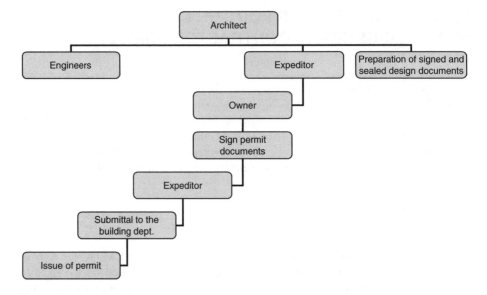

develop the construction documents they have to make sure that every code requirement has been addressed. The design team may only require the submittal of what is called a "building department set." These documents may not be 100% complete but contain sufficient information for the Building Department to analyze. In some cases, outside consultants, called expeditors are called in to review the construction documents for code compliance. The expeditors may also be used to obtain the permits from the municipality. The municipality usually wants to review, as a minimum, the following drawings:

1. Architectural
2. Structural
3. Heating, ventilation, and air conditioning (HVAC)
4. Ventilation index
5. Egress calculations
6. Equipment schedules
7. Sprinkler and hydraulic calculations
8. Plumbing
9. Electrical
10. Life safety
11. Furniture plan

Each of these documents has to be signed and sealed by a registered architect and an engineer (both registered in the state in which they are working). Each state has registration laws for construction design professionals. These state laws require the design team professionals to be practicing in their related field for a certain number of years and they must pass an exam. Reciprocity for registered professionals in one state to sign documents in another state is possible. The registration laws of the state will have to be reviewed to determine if reciprocity is allowed.

Depending upon the municipality in which you will be working will determine how the documents are submitted for review. In certain circumstances, the design team acts in the owner's behalf and submits the documents, but in other cases, the CM/GCs and sub-contractors submit the documents. This procedure is dependent upon the jurisdiction in which the PM is working.

The cost for filing documents is in the range of 1 to 2% of the proposed cost of construction. Estimates for the costs of the project as they relate to specific drawings to be reviewed will have to be developed, usually by the CM/GC. The municipal review process can take as little as two weeks and as long as six months. In all cases the local municipality has full control of the drawing review process. In other instances, the state may have to get involved with certain aspects of the construction process. This may include:

1. Cranes
2. Elevators (and other vertical transportation devices)
3. Change in grade
4. Energy conservation

The number of permits required will be determined by the local municipality, state, or federal agency with which you are working. Some of the permits that may be required are indicated in Exhibit 4-3.

1. Demolition
2. Base construction documents
3. Fire protection
4. Plumbing
5. Electrical
6. Life safety
7. Cranes and derricks
8. Hoists
9. Elevators (and other vertical transportation devices)
10. Street closings
11. Logistics plans
12. Life safety site plans
13. Overtime work
14. Sheds, bridges, and scaffolding
15. Local transportation authority
16. EPA
17. Asbestos or other hazardous materials
18. Brown fields
19. Federal Aviation Authority (FAA)
20. Army Corps of Engineers (for navigable waters)

Exhibit 4-3
Required permits.

Exhibit 4-4

PERMT/TCO/CO
tracking log.

	Project	Permit	TCO/CO	Contractor/ Subcontractor	Architect/ Engineer	Sign Offs	Permit Status	Missing Information
1								
2								
3								
4								
5								
6								
7								
8								
9								
10								
11								
12								
13								
14								
15								
16								
17								
18								
19								
20								

The Building Department will review all the documents for compliance with the local codes. Any rejections will have to be reviewed between the building department examiner and the design team professionals (or expeditor) whose drawings are being reviewed. Sometimes a letter is submitted enumerating the reasons for the rejection along with copies of marked-up drawings. In these particular cases the design team professional must revise the drawings and then submit them back to the Building Department. Once approval has been received and the remaining fee is paid, the permits are issued for the project and sent to the expeditor or CM/GC (depending upon the requirements of the jurisdiction). The CM/GC is to have on file with the Building Department adequate insurance, workers' compensation, and possibly bonding certificates. The local municipality should be consulted for the minimum amounts required for insurance, workers' compensation, and bonding.

In an urban environment, numerous permits are required, especially where public safety is of utmost importance. In order to make sure the PM has obtained all the required permits, a tracking log (as noted in Exhibit 4-4) should be kept.

After the permits are obtained, the Building Department usually wants copies of the permits to be displayed in a public area (see Exhibit 4-5).

INSPECTIONS

Municipalities have inspectors (usually trained in some form of construction discipline) who work for the local building department. The inspectors visit the various projects to make sure the CM/GC is contracting the project according to the documents approved by the Building Department. In certain cases inspectors are asked to come to the site by the CM/GCs for the following phases:

Exhibit 4-5
Permits displayed on construction site-5-7-08.JPG.

1. Foundation bearing conditions
2. Foundations
3. Plumbing roughing
4. Electrical roughing
5. Dry wall installation
6. Erection of structure
7. Asbestos removal
8. Final plumbing
9. Final electrical
10. Life safety
11. Boiler
12. Construction safety

When the inspectors review the project, they have the authority to stop a project and issue violations. Obviously if these conditions occur then the project schedule could be in jeopardy. Again, it behooves the PM to be familiar with all aspects of the project. The PM should walk with the inspectors when they come to the job site so that a good understanding of noncompliance issues can be reviewed. In addition, the approved Building Department drawings should be readily available at the job site along with the approved permits. It may be that a misunderstanding has occurred and a valid explanation is all that is required. This can put a stop to many frivolous problems. Major problems uncovered by the building inspector will have to be addressed expeditiously. Every effort has to be made to convince inspectors that any problems will be remedied that day. Then ask the inspector to come back the next day. A safety violation is another story and the inspectors may shut the project down immediately. This should never happen unless some extreme event has occurred that could not have been stopped with due diligence. See Chapter 5 for more details on safety concerns for a project.

TESTING

Some municipalities require that some form of controlled inspections take place for a variety of project components. This means that certain areas of the project have to be properly tested. Review Chapter 3 for details on testing procedures. The Building Department may require the following tests:

1. Borings
2. Test pits
3. Pilings
4. Underpinning
5. Soil bearing pressure
6. Welding
7. High-strength bolts
8. Smoke tests
9. Spray-on fire proofing
10. Concrete design mix
11. Concrete test cylinders
12. Sprinkler test (hydrostatic)
13. Emergency Lighting
14. Emergency Power

In addition, the local fire department may require that the following be tested or checked:

1. Smoke purge
2. Fire alarm system
3. Fuel storage
4. Elevator recall
5. Fire department key for elevator operation
6. Stair pressurization
7. Compartmentation
8. Storage of hazardous material and gases
9. Fire Pump

CERTIFICATE OF OCCUPANCY (C OF O) AND TEMPORARY CERTIFICATE OF OCCUPANCY (TCO)

Prior to tenant occupancy of any newly created project, the owner will require the municipality to issue either a certificate of occupancy (C of O) or a temporary certificate of occupancy (TCO). A TCO would be issued when a majority of the project has been completed and the project complies with all life safety requirements. By occupying the building, the building population would not be in any jeopardy if evacuation was required due to a fire. Depending upon the jurisdiction, the requirements may be different for obtaining the C of O or TCO. In most cases, the minimum requirements would be as follows:

1. Life safety systems have been tested and approved by the Building Department and/or Fire Department.
2. Make sure the stair locations and widths, door locations and corridor widths, and fireproofing meet the intent of the code.
3. Fire stairs have direct access to the outside or to be fire-rated assembly areas.
4. Inspectors have submitted their reports and no outstanding issues exist.
5. Building has no violations.
6. All controlled inspection reports have been signed off by the approved registered architects and engineers.
7. Project is complete and the space is ready for occupancy.
8. May require that all furniture be installed.
9. Subcontractors have submitted their documents indicating that certain Building Department requirements have been satisfied (usually electrical, plumbing, and sprinkler items).
10. Elevators are working properly and all safety devices are operational and they conform to the Fire Department requirements (i.e., elevators drop to the lobby upon a fire signal).
11. Operation of Fire Command Station.
12. Fire safety plan filed and approved.

A typical TCO checklist is shown in Exhibit 4-6.

LANDMARKS

A landmarks designation is an indication that the project or district in which you are working has significant architectural, cultural, or historical status. The city in which these designations occur wants to make sure that any proposed modifications to a project or the construction of a new project fall within the guidelines established to maintain

Exhibit 4-6

TCO requirement.

Contractor	Item No.	Activity	Target Date	Actual Completion
Electrical	1	Permanent Power		
		a. Utility will have vault completed by		
		b. Electrical contractor will require 5 days to sequence the installation of permanent power to the building from the time the power company brings power to the building.		
		c. Removing all temporary power		
	2	Wiring of smoke fans		
	3	Testing rotation of all fans and pumps with permanent power		
	4	City electrical inspections ongoing		
	5	All mechanical dampers completely wired to fire alarm system		
	6	Fire alarm		
		a. Complete fire alarm system installation including in-house test		
		b. Fire alarm shall include the following:		
		1. Smoke detectors		
		2. Heat detectors		
		3. Pull stations		
		4. Apartment smoke detectors		
		5. Speaker/strobes		
		6. Dust detectors		
		7. Fire alarm panel and remote enunciator		
		8. Interface with:		
		-elevators		
		-generator		
		-fire pump		
		-smoke control		
		9. Recall alternate		
		10. Connection to sprinkler flow and tamper switches		
		11. Firefighters' communication		
		12. One-way communication		
		13. In-house test		
	7	Fire alarm testing by city to begin		
	8	Fire alarm city testing completed by		
	9	Emergency Power including in-house generator test with loads and load bank to be completed by		
		a. Emergency power to include the following:		
		1. Life safety systems		
		2. Smoke control fans		
		3. Stair pressurization		
		4. Smoke control panel		
		5. Wire and supervision		
		6. Fire dampers		
		7. Elevator shaft pressurization		
		8. Garage smoke purge		
		9. Emergency lighting		
		10. Stair and exit lighting		
		11. One elevator per bank		
		12. Exit signs		
		13. Emergency lighting		
	10	City emergency test (TCO)		

the architectural integrity of the district. The federal government also has guidelines for the preservation of historical buildings.

If you are working in a landmarks building or in a landmarks district, then additional permit requirements need to be satisfied. In most cases, the professional architects and engineers have to submit drawings to the Landmarks Commission and Arts Commission and possibly to the local community board for approvals. This process can take from months to, in some cases, years to obtain the required approvals. In most cases, construction permits cannot be obtained until the Landmarks Commission or Federal Government has approved the scope of work proposed.

Contractor	Item No.	Activity	Target Date	Actual Completion
HVAC	1	Smoke fans less duct work installed		
	2	Smoke fans including all duct work and controls including cleaning of duct work complete		
	3	Stair pressurization including duct work and grills installation complete		
	4	Common area smoke fans including duct work and cleaning of duct work complete		
	5	All dampers installation complete		
	6	Elevator shaft pressurization installation complete		
	7	Duct work for elevator shaft pressurization complete		
	8	City inspection of mechanical system complete		
	9	In-house test of smoke evacuation system under normal permanent power including the following: a. Common areas b. Stair pressurization c. Elevator shaft pressurization d. Dampers e. Interface with fire alarm		
	10	Checking rotation and wiring of all fan systems with electrical contractor complete		
	11	Engineer of record smoke bomb placement drawing		
	12	All grills and diffusers installation complete		
	13	City smoke test		
Plumbing	1	Incoming water service		
	2	Back flow preventer		
	3	Tags on all valves		
	4	All condo floors have been inspected by city		
	5	8" and 2" backflow preventers were installed		
Fire protection	1	Installation of taper and flow switches		
	2	Tower risers including manifolds		
	3	Loft risers including manifolds		
	4	Fire pumps		
	5	Jockey pumps		
	6	City inspection of fire pumps		
	7	Sprinkler installation		
	8	Backflow preventer		
	9	Ground floor sprinkler		

Exhibit 4-6
(Continued)

The responsibility of the CM/GC usually involves the submittal of material that will be used for the landmarks project. The Landmarks Commission wants to make sure that the specified material (as agreed to by the Landmarks Commission) and the actual material are consistent with the intent of the architecture approved. The CM/GC should be aware of potential time delays in obtaining the necessary approvals for materials and their finishes.

SUMMARY

- Codes were developed to protect the public and occupants of buildings.
- Major code modifications take place after catastrophic events have occurred. This happened after the great Chicago fire of 1871, the San Francisco earthquake of 1906, and the September 11, 2001 attacks in New York City.
- The IBC has been accepted by 48 states and the District of Columbia. This code now incorporates the latest additions of the NFPA and the NEC.
- Permits are required to ensure that the professionals designing the project are conforming to the jurisdiction's building code.
- The CM/GCs have to ensure that they are conforming to certain provisions of the code.
- Testing of certain elements of a project is also required to ensure that the tests conform to industry standards.
- Prior to occupying a building, the owner must obtain a C of O or a TCO.
- If the CM/GC is working on a landmarks project, then they must be cognizant of potential delays for obtaining material and finishing approvals.

5 Safety
(Why we cannot police ourselves.)

HISTORY OF SAFETY ISSUES

The construction industry has the greatest number of deaths and accidents as compared to other occupations. The U.S. Department of Labor statistics (Fatal Occupational Injuries by Occupation and Events or Exposure) noted that 1239 construction workers died in 2006 (see Exhibit 5-1). This represents 21% of the total work force fatalities. To put it another way, 3.4 construction workers die every day of the year. The number of fatalities for 2006 was 4% higher then what occurred in 2005. Construction laborers constituted 27% of the total construction fatalities (see Exhibit 5-1A). Updated information on fatalities in the construction industry can be obtained via the U.S. government Website (www.osha.gov). See Exhibit 5-1B for the safety flowchart.

OCCUPATIONAL SAFETY HEALTH ADMINISTRATION (OSHA)

Due to the number of accidents that were occurring in the construction industry and lack of safety concerns by the construction industry (and other occupations), Congress passed the Occupational Safety Health Administration (OSHA) Law on December 29, 1970. In 1990, OSHA set up a separate construction and engineering division.

OSHA Regulations Section 1926 (Safety and Health Regulations for Construction) deals with safety issues relating to the construction industry. Section 1926 index is as noted in Exhibit 5-2.

OSHA also publishes a pocket guide called Workers Safety Series. Examples of some of these guides are noted in Exhibit 5-3 through Exhibit 5-25. Given that Spanish is the second most spoken language in the United States, and there are a large number of construction workers of Spanish heritage working in the United States, OSHA has made available safety signs in both English and Spanish to facilitate communications and safety. These guides should be handed out to all superintendents and subcontractor foremen at the first toolbox meeting. The toolbox meeting is held by each subcontractor formen and their workers to discuss safety and the performance of the construction work.

Exhibit 5-1

Department of
Labor fatality rate.

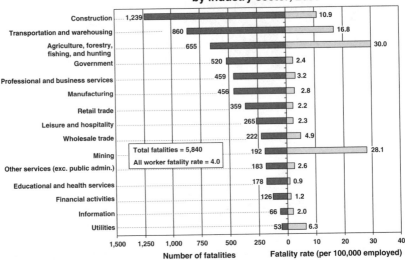

Number and rate of fatal occupational injuries,
by industry sector, 2006

Although the construction sector had the highest number of fatal injuries in 2006, the industry sectors with the highest fatality rates were agriculture, forestry, fishing, and hunting and mining.

SOURCE: U.S. Bureau of Labor Statistics, U.S. Department of Labor, 2008

Of extreme interest is the OSHA Quick Card, "Top Four Construction Hazards" (Exhibits 5-3 through 5-14 in both Spanish and English). This card enumerates the four greatest hazards that are found on a construction site:

1. Falls
2. Hits by equipment and vehicles
3. Cave-ins
4. Electrocution

Exhibit 5-1A

Department of
Labor fatality rate
by selected
occupations.

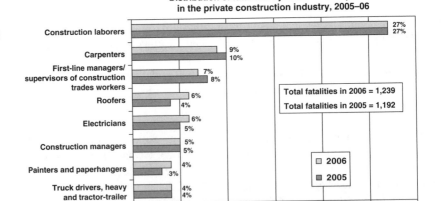

Distribution of fatalities by selected occupations
in the private construction industry, 2005–06

Fatal work injuries involving construction laborers accounted for more than one out of every four private construction fatalities in both 2005 and 2006.

SOURCE: U.S. Bureau of Labor Statistics, U.S. Department of Labor, 2008

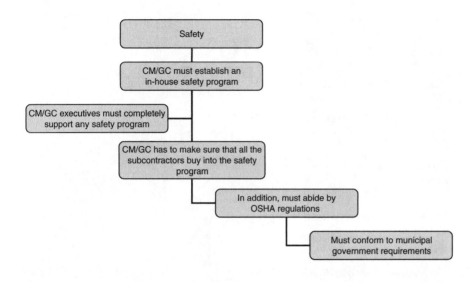

Exhibit 5-1B
Safety flowchart.

Subpart A General

Subpart B General Interpretation

Subpart C General Safety and Health Provisions

Subpart D Occupational Health and Environmental Controls

Subpart E Personal Protection and Life saving Equipment

Subpart F Fire Protection and Prevention

Subpart G Sign, Signal, and Barricades

Subpart H Materials Handling, Storage, Use, and Disposal

Subpart I Tools-Hand and Power

Subpart J Welding and Cutting

Subpart K Electrical

Subpart L Scaffolds

Subpart M Fall Protection

Subpart N Cranes, Derricks, Hoists, Elevators, and Conveyors

Subpart O Motor Vehicles, Mechanical Equipment, and Marine Operations

Subpart P Excavation

Subpart Q Concrete and Masonry Construction

Subpart R Steel Erection

Subpart S Tunnels, Shafts, Caissons, Cofferdams, and Compressed Air

Subpart T Demolition

Subpart U Blasting and use of Explosives

Subpart V Power Transmission and Distribution

Subpart W Rollover Protective Structures; Overhead Protection

Subpart X Stairway and Ladders

Subpart Y Commercial Diving Operations

Subpart Z Toxic and Hazardous Substance

Exhibit 5-2
Section 1926 of
OSHA regulations.

Exhibit 5-3

OSHA-Spanish top four construction hazards quick card.

OSHA DATOS RÁPIDOS

Los Cuatros Riegos Principales de la Construcción

Las cuatro causas más comunes de fatalidades en la construcción son: caídas, golpes, quedar atrapado/ pillado y electrocución.

Prevenga caídas

- Póngase y use el equipo personal de detención de caídas.
- Instale y mantenga una protección de perímetro.
- Póngale cubiertas a las aberturas en los pisos. Asegure y rotule todas las cubiertas.
- Use las escaleras y andamios de manera segura.

Prevenga golpes

- Nunca se coloque entre un objeto fijo y uno en movimiento.
- Use ropa de alta visibilidad cerca de equipos y vehículos.

Prevenga quedar atrapado o pillado

- Nunca entre a una zanja o excavación de 5 pies o más sin un sistema de protección adecuado en el lugar; algunas zanjas con menos de 5 pies de profundidad podrían también necesitar tal sistema.
- Asegúrese que la zanja o excavación esté protegida por un sistema de pendientes, apuntalamientos, bancos o cajas de zanja.

Prevenga electrocuciones

- Localice e identifique los servicios públicos antes de comenzar el trabajo.
- Esté atento a líneas eléctricas aéreas cuando esté operando cualquier equipo.
- Mantenga una distancia segura lejos de las líneas eléctricas; aprenda los requerimientos de distancia seguros.
- No opere herramientas eléctricas portátiles a menos que estén conectadas a tierra o sean de tipo doble aislamiento.
- Para protección, use un interruptor de circuito con pérdida a tierra.
- Esté alerta a riesgos eléctricos cuando trabaje con escaleras, andamios u otras plataformas.

Para información más completa:

OSHA Administración de Seguridad y Salud Ocupacional

Departamento del Trabajo de EE.UU.
www.osha.gov (800) 321-OSHA

Exhibit 5-4
OSHA-Spanish fall protection quick card.

OSHA DATOS RÁPIDOS

Consejos para Protección de Caídas

- Antes de comenzar a trabajar, identifique los riesgos potenciales contra tropiezos y caídas.

- Esté atento a los riesgos de caídas tales como inseguridades en pisos, cubiertas/orillas no protegidas, huecos, tragaluces, escaleras y aperturas en techos/orillas.

- Antes de usar el equipo de protección contra caídas inspecciónelo e identifique los defectos.

- Seleccione, póngase y use apropiadamente el equipo de protección contra caídas para la tárea asignada.

- Asegure y estabilize todas las escaleras antes de subirlas.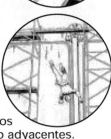

- Nunca se pare en el peldaño o escalón superior de una escalera.

- Use las barandas cuando usted suba o baje una escalera.

- Practique buena organización/limpieza. Mantenga los cables, cables de soldadura y mangas de aire fuera de los pasillos o áreas de trabajo adyacentes.

Para información más completa:

OSHA Administración de Seguridad y Salud Ocupacional
Departamento del Trabajo de EE.UU.
www.osha.gov (800) 321-OSHA

Exhibit 5-5

Spanish aerial lifts safety tips quick card.

OSHA DATOS RÁPIDOS

Protéjase
Equipo de Protección Personal en la Construcción

Protección Para los Ojos y la Cara
- Las gafas de seguridad o caretas se usan siempre que las operaciones en el trabajo puedan causar que objetos extraños entren a los ojos. Por ejemplo, cuando se esté soldando, cortando, puliendo, clavando (o cuando se esté trabajando con concreto y/o químicos peligrosos o expuesto a partículas que vuelan). Utilícelos cuando esté expuesto a cualquier riesgo eléctrico, incluyendo el trabajar en sistemas eléctricos energizados (vivos).
- Protectores para ojos y cara – se seleccionan en base a los riesgos anticipados.

Protección para los Pies
- Los trabajadores de la construcción deben utilizar zapatos o botas de trabajo con suelas resistentes a resbalones y perforaciones.
- El calzado con punta de metal es usado para prevenir que los dedos de los pies queden aplastados cuando se trabaja alrededor de equipo pesado u objetos que caen.

Protección para las Manos
- Los guantes deben ajustar cómodamente.
- Los trabajadores deben usar los guantes correctos para el trabajo que van a hacer (ejemplos: guantes de goma de alta resistencia para trabajos con concreto, guantes de soldar para soldaduras, guantes y mangas con aislamiento cuando se esté expuesto a riesgos eléctricos).

Protección para la Cabeza
- Use cascos de seguridad donde haya potencial de que objetos caigan desde arriba, de golpes en la cabeza por objetos fijos o contacto accidental de la cabeza con riesgos eléctricos.
- Cascos de seguridad – inspecciónelos rutinariamente para detectar abolladuras, grietas o deterioro. Reemplácelos después de que hayan recibido un golpe fuerte o descarga eléctrica. Manténgalos en buenas condiciones.

Protección para los Oídos
- Use tapones para oídos/orejeras en áreas de trabajo de alto ruido donde se usen sierras de cadena o equipo pesado. Limpie o reemplace los tapones para oídos regularmente.

Para información más completa:

OSHA Administración de Seguridad y Salud Ocupacional

Departamento del Trabajo de EE.UU.
www.osha.gov (800) 321-OSHA

Exhibit 5-6
OSHA-Spanish crane safety quick card.

OSHA DATOS RÁPIDOS

Jirafas
Consejos de Seguridad

Las jirafas incluyen a las plataformas elevadas de aguilón sostenido, como lo son las de puntal extensible con canasta (mejor conocidas como "cherry pickers") o los camiones canasta. Las principales causas de muertes son por caídas, electrocuciones y colapsos o volteos.

Prácticas de Trabajo Seguras

- Asegúrese que los trabajadores que operan jirafas están adecuadamente adiestrados en el uso seguro del equipo.
- Mantenga y opere las plataformas de trabajo elevadas de acuerdo con las instrucciones del manufacturero.
- Nunca invalide los dispositivos de seguridad hidráulicos, mecánicos o eléctricos.
- Nunca mueva el equipo con trabajadores en una plataforma elevada, a menos que sea permitido por el manufacturero.
- No permita a los trabajadores ponerse entre riesgos que estén por encima de la cabeza, como viguetas y vigas, y las barandas del canasto. El movimiento de la jirafa puede aplastar al trabajador.
- Mantenga una distancia mínima segura de las líneas eléctricas aéreas más cercanas de al menos 10 pies, ó 3 metros.
- Siempre trate a las líneas de energía eléctrica, alambres y otros conductores como si estuvieran energizados (vivos), aún si están fuera de servicio o parece que están aislados.
- Use un arnés de cuerpo o correa que restringe el movimiento con una cuerda de seguridad atada al aguilón o canasto para prevenir que el trabajador salga disparado o sea tirado del canasto.
- Ponga los frenos y use calzos cuando esté en un área inclinada.
- Use estabilizadores, si son provistos.
- No exceda la carga límite del equipo. Tome en cuenta el peso combinado del trabajador, herramientas y materiales.

Para información más completa:

OSHA
Administración de Seguridad y Salud Ocupacional
Departamento del Trabajo de EE.UU.
www.osha.gov (800) 321-OSHA

Exhibit 5-7

OSHA-Spanish
PPE quick card.

OSHA DATOS RÁPIDOS

Protéjase
Seguridad en Grúas

Muertes y lesiones serias pueden ocurrir si las grúas no son inspeccionadas y utilizadas correctamente. Muchas muertes pueden ocurrir cuando el aguilón de la grúa, el cable de carga o la carga contacta las líneas de energía eléctrica y descarga la electricidad hacia la tierra. Otros incidentes ocurren cuando los trabajadores son golpeados por la carga, son atrapados dentro del radio de movimiento de la grúa o fallan en montar/desmontar la grúa correctamente.

- Las grúas *solamente* pueden ser operadas por personal calificado y capacitado.
- Una persona competente designada tiene que inspeccionar todos los controles de la grúa para asegurarse que trabajan apropiada antes de su uso.
- Asegúrese que la grúa esté sobre una superficie firme/estable y nivelada.
- Mientras monte/desmonte la grúa no abra o remueva pernos a menos que las secciones estén bloqueadas y seguras (estables).
- Extienda completamente los estabilizadores y coloque una barricada en las áreas accesibles dentro del radio de movimiento de la grúa.
- Esté pendiente de las líneas eléctricas aéreas y mantenga al menos una distancia segura de 10 pies de las líneas.
- Inspeccione todos los aparejos antes de usarlos. No enrolle las líneas para levantar la carga alrededor de la carga.
- Asegúrese de usar la gráfica de carga correcta para la configuración actual de la grúa y su instalación, la carga total y la ruta de levantamiento.
- No exceda la capacidad según la gráfica de cargas mientras esté haciendo levantamientos.
- Eleve la carga unas pulgadas, aguántela, verifique la capacidad/balance y pruebe el sistema de frenos antes de entregar la carga.
- No mueva cargas sobre trabajadores.
- Asegúrese de obedecer las señales e instrucciones del manufacturero mientras esté operando grúas.

Para información más completa:

OSHA Administración de
Seguridad y Salud
Ocupacional
Departamento del Trabajo de EE.UU.
www.osha.gov (800) 321-OSHA

Exhibit 5-8
OSHA-Spanish electrical safety quick card.

OSHA DATOS RÁPIDOS

Seguridad Eléctrica

Los riesgos eléctricos pueden causar quemaduras, choques eléctricos y electrocución (muerte).

Consejos de Seguridad

- Presuma que todos los cables aéreos están energizados (vivos) a voltajes fatales. Nunca presuma que se puede tocar un cable de manera segura aún si está fuera de servicio o parece que está aislado.
- Nunca toque una línea de energía eléctrica que se haya caído. Llame a la compañía de servicio eléctrico para reportar líneas eléctricas caídas.
- Manténgase al menos 10 pies (3 metros) alejado de cables aéreos durante limpiezas y otras actividades. Si está trabajando a alturas o manejando objetos largos, antes de comenzar a trabajar evalúe el área para detectar la presencia de cables aéreos.
- Si un cable aéreo cae sobre su vehículo cuando esté guiando, manténgase dentro del vehículo y continúe guiando, alejándose del cable. Si el motor de su vehículo se detiene, no salga del vehículo. Adviértale a las personas que no toquen el vehículo o el cable. Llame, o pídale a alguien que llame, a la compañía local de servicio eléctrico y a servicios de emergencia.
- Nunca opere equipos eléctricos mientras esté parado sobre agua.
- Nunca repare cables o equipo eléctrico a menos que esté cualificado y autorizado.
- Antes de energizar el equipo eléctrico que se ha mojado, haga que un electricista cualificado lo inspeccione.
- Si está trabajando en áreas húmedas, inspeccione los cables y equipo eléctrico para asegurarse que están en buenas condiciones y sin defectos, y use un interruptor de circuito con pérdida a tierra (GFCI, por sus siglas en inglés).
- Siempre tenga cuidado cuando esté trabajando cerca de electricidad.

Para información más completa:

OSHA Administración de Seguridad y Salud Ocupacional
Departamento del Trabajo de EE.UU.
www.osha.gov (800) 321-OSHA

Exhibit 5-9

OSHA-Spanish lead in construction quick card.

OSHA DATOS RÁPIDOS

Protéjase
Plomo en la Construcción

El plomo es un elemento peligroso encontrado en muchos lugares de construcción. La exposición a plomo proviene de inhalar humos y polvo, y el plomo puede ser ingerido cuando las manos están contaminadas con polvo de plomo. Los trabajadores pueden llevarse el plomo a sus casas en las ropas, piel, cabello, herramientas y vehículos.

La exposición a plomo puede ocurrir en actividades de demolición, rescate, remoción, encapsulación renovación y limpieza.

Evite la Exposición
* Use equipo de protección personal adecuado (por ejemplo, guantes, ropa y respiradores aprobados).
* Lave las manos y cara después del trabajo y antes de comer.
* Nunca entre a áreas de comer con el equipo de protección puesto.
* Nunca use fuera del trabajo ropas y zapatos que fueron usados durante la exposición al plomo.
* Lave la ropa diariamente, use métodos de limpieza adecuados.
* Esté alerta a síntomas de exposición a plomo (por ejemplo, dolor abdominal severo, dolores de cabeza, pérdida de coordinación motora).

Use Respiradores
* Use respiradores apropiados, según le sea indicado.
* Haga que el usuario conduzca un cotejo-de-sellado cada vez que se ponga el respirador.
* Esté atento al programa de protección respiratoria de su compañía; conozca las limitaciones y riesgos potenciales de los respiradores.

Prevenga Futuras Exposiciones
* Procure una ventilación adecuada
  * En exteriores, manténgase en dirección contraria a la que vaya cualquier contaminante.
* Cuando sea posible, use equipo para recoger polvo.
* Use materiales y químicos sin plomo.
* Use métodos húmedos para disminuir el polvo.
* Use ventilación con extracción local para trabajos en áreas cerradas.

Para información más completa:

OSHA Administración de Seguridad y Salud Ocupacional
Departamento del Trabajo de EE.UU.
www.osha.gov (800) 321-OSHA

Exhibit 5-10
OSHA-Spanish demolition safety tips quick card.

Consejos de Seguridad para Demolición

Trabajos de demolición incluyen muchos de los mismos peligros que se presentan durante otras actividades de construcción. Sin embargo, demolición también incluye peligros adicionales debido a la variedad de otros factores. Algunos de estos incluyen: pinturas con contenido de plomo, objetos puntiagudos y materiales que contienen asbestos.

- Reforzarle o apuntale las paredes y piso de la estructuras las cuales han sido dañadas y en las cuales los empleados debén entrar.
- Inspeccione el equipo de protección personal (niciales en ingles - PPE) antes de usarlo.
- Seleccione, póngase y utilice un equipo de protección personal apropiado para la taréa.
- Inspeccione los escalones, pasillos y escaleras; ilumine todas las escaleras.
- Apague y tape todos los servicios de luz, gas, agua, vapor, alcantarillados y otros servicios de linea; notifique a la compañia de servicios apropiada.
- Proteja todas las aperturas de paredes hasta una altura de 42 pulgadas; cubra y asegure las aperturas de pisos con materiales que sean capaces de soportar la carga que será posiblemente aplicada.
- La aperturas en el piso usadas para la disposición de materiales no debén ser más de 25% del área total del piso.
- Use vertederos de escombro encerrado con portones en el final de la descarga. para dejar caer material de la demolición al suelo o en un contenedor de la basura.
- Demolición de paredes exteriores y piso debén comenzar en la parte de arriba de la estructura y proceder hacia abajo.
- Los miembros estructurales y los que soportan la cargas en cualquier piso no debén ser cortados o removidos hasta que todos los pisos encima de ese piso hayan sido removidos.
- Todas las cornizas de techos o otras piedras ornamentales debén ser removidas antes de tirar las paredes hacia abajo.
- Empleados no se deben permitir trabajar donde existen peligros de colapso de estructura hasta que sea correjido con puntales, refuerzos o otros medios efectivos.

Para información más completa:

OSHA Administración de Seguridad y Salud Ocupacional
Departamento del Trabajo de EE.UU.
www.osha.gov (800) 321-OSHA

Exhibit 5-11

OSHA-Spanish work zone safety tips quick card.

OSHA DATOS RÁPIDOS

Seguridad del Tráfico en las Zonas de Trabajo

El que los trabajadores sean golpeados por vehículos o equipo móvil conduce a muchas lesiones fatales en las zonas de trabajo. Las zonas de trabajo necesitan controles de tráfico identificados por letreros, conos, barriles y barreras.

Los conductores, trabajadores a pie y peatones tienen que ser capaces de ver y entender las rutas apropiadas. Los gerentes de proyectos de construcción establecen planes de control de tráfico dentro de los lugares de trabajo de construcción/demolición.

- Los dispositivos de control de tráfico, señales y tableros de mensajes instruyen a los conductores a seguir rutas que se alejan del área donde se realiza el trabajo.
- Los dispositivos de control de tráfico aprobados, incluyendo conos, barriles, barricadas y postes de demarcación, también se usan dentro de las zonas de trabajo.

Protección de la Zona de Trabajo: Los atenuadores como concreto, agua, arena, barreras que colapsan, amortiguadores de choques, y atenuadores montados-en-camión pueden ayudar a limitar la entrada no autorizada de conductores a la zona de trabajo de construcción.

Señalar mediante Banderas: Los trabajadores encargados de dar señales con banderas deben usar ropa de alta visibilidad con fondo fluorescente, y hecha de material que refleje. Esto permite que se puedan ver a los trabajadores a una distancia de al menos 1,000 pies en cualquier dirección. Coteje la etiqueta o empaque para asegurarse que la ropa sea de rendimiento clase 2 ó 3. Los conductores deben ser advertidos por medio de letreros de que va a haber trabajadores encargados de dar señales con banderas. Estos trabajadores deben usar letreros de PARAR/REDUCIR VELOCIDAD, letreros con luces o banderas (solamente durante emergencias).

Iluminación: Las estaciones de los trabajadores encargados de dar señales con banderas deben estar iluminadas. La iluminación para los trabajadores a pie y operadores de equipo debe ser de 5 bujías-pie o más. Donde la iluminación disponible no sea suficiente, luces de bengala o iluminación química deben ser usadas. Los reflejos deben ser controlados o eliminados.

Capacitación: Los trabajadores encargados de dar señales con banderas tienen que estar adiestrados/certificados, y usar métodos de señalización autorizados.

Conducir: Los cinturones de seguridad y la protección contra vuelcos deben ser usados en equipos y vehículos según sea recomendado por el manufacturero.

Para información más completa:

OSHA
Administración de
Seguridad y Salud
Ocupacional
Departamento del Trabajo de EE.UU.
www.osha.gov (800) 321-OSHA

Exhibit 5-12

OSHA-Spanish supported scaffold safety tips quick card.

Andamio de Soporte
Consejos de Seguridad

Los andamios de soporte consisten de una o más plataformas sostenidas por vigas voladizas (estabilizadoras), abrazaderas, postes, patas, montantes rectos, marcos o soportes rígidos similares.

Para la prevención/protección contra caídas, se requieren barandas o sistemas personales de prevención de caídas para los empleados sobre plataformas a 10 o más pies de alto.

Las plataformas y entarimados de trabajo tienen que estar entablados cercano a las barandas.

El entablado (Los tablones) ha (han) de estar solapado (s) sobre un soporte al menos 6 pulgadas, pero no más de 12 pulgadas.

Las patas, postes, marcos y montantes rectos tienen que estar sobre placas de soporte y durmientes, o sobre cimientes confiables, y estar aplomados y reforzados.

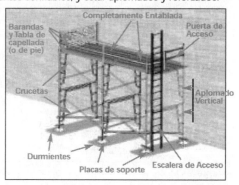

La capacitación para los usuarios de andamios tiene que incluir:
- Los riesgos específicos del tipo de andamio que se está usando,
- La carga máxima prevista y capacidad del andamio,
- El reconocimiento y reporte de defectos,
- Los riesgos de caídas,
- Los riesgos eléctricos, incluyendo las líneas aéreas,
- Riesgos causados por objetos que caen,
- Otros riesgos que se puedan encontrar.

Para información más completa:

 Administración de Seguridad y Salud Ocupacional
Departamento del Trabajo de EE.UU.
www.osha.gov (800) 321-OSHA

Exhibit 5-13A

OSHA-Spanish supported scaffold inspection tips quick card.

OSHA DATOS RÁPIDOS

Andamio de Soporte
Consejos para una Inspección

Inspeccione los andamios y las partes del andamio diaria-mente, antes de comenzar cada turno de trabajo y despúes de cualquier evento que pueda haber causado daño al andamio.

- Verifique a ver si las líneas electricás aéreas que están cerca de los andamios están desenergizadas o que los andamios estén a un mínimo de 10 pies de distancia de las líneas eléctricas aéreas que están energizadas.
- Asegúrese que las herramientas y los materiales estén a un mínimo de 10 pies de separación de las líneas eléctric-as aéreas energizadas.
- Verifique que el andamio es el tipo correcto para las car-gas, materiales, empleados y condiciones atmosféricas.
- Verifique si las zapatas del andamio están niveladas, sóli-das, rígidas y son capaces de soportar el andamio cargado.

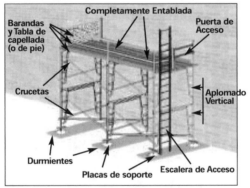

Completamente Entablada
Barandas y Tabla de capellada (o de pie)
Puerta de Acceso
Crucetas
Aplomado Vertical
Durmientes
Placas de soporte
Escalera de Acceso

- Verifique las patas, rectos, marcos y montantes para ver si ellos están en la placa de soporte y los durmiente.
- Verifique si los componentes de metal tienen dobles, agri-etaduras, huecos, moho, salpicaduras de soldaduras, que-braduras, soldaduras rotas y partes no compatibles.
- Verifique a ver si la plataforma está a 14 pulgadas o más de separación de la pared o 18 pulgadas o menos de sep-aración si se está haciendo trabajo de empañetamiento/estuco.
- Verifique que haya un acceso seguro. No use las crucetas como una escalera de acceso o salida.

 **Administración de Seguridad y Salud Ocupacional**
Departamento del Trabajo de EE.UU.
www.osha.gov (800) 321-OSHA

OSHA 3319-09-06

Exhibit 5-13B
OSHA-Spanish supported scaffold inspection tips quick card.

OSHA DATOS RÁPIDOS

- Verifique si los tablones de madera tienen quebraduras, rajaduras mayores de $1/4$ de pulgada de ancho, rajaduras en los extremos que sean largas, varios nudos grandes y flojos, torceduras o deformaciones mayores de $1/4$ de pulgada, tablas y extremos con ranuras, moho, laminado(s) separado(s) o inclinación de grano mayor de 1 en 12 pulgadas desde el borde largo y que sean tablas de calidad apta para los andamios o equivalentes.
- Si los tablónes deflecionan más de 1/60 del ancho del tablón o 2 pulgadas en un tablón de madera de 10 pies, el tablón se ha dañado y no debe ser utilizado.
- Verifique a ver si los tablones están uno al lado del otro, con espacios no mayor de una pulgada alrededor de los montantes rectos.
- Verifique a ver si los tablones de 10 pies o menos están de 6 a 12 pulgadas sobre la línea de soporte del centro y que los tablones de 10 pies o más no estén a más de 18 pulgadas al otro lado del extremo.
- Verifique que hayan barrandas y largueros intermedios en la plataforma donde se está realizando el trabajo.
- Verifique los empleados que trabajan debajo de la plataforma y proveales con protección contra objetos cayendo o barricadas alrededor. Asegúrese que los cascos de seguridad se usan.
- Use refuerzos, amarres y líneas de retención de vientos según descrito por el fabricante del andamio, en cada extremo del andamio, vertical y horizontalmente, para evitar la inclinación de éste.

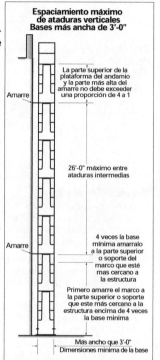

Espaciamiento máximo de ataduras verticales Bases más ancha de 3'-0"

La parte superior de la plataforma del andamio y la parte más alta del amarre no debe exceeder una proporción de 4 a 1

26'-0" máximo entre ataduras intermedias

4 veces la base mínima amarralo a la parte superior o soporte del marco que esté mas cercano a la estructura

Primero amarrar el marco a la parte superior o soporte que este más cercano a la estructura encima de 4 veces la base mínima

Más ancho que 3'-0" Dimensiones mínima de la base

OSHA Administración de Seguridad y Salud Ocupacional
Departamento del Trabajo de EE.UU.
www.osha.gov (800) 321-OSHA

Exhibit 5-14

OSHA-English top four construction hazards quick card.

OSHA QUICK CARD™

Top Four
Construction Hazards

The top four causes of construction fatalities are: Falls, Struck-By, Caught-In/Between and Electrocutions.

Prevent Falls

- Wear and use personal fall arrest equipment.
- Install and maintain perimeter protection.
- Cover and secure floor openings and label floor opening covers.
- Use ladders and scaffolds safely.

Prevent Struck-By

- Never position yourself between moving and fixed objects.
- Wear high-visibility clothes near equipment/vehicles.

Prevent Caught-In/Between

- Never enter an unprotected trench or excavation 5 feet or deeper without an adequate protective system in place; some trenches under 5 feet deep may also need such a system.
- Make sure the trench or excavation is protected either by sloping, shoring, benching or trench shield systems.

Prevent Electrocutions

- Locate and identify utilities before starting work.
- Look for overhead power lines when operating any equipment.
- Maintain a safe distance away from power lines; learn the safe distance requirements.
- Do not operate portable electric tools unless they are grounded or double insulated.
- Use ground-fault circuit interrupters for protection.
- Be alert to electrical hazards when working with ladders, scaffolds or other platforms.

For more complete information:

OSHA Occupational Safety and Health Administration
U.S. Department of Labor
www.osha.gov   (800) 321-OSHA

OSHA 3216-6N-06

Exhibit 5-15
OSHA-English fall protection tips quick card.

OSHA QUICK CARD™

Fall Protection Tips

- Identify all potential tripping and fall hazards before work starts.

- Look for fall hazards such as unprotected floor openings/edges, shafts, skylights, stairwells, and roof openings/edges.

- Inspect fall protection equipment for defects before use.

- Select, wear, and use fall protection equipment appropriate for the task.

- Secure and stabilize all ladders before climbing them.

- Never stand on the top rung/step of a ladder.

- Use handrails when you go up or down stairs.

- Practice good housekeeping. Keep cords, welding leads and air hoses out of walkways or adjacent work areas.

For more complete information:

OSHA Occupational
Safety and Health
Administration
U.S. Department of Labor
www.osha.gov   (800) 321-OSHA

OSHA 3257-11R-05

Exhibit 5-16

OSHA-English
aerial lifts safety
tips quick card.

OSHA QUICK CARD™

Aerial Lifts
Safety Tips

Aerial lifts include boom-supported aerial platforms, such as cherry pickers or bucket trucks. The major causes of fatalities are falls, electrocutions, and collapses or tip overs.

Safe Work Practices

- Ensure that workers who operate aerial lifts are properly trained in the safe use of the equipment.
- Maintain and operate elevating work platforms in accordance with the manufacturer's instructions.
- Never override hydraulic, mechanical, or electrical safety devices.
- Never move the equipment with workers in an elevated platform unless this is permitted by the manufacturer.
- Do not allow workers to position themselves between overhead hazards, such as joists and beams, and the rails of the basket. Movement of the lift could crush the worker(s).
- Maintain a minimum clearance of at least 10 feet, or 3 meters, away from the nearest overhead lines.
- Always treat powerlines, wires and other conductors as energized, even if they are down or appear to be insulated.
- Use a body harness or restraining belt with a lanyard attached to the boom or basket to prevent the worker(s) from being ejected or pulled from the basket.
- Set the brakes, and use wheel chocks when on an incline.
- Use outriggers, if provided.
- Do not exceed the load limits of the equipment. Allow for the combined weight of the worker, tools, and materials.

For more complete information:

OSHA Occupational
Safety and Health
Administration
U.S. Department of Labor
www.osha.gov   (800) 321-OSHA

OSHA 3267-09N-05

Exhibit 5-17
OSHA-English crane safety quick card.

OSHA QUICK CARD™

Protect Yourself
Crane Safety

Fatalities and serious injuries can occur if cranes are not inspected and used properly. Many fatalities can occur when the crane boom, load line or load contacts power lines and shorts electricity to ground. Other incidents happen when workers are struck by the load, are caught inside the swing radius or fail to assemble/disassemble the crane properly.

- Cranes are to be operated *only* by qualified and trained personnel.
- A designated competent person must inspect the crane and all crane controls before use.
- Be sure the crane is on a firm/stable surface and level.
- During assembly/disassembly do not unlock or remove pins unless sections are blocked and secure (stable).
- Fully extend outriggers and barricade accessible areas inside the crane's swing radius.
- Watch for overhead electric power lines and maintain at least a 10-foot safe working clearance from the lines.
- Inspect all rigging prior to use; do not wrap hoist lines around the load.
- Be sure to use the correct load chart for the crane's current configuration and setup, the load weight and lift path.
- Do not exceed the load chart capacity while making lifts.
- Raise load a few inches, hold, verify capacity/balance, and test brake system before delivering load.
- Do not move loads over workers.
- Be sure to follow signals and manufacturer instructions while operating cranes.

For more complete information:

OSHA Occupational Safety and Health Administration
U.S. Department of Labor
www.osha.gov (800) 321-OSHA

OSHA 3259-09R-05

Exhibit 5-18

OSHA-English
PPE quick card.

OSHA QUICK CARD™

Protect Yourself
Construction Personal Protective Equipment (PPE)

Eye and Face Protection
- Safety glasses or face shields are worn any time work operations can cause foreign objects to get in the eye. For example, during welding, cutting, grinding, nailing (or when working with concrete and/or harmful chemicals or when exposed to flying particles). Wear when exposed to any electrical hazards, including working on energized electrical systems.
- Eye and face protectors – select based on anticipated hazards.

Foot Protection
- Construction workers should wear work shoes or boots with slip-resistant and puncture-resistant soles.
- Safety-toed footwear is worn to prevent crushed toes when working around heavy equipment or falling objects.

Hand Protection
- Gloves should fit snugly.
- Workers should wear the right gloves for the job (examples: heavy-duty rubber gloves for concrete work; welding gloves for welding; insulated gloves and sleeves when exposed to electrical hazards).

Head Protection
- Wear hard hats where there is a potential for objects falling from above, bumps to the head from fixed objects, or of accidental head contact with electrical hazards.
- Hard hats – routinely inspect them for dents, cracks or deterioration; replace after a heavy blow or electrical shock; maintain in good condition.

Hearing Protection
- Use earplugs/earmuffs in high noise work areas where chainsaws or heavy equipment are used; clean or replace earplugs regularly.

For more complete information:

OSHA Occupational Safety and Health Administration
U.S. Department of Labor
www.osha.gov   (800) 321-OSHA

OSHA 3260-09N-05

Exhibit 5-19
OSHA-English
electrical safety
quick card.

OSHA QUICK CARD™

Electrical Safety

Electrical hazards can cause burns, shocks and electrocution (death).

Safety Tips

- Assume that all overhead wires are energized at lethal voltages. Never assume that a wire is safe to touch even if it is down or appears to be insulated.

- Never touch a fallen overhead power line. Call the electric utility company to report fallen electrical lines.

- Stay at least 10 feet (3 meters) away from overhead wires during cleanup and other activities. If working at heights or handling long objects, survey the area before starting work for the presence of overhead wires.

- If an overhead wire falls across your vehicle while you are driving, stay inside the vehicle and continue to drive away from the line. If the engine stalls, do not leave your vehicle. Warn people not to touch the vehicle or the wire. Call or ask someone to call the local electric utility company and emergency services.

- Never operate electrical equipment while you are standing in water.

- Never repair electrical cords or equipment unless qualified and authorized.

- Have a qualified electrician inspect electrical equipment that has gotten wet before energizing it.

- If working in damp locations, inspect electric cords and equipment to ensure that they are in good condition and free of defects, and use a ground-fault circuit interrupter (GFCI).

- Always use caution when working near electricity.

For more complete information:

OSHA Occupational
Safety and Health
Administration
U.S. Department of Labor
www.osha.gov   (800) 321-OSHA

OSHA 3298-09N-05

Exhibit 5-20

OSHA-English lead in construction quick card.

OSHA QUICK CARD™

Protect Yourself
Lead in Construction

Lead is a common hazardous element found at many construction sites. Lead exposure comes from inhaling fumes and dust, and lead can be ingested when hands are contaminated by lead dust. Lead can be taken home on workers' clothes, skin, hair, tools and in vehicles.

Lead exposure may take place in demolition, salvage, removal, encapsulation, renovation and cleanup activities.

Avoid Exposure
- Use proper personal protective equipment (e.g., gloves, clothing and approved respirators).
- Wash hands and face after work and before eating.
- Never enter eating areas wearing protective equipment.
- Never wear clothes and shoes that were worn during lead exposure away from work.
- Launder clothing daily; use proper cleaning methods.
- Be alert to symptoms of lead exposure (e.g., severe abdominal pain, headaches, loss of motor coordination).

Use Respirators
- Wear appropriate respirators as directed.
- Conduct a user seal check each time a respirator is donned.
- Be aware of your company's respiratory protection program; understand the limitations and potential hazards of respirators.

Prevent Further Exposure
- Ensure adequate ventilation.
  - When outdoors, stand upwind of any plume.
- Use dust collecting equipment, when possible.
- Use lead-free materials and chemicals.
- Use wet methods to decrease dust.
- Use local exhaust ventilation for enclosed work areas.

For more complete information:

OSHA Occupational Safety and Health Administration
U.S. Department of Labor
www.osha.gov   (800) 321-OSHA

OSHA 3291-10-05

OSHA QUICK CARD™

Demolition Safety Tips

Demolition work involves many of the same hazards that arise during other construction activities. However, demolition also involves additional hazards due to a variety of other factors. Some of these include: lead-based paint, sharp or protruding objects and asbestos-containing material.

- Brace or shore up the walls and floors of structures which have been damaged and which employees must enter.
- Inspect personal protective equipment (PPE) before use.
- Select, wear and use appropriate PPE for the task.
- Inspect all stairs, passageways, and ladders; illuminate all stairways.
- Shut off or cap all electric, gas, water, steam, sewer, and other service lines; notify appropriate utility companies.
- Guard wall openings to a height of 42 inches; cover and secure floor openings with material able to withstand the loads likely to be imposed.
- Floor openings used for material disposal must not be more than 25% of the total floor area.
- Use enclosed chutes with gates on the discharge end to drop demolition material to the ground or into debris containers.
- Demolition of exterior walls and floors must begin at the top of the structure and proceed downward.
- Structural or load-supporting members on any floor must not be cut or removed until all stories above that floor have been removed.
- All roof cornices or other ornamental stonework must be removed prior to pulling walls down.
- Employees must not be permitted to work where structural collapse hazards exist until they are corrected by shoring, bracing, or other effective means.

For more complete information:

OSHA

Occupational
Safety and Health
Administration

U.S. Department of Labor
www.osha.gov   (800) 321-OSHA

OSHA 3290-10N-05

Exhibit 5-21
OSHA-English demolition safety tips quick card.

Exhibit 5-22

OSHA-English
work zone safety
tips quick card.

OSHA QUICK CARD™

Work Zone Traffic Safety

Employees being struck by vehicles or mobile equipment lead to many work zone fatalities or injuries. Work zones need traffic controls identified by signs, cones, barrels and barriers.

Drivers, employees on foot, and pedestrians must be able to see and understand the proper routes. Construction project managers determine traffic control plans within construction/demolition worksites.

- Traffic control devices, signals, and message boards instruct drivers to follow paths away from where work is being done.

- Approved traffic control devices, including cones, barrels, barricades, and delineator posts are also used inside work zones.

Work Zone Protections: Various concrete, water, sand, collapsible barriers, crash cushions, and truck-mounted attenuators can help limit motorist intrusions into construction work zones.

Flagging: Flaggers should wear high visibility clothing with a fluorescent background and made of retroreflective material. This makes employees visible for at least 1,000 feet in any direction. Check the label or packaging to ensure that the garments are performance class 2 or 3. Drivers should be warned with signs that there will be flaggers ahead. Flaggers should use STOP/SLOW paddles, paddles with lights, or flags (only in emergencies).

Lighting: Flagger stations should be illuminated. Lighting for employees on foot and for equipment operators should be at least 5 foot-candles or greater. Where available lighting is not sufficient, flares or chemical lighting should be used. Glare should be controlled or eliminated.

Training: Flaggers must be trained/certified and use authorized signaling methods.

Driving: Seat belts and rollover protection should be used on equipment and vehicles as the manufacturer recommends.

For more complete information:

OSHA
Occupational
Safety and Health
Administration
U.S. Department of Labor
www.osha.gov   (800) 321-OSHA

OSHA 3284-05R-07

OSHA QUICK CARD™

Supported Scaffold
Safety Tips

Supported scaffolds consist of one or more platforms supported by outrigger beams, brackets, poles, legs, uprights, posts, frames, or similar rigid support.

Guardrails or personal fall arrest systems for fall prevention/protection are required for workers on platforms 10 feet or higher.

Working platforms/decks must be planked close to the guardrails.

Planks are to be overlapped on a support at least 6 inches, but not more than 12 inches.

Legs, posts, frames, poles, and uprights must be on base plates and mud sills, or a firm foundation; and, be plumb and braced.

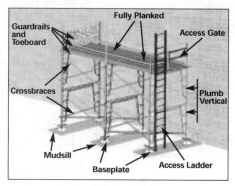

Scaffold user training must include:
• The hazards of type of scaffold being used;
• Maximum intended load and capacity;
• Recognizing and reporting defects;
• Fall hazards;
• Electrical hazards including overhead lines;
• Falling object hazards;
• Other hazards that may be encountered.

OSHA Occupational Safety and Health Administration
U.S. Department of Labor
www.osha.gov   (800) 321-OSHA

OSHA 3242-08-05

Exhibit 5-23
OSHA-English supported scaffold safety tips quick card.

Exhibit 5-24A

OSHA-English supported scaffold inspection tips quick card.

OSHA QUICK CARD™

Supported Scaffold Inspection Tips

Inspect scaffolds and scaffold parts daily, before each work shift, and after any event that may have caused damage.

- Check to see if powerlines near scaffolds are deenergized or that the scaffolds are at least 10 feet away from energized powerlines.
- Make sure that tools and materials are at least 10 feet away from energized powerlines.
- Verify that the scaffold is the correct type for the loads, materials, employees, and weather conditions.
- Check footings to see if they are level, sound, rigid, and capable of supporting the loaded scaffold.

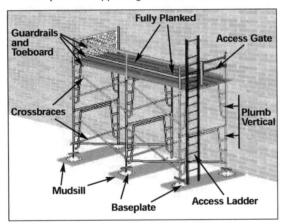

- Check legs, posts, frames, and uprights to see if they are on baseplates and mudsills.
- Check metal components for bends, cracks, holes, rust, welding splatter, pits, broken welds, and non-compatible parts.
- Check for safe access. Do not use the crossbraces as a ladder for access or exit.

Occupational Safety and Health Administration
U.S. Department of Labor
www.osha.gov   (800) 321-OSHA

OSHA 3318-09-06

Exhibit 5-24B
OSHA-English supported scaffold inspection tips quick card.

OSHA QUICK CARD™

- Check wooden planks for cracks, splits greater than $1/4$ inch, end splits that are long, many large loose knots, warps greater than $1/4$ inch, boards and ends with gouges, mold, separated laminate(s), and grain sloping greater than 1 in 12 inches from the long edge and are scaffold grade lumber or equivalent.
- If the planks deflect $1/60$ of the span or 2 inches in a 10-foot wooden plank, the plank has been damaged and must not be used.
- Check to see if the planks are close together, with spaces no more than 1 inch around uprights.
- Check to see if 10-foot or shorter planks are 6 to 12 inches over the center line of the support, and that 10-foot or longer planks are no more than 18 inches over the end.
- Check to see if the platform is 14 inches or more away from the wall or 18 inches or less away if plastering/stucco.
- Check for guardrails and midrails on platforms where work is being done.
- Check for employees under the platform and provide falling object protection or barricade the area. Make sure that hard hats are worn.
- Use braces, tie-ins and guying as described by the scaffold's manufacturer at each end, vertically and horizontally to prevent tipping.

Maximum Vertical Tie Spacing Wider Than 3'- 0" Bases

Tie. Top of scaffold platform and uppermost tie not to exceed 4 to 1 ratio

26'- 0" maximum between intermediate ties

Tie. 4 times minimum base tie at closest frame header or bearer

First tie closest frame header or bearer above 4 times the minimum base dimension

Wider than 3'- 0" minimum base dimension

OSHA Occupational Safety and Health Administration
U.S. Department of Labor
www.osha.gov (800) 321-OSHA

Exhibit 5-25

OSHA-pocket quide.

OSHA POCKET GUIDE

Occupational Safety and Health Administration

www.osha.gov

Worker Safety Series
Construction

OSHA 3252-05N 2005

In general, OSHA visits construction sites on a very limited basis. This is true for most of the urban centers; one of the few exceptions is New York City, which has its own safety organization. This is primarily due to the limited number of people employed by OSHA. However, when a construction accident does occur, OSHA investigates the cause of the accident and submits reports to the contractor and the local municipality. OSHA also has the jurisdiction to issue fines.

For further information on OSHA and for obtaining the complete set of OSHA safety pocket cards, visit their Website: http://www.osha.gov.

NEW YORK CITY SAFETY PROGRAMS

According to the New York Building Congress, New York City's construction spending exceeded $26 billion in 2007—more total construction than any other U.S. city. Due to this volume of construction and the safety concerns of the public, New York City has one of the most safety conscientious building departments in the United States. New York City has set up a separate division called the B.E.S.T. (Building Enforcement Safety Team) squad. The B.E.S.T. squad has the responsibility for inspecting high rise projects (greater than 15 stories, above 200 feet in height, and more then 100,000 square feet) construction, making sure the work conforms to the safety requirements of the New York City Building Code (International Building Code was approved in 2007 and will go into effect in July 2008).

Even with New York City's oversight, 29 construction workers died in 2006. This was a 60% increase over the 18 deaths reported in 2005. Of the 29 deaths in 2006, 17 were attributed to falls from buildings and scaffolds. Due to these deaths, New York City Building Department has set up a Scaffold Worker Safety Task Force. This task force visits sites and provides safety classes to subcontractors who use scaffolds.

During 2007, New York City has set up other safety initiatives:

1. Superintendents safety requirements
2. Safety week (lectures, seminars, demonstrations)
3. Low rise safety program (buildings greater than 7 stories but less than 14 stories)
4. Increased safety regulations for tower cranes
5. Excavation/earthwork regulations

Additional information on New York City's safety programs can be found on the Building Department's Website: www.nyc.gov/buildings.

TRAGIC ACCIDENTS

On October 30, 2003, a tragic accident occurred in Atlantic City, New Jersey. Trades people were placing concrete for the eighth level of a garage that would contain 2400 automobile parking spaces at the Tropicana Hotel. Without warning, the garage collapsed and killed 4 workers and injured 30 others. OSHA was involved in the investigation of the collapse and found that the engineer's design for slab reinforcing bar connections to the main structural wall were not followed and proper shoring was not installed. There were some indications that the contractors were trying to expedite the garage construction in order to meet a March 2004 hotel opening date. Based on the events and subsequent investigations by OSHA, numerous agencies, and consulting firms, the following were the adverse results of this tragic accident:

1. Four workers were killed.
2. Thirty workers were injured.
3. Four construction firms were fined $119,000 by OSHA (maximum under the current guidelines at that time).
4. Lawsuits were settled for $101 million (one of the largest settlements for a construction "accident").
5. The project was delayed.
6. The garage was reconstructed.

This is a tragic example of what can happen when proper safety precautions are not addressed and the contractors do not follow the engineer's drawings.

On March 15, 2008, a crane collapsed in a dense neighborhood of New York City killing seven people (five were from the crane rigging company, one was the crane operator, and one was a woman staying in one of the buildings that was hit by the crane's tower). The collapse occurred when the riggers were jumping the tower crane and placing the support collar onto the building structure. The collar broke away from the riggers (supposedly due to a faulty $50 nylon sling), slid down the crane's support tower, and caused the other support collars to break away from the building. The tower crane became unstable and fell into several apartment buildings just south of where the original construction was taking place. Exhibit 5-25A shows part of the crane leaning on an apartment building.

Exhibit 5-25A
Crane collapse.

The event that caused this tragic accident was that the riggers were supposedly using defective rigging equipment. The crane's collapse caused the death of 7 people and injury to an additional 25 people. In addition, a high-density urban area (affecting 300 apartments) was shut down several months, affecting businesses and preventing residents from occupying their apartments. The filing of lawsuits, loss of business, fines, and the project being shut down are just some of the adverse secondary losses that have occurred with the crane's collapse. In addition, after further review of the zoning for the building, officials determined that the total square footage being constructed was in violation of current zoning regulations.

On May 30, 2008, another crane accident occurred in New York City, killing the crane operator and another worker on the ground. Initial indications were that mechanical failure occurred at the turntable below the cab, which caused the cab to fall from the mast and land on a busy New York City street. The falling cab and boom caused extensive damage to several apartment buildings surrounding the construction site. As in the March 15, 2008 incident, not only is a construction project stopped but also a dense urban neighborhood is being adversely affected. The New York City Department of Buildings issued a stop work order to all construction sites using jumping cranes.

The March 10, 2008 *Engineering News Record* article "Cranes, Stalled Federal Rules Prompt State Action" reviewed the construction industry's lack of crane inspection and proper crane operator's certification. A death in Miami-Dade County, Florida caused by a crane collapse has prompted the Miami-Dade County building officials to create an ordinance for better crane safety. On March 24, 2008, another crane collapsed in Miami Beach, Florida and caused two deaths and five injuries. This collapse seems to have been caused by heavy winds when the crane was being jumped. It seems the riggers were rushing to jump the crane after several days of delays caused by high winds.

With the number of cranes operating in major urban centers, it behooves all project managers (PM) to review and evaluate the safety requirements, rigging equipment used, connections inspected, and operator certification for the cranes. Just as the crane collapses in New York City on March 15, 2008 and May 30, 2008 caused several deaths and extensive damage to dense neighborhoods, additional and more extensive safety care has to be exercised by all participating members of the construction process in an urban environment.

SAFETY RECORD

Based on information recorded in newspapers and construction magazines, there seems to be a direct correlation between the volume of construction work in a major city and an increase in the number of construction injuries and deaths that occur. The construction boom in Las Vegas, Nevada is a prime example of what is happening with safety on the job site.

It seems with expedited schedules, construction workers tend to take greater risks. During 2007, 13 construction workers died from various construction accidents in Las

Vegas. With contractors trying to meet schedules, it seems that safety takes a back seat. However, smart contractors realize that safe job sites not only create larger profits, but also expedite projects in the end.

INJURY AND DEATH IMPACT ON THE CONSTRUCTION PROCESS

The deaths and injuries sustained in the construction industry have a profound impact on the contractors whose workers are injured on the job site. A list of items that affect the Construction Manager/General Contractor (CM/GC) is indicated in Exhibit 5-26.

SAFETY AND COST IMPLICATIONS

CM/GCs may not always consider the actual cost of the items enumerated in Exhibit 5-26 when evaluating the cost of a project and the potential profit (or loss) of a particular job. By evaluating some key elements of potential costs, the contractor can have a better understanding of the cost impact of injuries at the job site.

Exhibit 5-26

Death and injury impact on the CM/GC.

1. Injury to valuable workers
2. Tragic deaths
3. Liability insurance increases
4. Builders all risk insurance increases
5. Workers' compensation rate increases
6. Delay of the project
7. Investigation by local authorities
8. Investigation by OSHA
9. Potential lawsuits
10. Depositions that also require extensive time away from the project
11. Fines and penalties
12. Loss of productivity
13. Investigation reports required, which take valuable time away from the PM and superintendent
14. Retention of new workers to take the place of injured workers (lost time for getting up to "speed" on the work involved)
15. Training of new workers
16. Reschedule of work
17. Paying of wages while an injured worker is out on leave
18. Any repair or replacement caused by the accident
19. Increase in bonding premiums or total rejection by the bonding companies

STANFORD ACCIDENT COST ACCOUNTING SYSTEM

Michael Robert Robinson developed the Stanford Accident Cost Accounting System. The matrix is noted in Exhibit 5-27. The matrix shows:

1. Body part injured
2. Nature of the injury (sprain, fracture, bruises, etc.)
3. Down time required of the worker
4. Cost associated with the injury
5. Information based on historical data

Contractors can use this matrix to approximate the cost of an accident that may occur on the job site. The negative cost information would then be placed as a liability on the balance sheet for the project. These negative costs can have a profound impact on the bottom line cost of a project.

WORKERS' COMPENSATION

Workers' compensation rates for various construction trades throughout the United States are indicated in Exhibit 5-28 from *Engineering News Record-McGraw-Hill Companies*. The rates are given per $100 in wages. Therefore, for New York, an iron worker's workers' compensation would be $16 for every $100 of wages. If we review the wage rate for iron worker as noted in Exhibit 5-29 we see that workers' compensation constitutes approximately 12% of the wage.

The workers' compensation rates noted in Exhibit 5-28 are for "average" accident rates. If an iron worker works for a company where the accident history is higher than the average, then workers' compensation rates start to escalate. If you look at Exhibit 5-29, we now see that the workers' compensation rates now constitute 17% of the total wage. In addition, the total wage rate has gone up $9.00 per hour. This could make a big difference when one is bidding a job.

The adjustment to the rate is called the experience modification rating (EMR). The EMR is based on a three-year accident history of the company being evaluated and the number of accidents (and severity) that occurred over that period.

LIABILITY INSURANCE

The insurance companies look over the contractor's safety record for the past five years. They look at the number of accidents and the severity. In addition, they want to make sure that all cases have been resolved. If the safety record has not improved, then the premiums will increase or the insurance company will not renew the policy.

Accident Cost Matrix (Approximate costs in 2008 dollars)

Workers' Compensation Rates for Wrecking Iron or Steel Structure Workers, CA

Body part		Amputation	Strain, sprain, crush, mash, smash Non-lost time	Strain, sprain, crush, mash, smash Lost time	Fracture Non-lost time	Fracture Lost time	Cut, puncture, laceration Non-lost time	Cut, puncture, laceration Lost time	Burn Non-lost time	Burn Lost time	Bruise, abrasion Non-lost time	Bruise, abrasion Lost time	Other Non-lost time	Other Lost time
Head, face		NA	NA	NA	2,538	29,007	906	10,878	1,178	27,194	906	3,626	1,269	21,755
Eye(s)	(1)	163,163	NA	NA	NA	NA	906	10,878	725	18,129	1,088	3,626	1,088	18,129
	(2)	525,747												
Neck and shoulders		NA	1,269	25,381	5,439	29,007	906	10,878	1,178	18,129	906	7,252	1,088	25,381
Arm(s) and elbow(s)	(1)	688,910	1,269	14,503	3,626	21,755	906	10,878	906	18,129	1,088	10,878	906	21,755
	(2)	906,460												
Wrist(s) and hand(s)	(1)	188,544	997	9,065	2,538	32,633	906	10,878	1,178	18,129	906	14,503	1,269	29,007
	(2)	906,460												
Thumb(s) and finger(s)	(1)	29,0037 ea up to 135,051	997	9,065	1,269	18,129	906	10,878	725	18,129	725	10,878	725	18,129
Back		NA	7,252	36,258	NA	362,584	906	10,878	1,178	27,194	1,160	18,129	1,269	36,258
Chest and lower trunk		NA	1,632	14,503	NA	NA	906	29,007	1,178	18,129	906	10,878	906	32,633
Ribs		NA	1,269	3,626	1,632	14,503	NA	NA	1,178	18,129	1,269	10,878	1,088	32,633
Hip		NA	NA	12,690	1,632	43,510	725	10,878	1,178	18,129	1,632	18,129	1,632	14,503
Leg(s) and knee(s)	(1)	322,700	1,360	14,503	1,813	54,388	906	10,878	1,178	18,129	906	10,878	1,088	29,007
	(2)	1,015,235												
Foot (feet) and ankle(s)	(1)	159,537	1,088	9,065	1,632	32,633	725	9,065	1,088	10,878	1,015	3,626	1,269	7,252
	(2)	326,326												
Toe(s)		25,381 ea up to 145,034	1,088	5,439	725	9,065	906	10,878	1,178	7,252	725	3,626	798	7,252
Hernia rupture													725	29,007
Heart attack														108,775
Hearing loss														36,258
Death														326,326

Adapted from Table 2.1: Stanford Accident Cost Matrix (Costs in 1993 Dollars)

Exhibit 5-27

Stanford accident cost accounting.

Exhibit 5-28

Compensation insurance base rates.

Compensation Insurance Base Rates

EFFECTIVE SEPT. 9, 2007

CLASSIFICATION OF WORK	ALA.	ALASKA	ARIZ.	ARK.	CALIF.	COLO.	CONN.	DEL.	D.C.	FLA.	GA.	HAWAII	IDAHO	ILL.	IND.	IOWA	KAN.	KY.	LA.	MAINE	MD.	MASS.	MICH.
CARPENTRY																							
Detached One and Two-Family Dwellings	35.90	21.95	9.36	11.96	18.11	15.86	16.37	15.53	8.89	27.71	27.89	11.74	13.20	22.21	7.06	9.07	12.46	25.81	35.07	17.56	13.46	9.03	13.88
Dwellings-Three Stories or Less	27.56	18.10	9.10	9.74	5.59	15.45	13.87	20.72	7.89	17.61	22.77	11.32	9.62	23.27	7.80	13.19	10.40	10.10	26.60	11.85	14.57	9.03	13.88
NOC	26.79	17.20	12.36	10.61	18.11	11.74	25.96	15.53	8.39	18.25	16.95	15.76	11.41	22.28	7.52	10.60	11.69	27.08	33.60	30.16	12.90	16.48	16.66
Installation of Cabinet Work Interior Trim	16.01	12.99	5.52	4.85	—	7.39	13.65	11.58	9.35	13.01	11.44	9.00	5.65	15.49	4.97	7.87	9.03	11.51	21.06	11.58	8.78	6.88	8.74
CONCRETE																							
Work-Floor, yard or sidewalks	8.90	15.09	2.75	4.13	7.39	7.81	14.17	15.64	6.96	19.07	7.58	8.11	6.35	11.95	4.44	9.14	7.93	9.19	10.53	11.36	8.10	7.22	9.16
Construction connection with Bridges or Culverts	32.30	17.04	7.43	10.34	8.61	11.76	18.32	17.04	18.77	20.67	19.69	15.69	12.97	36.23	4.71	19.06	11.36	21.51	16.76	17.60	18.39	8.61	11.15
Work-incidental to the Construction of Private Residence	8.90	15.09	2.75	4.13	7.39	7.81	14.17	15.64	6.96	19.07	7.58	8.11	6.35	11.95	4.44	9.14	7.93	9.19	10.53	11.36	8.10	7.22	9.16
Construction NOC*	10.72	14.66	6.39	7.93	6.79	12.70	18.37	15.64	7.31	18.38	9.86	10.23	10.52	29.10	5.08	11.45	8.31	17.79	16.28	25.79	13.60	22.32	14.73
ELECTRICAL WIRING WITHIN BUILDING	7.73	11.41	4.01	3.28	4.66	4.76	7.03	8.26	4.93	7.48	6.68	5.40	4.32	10.50	2.77	4.45	3.79	8.19	9.40	5.61	8.61	4.18	3.51
EXCAVATION																							
Rock Excavation & Driver	10.44	14.94	4.22	5.02	7.88	9.53	10.00	12.82	9.42	9.33	11.57	6.76	11.66	11.20	4.67	5.84	5.11	11.27	14.26	10.93	8.94	6.08	7.85
Grading of Land NOC	10.44	14.94	4.22	5.02	7.88	9.53	10.00	12.82	9.42	9.33	11.57	6.76	6.17	11.20	4.67	5.84	5.11	11.27	14.26	10.93	8.94	6.08	7.85
GLAZIER - AWAY FROM SHOP	14.31	28.84	6.28	6.40	10.67	9.01	19.49	14.75	13.93	14.41	10.84	9.42	9.27	21.43	5.65	11.14	9.35	10.21	21.89	23.03	15.17	9.25	12.07
INSULATION WORK	11.33	28.68	10.29	10.66	7.71	12.80	12.59	18.07	7.50	11.19	15.27	7.12	7.82	18.38	6.83	7.92	8.32	11.17	20.61	14.94	11.65	14.59	9.83
LATHING AND DRIVING	8.68	10.24	4.56	3.84	7.02	5.55	23.28	17.89	9.95	9.35	10.05	4.75	7.38	13.34	2.25	4.92	4.15	8.17	11.62	8.34	7.83	5.89	14.73
MASONRY NOC	17.88	40.28	5.79	6.49	10.44	13.16	19.92	17.04	11.66	13.96	13.39	11.35	8.34	20.38	5.20	9.36	8.10	14.30	16.60	21.19	13.45	15.67	11.57
PAINTING OR PAPERHANGING NOC AND SHOP	17.12	28.01	5.94	7.50	9.68	9.47	14.08	21.27	6.49	13.51	12.88	9.37	7.94	12.64	4.82	7.30	6.71	17.59	18.25	16.43	8.71	6.85	10.98
PILE DRIVING	30.60	34.04	10.69	10.67	8.95	13.88	24.27	25.97	10.95	38.24	18.23	14.00	14.23	44.00	8.56	10.55	12.93	22.78	30.86	18.28	15.35	15.92	34.98
PLASTERING OR STUCCO WORK ON OUTSIDE OF BUILDING	17.88	40.28	5.79	6.49	10.44	13.16	19.92	17.04	11.66	13.96	13.39	11.35	8.34	20.38	5.20	9.36	8.10	14.30	16.60	21.19	13.45	15.67	11.57
PLUMBING - NOC	8.13	11.51	3.97	3.35	7.41	7.78	8.84	10.67	9.13	8.13	7.41	5.07	5.39	10.88	2.88	5.27	5.81	8.58	8.27	11.11	8.71	4.80	5.38
ROOFING ALL KINDS	47.75	31.31	12.30	15.07	31.00	22.04	38.38	36.63	16.65	28.05	30.32	22.62	31.77	34.47	11.48	17.75	17.38	25.68	51.78	32.85	34.57	47.57	21.91
SHEET METAL WORK - SHOP AND OUTSIDE NOC	15.51	11.14	7.38	6.95	10.85	10.27	13.34	9.59	6.92	12.68	14.12	8.82	8.89	16.69	4.49	7.52	7.93	18.96	16.92	11.85	11.28	6.72	8.68
STEEL OR IRON ERECTION																							
Doors and Door Frame, or Sash Erection -Metal	10.31	10.87	9.60	4.39	8.40	7.72	16.27	11.07	9.98	10.60	11.08	5.97	6.57	16.48	4.11	6.79	6.44	7.35	9.35	11.28	14.26	11.04	7.46
Construction of Dwellings Not Over Two Stories	66.77	40.12	20.51	22.67	24.96	5.64	64.19	37.36	56.49	58.10	57.59	34.76	30.10	76.06	18.49	30.15	28.80	31.77	69.80	59.06	63.22	34.06	34.65
Interior Cap Work Reference Carpentry - Interior	16.01	12.99	5.52	4.85	18.11	7.39	13.65	11.58	9.35	13.01	11.44	9.00	5.65	15.49	4.97	7.87	9.03	11.51	21.06	11.58	8.78	6.88	8.74
NOC	23.79	29.43	11.91	16.79	12.21	14.59	21.83	37.36	14.17	23.49	28.08	15.25	11.34	23.94	8.06	40.70	13.57	17.16	12.21	33.20	18.85	34.06	19.86
Frame Structures	50.94	48.91	22.01	20.54	10.30	32.69	40.33	37.36	33.95	35.41	33.83	27.99	31.36	74.47	15.90	36.84	28.24	37.30	35.28	43.10	41.80	43.69	34.98
Frame Structures Not Over Stories in Height	101.65	53.51	25.09	23.67	24.96	63.62	64.82	37.36	34.47	56.25	53.70	30.85	32.83	65.27	24.47	27.85	40.88	101.01	86.70	53.60	56.06	34.06	34.65
TILE WORK CERAMIC , STONE, MOSAIC, OR TERRAZZO	8.82	8.14	2.29	3.94	5.18	7.92	8.27	12.85	16.81	8.34	6.85	8.97	8.50	15.48	3.11	6.84	7.11	12.45	9.37	8.45	6.48	8.78	8.10
TIMEKEEPERS - CONSTRUCTION OR ERECTION	12.53	8.39	5.07	7.20	1.75	6.82	6.46	NAB	5.38	11.46	8.67	6.98	8.28	7.06	3.27	5.11	9.12	8.63	12.32	7.23	9.08	5.89	6.85
WATERPROOFING																							
Brush or Hand Pressured Caulking	17.12	28.01	5.94	7.50	9.68	9.47	14.08	21.27	6.49	13.51	12.88	9.37	7.94	12.64	4.82	7.30	6.71	17.59	18.25	16.43	8.71	6.85	10.98
Trowel Exterior of Buildings	17.88	40.28	5.79	6.49	10.44	13.16	19.92	17.04	11.66	13.96	13.39	11.35	8.34	20.38	5.20	9.36	8.10	14.30	16.60	21.19	13.45	15.67	11.57
Trowel Interior of Buildings	17.10	39.09	5.06	10.42	7.39	8.28	14.92	17.89	10.06	25.08	12.62	10.21	11.05	18.95	3.87	8.57	7.35	11.40	17.10	14.91	18.36	6.62	12.83
WRECKING BUILDING OR STRUCTURES																							
Concrete or concrete- Encased Building or Structures	10.72	14.66	6.39	7.93	6.79	12.70	18.37	15.64	7.31	18.38	9.86	10.23	10.52	29.10	5.08	11.45	8.31	17.79	16.28	25.79	13.60	22.32	14.73
Iron or Steel Buildings or Structures	23.79	29.43	11.91	16.79	12.21	14.59	21.83	37.36	14.17	23.49	28.08	15.25	11.34	23.94	8.06	40.70	13.57	17.16	12.21	33.20	18.85	34.06	19.86

CLASSIFICATION OF WORK	MINN.	MISS.	MO.	MONT.	NEB.	N.H.	N.J.	N.M.	N.Y.	N.C.	OKLA.	ORE.	PENN.	R.I.	S.C.	S.D.	TENN.	TEXAS	UTAH	VT.	VA.	WISC.
CARPENTRY																						
Detached One and Two-Family Dwellings	19.61	17.78	14.42	19.79	12.57	22.84	13.69	14.26	11.47	19.19	18.15	22.34	13.73	9.99	26.04	10.98	24.88	13.34	13.47	12.89	11.94	15.49
Dwellings-Three Stories or Less	20.63	14.72	16.71	26.11	13.54	31.80	13.69	11.72	14.21	13.64	12.64	15.10	15.43	10.65	21.46	22.87	15.00	13.34	11.92	17.01	9.77	10.98
NOC	41.98	18.86	11.13	34.43	13.27	20.53	13.69	12.45	12.39	13.43	13.47	15.35	13.73	14.12	18.79	24.36	17.95	13.34	8.85	13.55	7.09	14.88
Installation of Cabinet Work Interior Trim	20.87	8.72	8.89	17.01	7.91	11.05	8.84	3.91	6.29	11.67	12.03	7.92	10.62	8.31	14.19	6.70	11.43	10.20	6.94	8.85	6.04	12.78
CONCRETE																						
Work-Floor, yard or sidewalks	13.17	6.67	11.46	17.04	7.41	13.71	18.73	6.85	17.19	5.94	10.74	8.23	16.51	11.52	10.43	8.96	7.20	10.14	9.35	15.95	5.83	15.82
Construction connection with Bridges or Culverts	25.93	20.32	15.74	37.83	11.36	15.58	18.98	8.95	12.32	24.02	11.34	22.93	14.02	15.66	18.34	9.21	16.10	10.14	9.38	20.17	7.35	6.64
Work-incidental to the Construction of Private Residence	13.17	6.67	11.46	17.04	7.41	13.71	18.73	6.85	17.19	5.94	10.74	8.23	16.51	11.52	10.43	8.96	7.20	10.14	9.35	15.95	5.83	15.82
Construction NOC*	10.97	10.28	15.05	18.86	16.94	28.69	15.18	11.18	17.19	14.60	14.41	13.92	16.51	13.84	13.63	21.04	13.08	10.14	6.88	18.71	8.58	12.73
ELECTRICAL WIRING WITHIN BUILDING	7.97	4.88	6.08	9.75	5.94	6.77	4.42	3.91	6.69	9.11	7.46	5.03	6.99	5.17	9.92	4.55	6.35	7.58	4.09	6.03	4.54	5.46
EXCAVATION																						
Rock Excavation & Driver	11.40	8.88	7.30	23.71	10.39	16.94	7.83	6.74	8.37	10.06	24.36	9.13	9.72	8.98	11.99	9.06	10.40	9.59	7.79	12.46	10.86	7.75
Grading of Land NOC and Driver	11.40	8.88	7.30	23.71	10.39	16.94	7.83	6.74	8.37	10.06	11.25	9.13	9.72	8.98	11.99	9.06	10.40	9.59	7.79	12.46	5.63	7.75
GLAZIER - AWAY FROM SHOP	17.07	10.85	8.86	16.27	14.75	10.94	7.43	11.07	10.18	9.93	12.47	14.91	11.44	15.08	14.56	10.01	12.59	11.38	11.88	16.53	7.03	11.03
INSULATION WORK	12.49	12.86	12.08	48.86	19.57	21.32	13.14	8.06	9.91	13.66	12.16	11.56	13.30	19.72	13.34	9.92	10.74	13.51	8.55	17.23	8.73	14.06
LATHING AND DRIVING	10.97	5.72	6.17	16.88	6.61	8.43	11.63	5.37	11.27	15.63	8.23	7.45	13.75	6.84	10.28	6.34	7.67	7.33	6.36	7.25	12.94	7.05
MASONRY NOC	19.85	9.13	11.95	20.98	12.17	23.02	12.42	9.01	15.70	9.72	11.34	13.76	14.02	14.04	11.68	9.31	13.77	13.15	10.50	14.29	7.78	16.00
PAINTING OR PAPERHANGING NOC AND SHOP	16.29	11.40	9.97	14.66	12.41	13.86	11.61	8.40	10.41	10.54	10.66	10.40	15.70	11.05	15.29	7.84	9.79	10.09	7.41	11.31	7.35	12.96
PILE DRIVING	18.40	23.54	15.23	38.41	13.91	19.66	15.28	13.68	15.68	14.78	15.36	15.83	18.79	32.64	17.74	17.55	19.24	25.84	10.01	16.56	9.59	18.78
PLASTERING OR STUCCO WORK ON OUTSIDE OF BUILDING	19.85	9.13	11.95	20.98	12.17	23.02	12.42	9.01	15.70	9.72	11.34	13.76	14.02	14.04	11.68	9.31	13.77	13.15	10.50	14.29	7.78	16.00
PLUMBING - NOC	10.48	7.53	8.00	13.35	6.34	9.75	6.25	5.47	7.33	7.69	9.31	5.38	8.65	4.64	10.41	9.67	7.14	7.40	4.76	8.15	4.94	6.77
ROOFING ALL KINDS	54.28	32.16	23.87	57.36	20.36	50.28	35.61	20.57	33.39	21.46	23.21	25.75	32.55	19.06	43.47	17.07	21.02	22.70	21.68	39.48	16.90	43.79
SHEET METAL WORK - SHOP AND OUTSIDE NOC	15.11	12.78	10.50	16.11	11.07	12.59	6.75	7.19	12.79	12.12	11.18	9.22	16.51	7.28	11.76	10.65	11.12	14.21	9.77	9.67	6.64	7.44
STEEL OR IRON ERECTION																						
Doors and Door Frame, or Sash Erection -Metal	10.46	12.37	9.00	13.27	8.89	12.32	22.92	8.73	9.59	8.41	9.82	7.76	9.21	7.54	12.53	7.40	7.17	10.68	6.91	11.54	7.40	9.27
Construction of Dwellings Not Over Two Stories	146.62	87.57	39.85	78.68	39.87	64.54	22.92	28.62	31.19	40.91	74.61	21.78	25.66	39.97	75.07	77.26	40.02	18.16	23.16	55.12	24.79	28.03
Interior Cap Work Reference Carpentry - Interior	20.87	8.72	8.89	17.01	7.91	11.05	8.84	3.91	6.29	11.67	12.03	7.92	10.62	8.31	14.19	6.70	11.43	10.20	6.94	8.85	6.04	12.78
NOC	22.71	20.67	28.53	23.99	17.46	17.45	13.33	19.55	15.20	18.91	20.32	18.38	25.66	15.15	28.28	33.74	17.21	16.40	16.28	46.39	15.83	36.80
Frame Structures	137.37	27.74	27.52	41.86	34.04	44.18	26.18	31.59	21.77	64.72	32.85	26.39	25.66	27.66	29.58	65.93	26.29	31.16	23.64	48.72	29.86	31.67
Frame Structures Not Over Stories in Height	146.62	31.70	84.84	119.15	35.23	49.62	41.55	72.55	50.55	60.08	75.48	21.78	25.66	42.52	44.56	71.60	54.74	16.40	25.14	44.68	35.75	50.62
TILE WORK CERAMIC , STONE, MOSAIC, OR TERRAZZO	13.24	8.15	8.90	11.49	6.90	10.48	6.06	4.07	8.72	6.38	6.86	10.81	9.50	6.56	8.75	7.20	9.37	8.44	5.71	8.78	4.22	14.65
TIMEKEEPERS - CONSTRUCTION OR ERECTION	3.75	7.81	7.43	10.13	9.21	7.18	8.82	5.92	5.98	6.71	14.83	9.51	NAB	3.63	9.50	5.73	7.30	NAB	19.64	6.94	3.87	8.37
WATERPROOFING																						
Brush or Hand Pressured Caulking	16.29	11.40	9.97	14.66	12.41	13.86	11.61	8.40	10.41	10.54	10.66	10.40	15.70	11.05	15.29	7.84	9.79	10.09	7.41	11.31	7.35	12.96
Trowel Exterior of Buildings	19.85	9.13	11.95	20.98	12.17	23.02	12.42	9.01	15.70	9.72	11.34	13.76	14.02	14.04	11.68	9.31	13.77	13.15	10.50	14.29	7.78	16.00
Trowel Interior of Buildings	20.05	25.29	13.55	15.28	11.47	14.31	11.63	7.58	9.04	13.69	13.46	12.46	13.75	11.16	14.54	10.24	13.37	10.09	6.65	10.66	9.07	10.72
WRECKING BUILDING OR STRUCTURES																						
Concrete or concrete- Encased Building or Structures	10.97	10.28	15.05	18.86	16.94	28.69	15.18	11.18	17.19	14.60	14.41	13.92	16.51	13.84	13.63	21.04	13.08	10.14	6.88	18.71	8.58	12.73
Iron or Steel Buildings or Structures	22.71	20.67	28.53	23.99	17.46	17.45	13.33	19.55	15.20	18.91	20.32	18.38	25.66	15.15	28.28	33.74	17.21	16.40	16.28	46.39	15.83	36.80

COMPILED BY AON. RISK SERVICES, INSURANCE BROKERS, NEW YORK CITY. RATES ARE TYPICAL WORKERS' COMPENSATION INSURANCE RATES. NORTH DAKOTA, OHIO, WASHINGTON, WEST VIRGINIA AND WYOMING. MINNESOTA PER $100 PAYROLL AND THOSE APPROVED AND IN USE AS OF SEPT.14, 2007. THESE RATES ARE TYPICAL CURRENT RATES SUBJECT TO CHANGE ACCORDING TO EXPERIENCE RATING. LISTING DOES NOT INCLUDE MAINE OR MONOPOLISTIC-FUND STATES, NAMELY: NEVADA, IS AN ASSIGNED-RISK STATE; RATES ARE VARIABLE. IF SPECIALTY RATES ARE LEFT BLANK, REFER TO COMPANY. *NOC=NOT OTHERWISE CLASSIFIED.

Some insurance companies send out safety teams that review and evaluate the CM/GC's safety procedures. In addition, they walk the site to determine if any major safety violations have been noticed.

Premium increases could be substantial if the safety record has not improved. The CM/GC should check with the insurance provider to determine what kind of impact an accident will have on premium rates. This may give the CM/GC incentives for stepping up any safety program that is now established.

Exhibit 5-29A

Safety net stages.

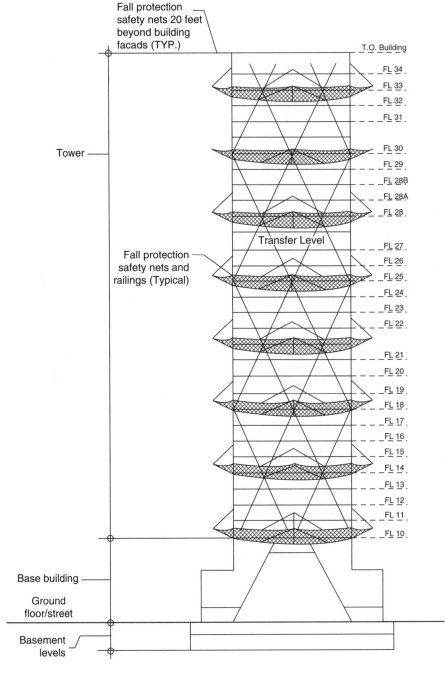

Fall protection safety nets 20 feet beyond building facads (TYP.)

T.O. Building

FL 34
FL 33
FL 32
FL 31

Tower

FL 30
FL 29
FL 28B
FL 28A
FL 28

Transfer Level

FL 27
FL 26

Fall protection safety nets and railings (Typical)

FL 25
FL 24
FL 23
FL 22

FL 21
FL 20

FL 19
FL 18
FL 17
FL 16
FL 15
FL 14
FL 13
FL 12
FL 11
FL 10

Base building

Ground floor/street

Basement levels

SAFETY NET STAGES

Exhibit 5-29B
Safety netting.

REDUCING ACCIDENTS ON THE URBAN JOB SITE

In order to maintain a safe environment on the job site, the following critical elements must be established:

1. Review the drawings for potential safety concerns.
2. Upper management must be totally committed to safety.
3. Learn to communicate safety to all workers on the job site.
4. Provide a safety manager and/or director on the job site.
5. Make sure that the site is clean with no debris scattered around.
6. The safety manager and PM must constantly walk the job site.
7. Enforce OSHA standards.
8. Have weekly toolbox meetings with all trade foremen.
   - Review any current problems.
   - Request ways to improve safety on the job site.
   - The meeting should be of short duration but cover critical issues.
9. Make sure all trades are using safety equipment prescribed by OSHA.
10. Listen to all the workers and make safety modifications as suggested.
11. Have special safety meetings with all "new" workers. (Approximately 25% of the accidents occur with workers working less than one month on the job site.)
12. Review all building codes to make sure all safety requirements are being met.
13. Listen to weather reports to make sure all equipment and material will be secured in case of high winds.
14. Make sure that certain trades do not work during inclement weather (i.e., iron workers).

15. Have strong language in the subcontractor's contract regarding safety issues. This should include the use of hard hats, safety harness, safety nets, attending safety meetings, etc. Disciplinary action for subcontractor trades people who do not conform to the safety standards must be established. See Exhibits 5-29A and 5-29B for the location for installing nets and a photograph of the nets installed.

16. A safety plan has to be prepared which would include location of first aid stations, evacuation plan in case of an emergency, communications set up, and safety horns.

17. Install safety signs around the site.

18. Prepare a safety manual that will be handed out to all subcontractors.

19. Invite insurance safety groups to visit the site and make recommendations.

20. Do not rush jobs and overwork the trades people.

21. Drugs and alcohol must be prohibited from the job site.

22. Smoking must be prohibited from the job site to eliminate potential fires.

WHEN AN ACCIDENT OCCURS ON THE JOB SITE

In the event that a major accident does occur on the job site, the CM/GC must be prepared to act immediately. Depending on the seriousness of the injury, the proper people must be contacted. In order to expedite the process, a contact list similar to Exhibit 5-30 must be available. It may be necessary to contact all the organizations listed so that the required assistance arrives at the site as quickly as possible. Any delays could cause catastrophic results. The CM/GC should also develop an on-site SWAT team that will be trained to act in an emergency situation. Their responsibility would be to give directions to all the trades people so that they would be out of harm's way. In addition, they would comfort any of the accident victims until the professionals arrive at the site.

Once the accident victims have been seen to and removed from the site, the paperwork process begins. OSHA requires that forms noted in Exhibits 5-31, 5-32, and 5-33 be filed. In addition, the form noted in Exhibit 5-34 should be filled out for the

Exhibit 5-30
Emergency call numbers.

Hospital	Name	Telephone	Cell Number	Address
Hospital				
Doctor				
Police Emergency Number				
Fire Emergency Number				
Building Department				
OSHA Office				
CM/GC Safety Director				
Local First Aid Center, Site Safety Manager				

OSHA's Form 300 (Rev. 01/2004)

Log of Work-Related Injuries and Illnesses

Attention: This form contains information relating to employee health and must be used in a manner that protects the confidentiality of employees to the extent possible while the information is being used for occupational safety and health purposes.

U.S. Department of Labor
Occupational Safety and Health Administration

Year 20 ____

Form approved OMB no. 1218-0176

You must record information about every work-related death and about every work-related injury or illness that involves loss of consciousness, restricted work activity or job transfer, days away from work, or medical treatment beyond first aid. You must also record significant work-related injuries and illnesses that are diagnosed by a physician or licensed health care professional. You must also record work-related injuries and illnesses that meet any of the specific recording criteria listed in 29 CFR Part 1904.8 through 1904.12. Feel free to use two lines for a single case if you need to. You must complete an injury and illness incident Report (OSHA Form 301) or equivalent form for each injury or illness recorded on this form. If you're not sure whether a case is recordable, call your local OSHA office for help.

Establishment name ____

City ____   State ____

Identify the person

(A) Case no.	(B) Employee's name	(C) Job title (e.g., Welder)

Describe the case

(D) Date of injury or onset of illness	(E) Where the event occurred (e.g., Loading dock north end)	(F) Describe injury or illness, parts of body affected, and object/substance that directly injured or made person ill (e.g., Second degree burns on right forearm from acetylene torch)

Classify the case

CHECK ONLY ONE box for each case based on the most serious outcome for that case:

Death (G)	Days away from work (H)	Remained at Work — Job transfer or restriction (I)	Remained at Work — Other recordable cases (J)

Enter the number of days the injured or ill worker was:

Away from work (K)	On job transfer or restriction (L)
days	days

Check the "Injury" column or choose one type of illness:

(M)

Injury (1)	Skin disorder (2)	Respiratory condition (3)	Poisoning (4)	Hearing loss (5)	All other illnesses (6)

Page totals ▶

Be sure to transfer these totals to the Summary page (Form 300A) before you post it.

Injury (1)	Skin disorder (2)	Respiratory condition (3)	Poisoning (4)	Hearing loss (5)	All other illnesses (6)

Page ____ of ____

Public reporting burden for this collection of information is estimated to average 14 minutes per response, including time to review the instructions, search and gather the data needed, and complete and review the collection of information. Persons are not required to respond to the collection of information unless it displays a currently valid OMB control number. If you have any comments about these estimates or any other aspects of this data collection, contact: US Department of Labor, OSHA Office of Statistical Analysis, Room N-3644, 200 Constitution Avenue, NW, Washington, DC 20210. Do not send the completed forms to this office.

Exhibit 5-31
OSHA form 300.

OSHA's Form 300A (Rev. 01/2004)

Summary of Work-Related Injuries and Illnesses

Year 20____

U.S. Department of Labor
Occupational Safety and Health Administration

Form approved OMB no. 1218-0176

All establishments covered by Part 1904 must complete this Summary page, even if no work-related injuries or illnesses occurred during the year. Remember to review the Log to verify that the entries are complete and accurate before completing this summary.

Using the Log, count the individual entries you made for each category. Then write the totals below, making sure you've added the entries from every page of the Log. If you had no cases, write "0."

Employees, former employees, and their representatives have the right to review the OSHA Form 300 in its entirety. They also have limited access to the OSHA Form 301 or its equivalent. See 29 CFR Part 1904.35, in OSHA's recordkeeping rule, for further details on the access provisions for these forms.

Number of Cases

Total number of deaths	Total number of cases with days away from work	Total number of cases with job transfer or restriction	Total number of other recordable cases
(G)	(H)	(I)	(J)

Number of Days

Total number of days away from work	Total number of days of job transfer or restriction
(K)	(L)

Injury and Illness Types

Total number of . . .
(M)
(1) Injuries _____
(2) Skin disorders _____
(3) Respiratory conditions _____

(4) Poisonings _____
(5) Hearing loss _____
(6) All other illnesses _____

Post this Summary page from February 1 to April 30 of the year following the year covered by the form.

Public reporting burden for this collection of information is estimated to average 58 minutes per response, including time to review the instructions, search and gather the data needed, and complete and review the collection of information. Persons are not required to respond to the collection of information unless it displays a currently valid OMB control number. If you have any comments about these estimates or any other aspects of this data collection, contact: US Department of Labor, OSHA Office of Statistical Analysis, Room N-3644, 200 Constitution Avenue, NW, Washington, DC 20210. Do not send the completed forms to this office.

Establishment information

Your establishment name _____

Street _____

City _____ State _____ ZIP _____

Industry description (e.g., Manufacture of motor truck trailers) _____

Standard Industrial Classification (SIC), if known (e.g., 3715) _____

OR

North American Industrial Classification (NAICS), if known (e.g., 336212) _____

Employment information (If you don't have these figures, see the Worksheet on the back of this page to estimate.)

Annual average number of employees _____

Total hours worked by all employees last year _____

Sign here

Knowingly falsifying this document may result in a fine.

I certify that I have examined this document and that to the best of my knowledge the entries are true, accurate, and complete.

Company executive _____ Title _____

Phone _____ Date ___/___/___

Exhibit 5-32
OSHA form 300A.

OSHA's Form 301

Injury and Illness Incident Report

U.S. Department of Labor
Occupational Safety and Health Administration

Form approved OMB no. 1218-0176

Attention: This form contains information relating to employee health and must be used in a manner that protects the confidentiality of employees to the extent possible while the information is being used for occupational safety and health purposes.

This *Injury and Illness Incident Report* is one of the first forms you must fill out when a recordable work-related injury or illness has occurred. Together with the *Log of Work-Related Injuries and Illnesses* and the accompanying *Summary*, these forms help the employer and OSHA develop a picture of the extent and severity of work-related incidents.

Within 7 calendar days after you receive information that a recordable work-related injury or illness has occurred, you must fill out this form or an equivalent. Some state workers' compensation, insurance, or other reports may be acceptable substitutes. To be considered an equivalent form, any substitute must contain all the information asked for on this form.

According to Public Law 91-596 and 29 CFR 1904, OSHA's recordkeeping rule, you must keep this form on file for 5 years following the year to which it pertains.

If you need additional copies of this form, you may photocopy and use as many as you need.

Information about the employee

1) Full name _____

2) Street _____
   City _____ State _____ ZIP _____

3) Date of birth ___/___/___

4) Date hired ___/___/___

5) ☐ Male
   ☐ Female

Information about the physician or other health care professional

6) Name of physician or other health care professional _____

7) If treatment was given away from the worksite, where was it given?
   Facility _____
   Street _____
   City _____ State _____ ZIP _____

8) Was employee treated in an emergency room?
   ☐ Yes
   ☐ No

9) Was employee hospitalized overnight as an in-patient?
   ☐ Yes
   ☐ No

Information about the case

10) Case number from the *Log* _____ (*Transfer the case number from the Log after you record the case.*)

11) Date of injury or illness ___/___/___

12) Time employee began work _____ AM / PM

13) Time of event _____ AM / PM ☐ Check if time cannot be determined

14) **What was the employee doing just before the incident occurred?** Describe the activity, as well as the tools, equipment, or material the employee was using. Be specific. *Examples:* "climbing a ladder while carrying roofing materials"; "spraying chlorine from hand sprayer"; "daily computer key-entry."

15) **What happened?** Tell us how the injury occurred. *Examples:* "When ladder slipped on wet floor, worker fell 20 feet"; "Worker was sprayed with chlorine when gasket broke during replacement"; "Worker developed soreness in wrist over time."

16) **What was the injury or illness?** Tell us the part of the body that was affected and how it was affected; be more specific than "hurt," "pain," or "sore." *Examples:* "strained back"; "chemical burn, hand"; "carpal tunnel syndrome."

17) **What object or substance directly harmed the employee?** *Examples:* "concrete floor"; "chlorine"; "radial arm saw." *If this question does not apply to the incident, leave it blank.*

18) *If the employee died, when did death occur?* Date of death ___/___/___

Completed by _____

Title _____

Phone (___) ___ - ___ Date ___/___/___

Public reporting burden for this collection of information is estimated to average 22 minutes per response, including time for reviewing instructions, searching existing data sources, gathering and maintaining the data needed, and completing and reviewing the collection of information. Persons are not required to respond to the collection of information unless it displays a current valid OMB control number. If you have any comments about this estimate or any other aspects of this data collection, including suggestions for reducing this burden, contact: US Department of Labor, OSHA Office of Statistical Analysis, Room N-3644, 200 Constitution Avenue, NW, Washington, DC 20210. Do not send the completed forms to this office.

Exhibit 5-33
OSHA form 301.

Exhibit 5-34

Accident report.

Company Name
Address
City, State, ZIP
Phone Number

Date: _____
Owner: _____
Contractor: _____
Project Name: _____

Name of injured: _____   Social Security # _____

Home address of injured: _____

Company: _____   Age: _____   M___   F___

Date of injury: _____   Time: _____   AM___   PM___

Occupation: _____   How long? _____

Type and nature of injury: _____

What was person doing at time of injury? _____

Where and how did the accident occur? _____

_____

_____

Specify machine, tool, substance, or object that directly injured employee: _____

_____

Was medical treatment sought?   Yes _____   No _____

Where and by whom? _____

Was person unable to work after injury? _____

If yes, for how long was he absent from job? _____

List names and addresses of witnesses: _____

_____

_____

This report filed by: _____   Date: _____

Corrective action taken: _____

Describe any unsafe acts or conditions contributing to accident: _____

_____

Explain specifically the corrective action taken: _____

_____

CM/GC's records. The evaluation procedures do not stop with filling out the forms. The PM, Safety Director, superintendent, and subcontractor must now analyze the cause of the accident. In addition, OSHA and the local and state building departments may perform their own investigation of the accident. Once the cause is found, new procedures must be implemented immediately to eliminate any potential similar accidents. In addition, all the trades people on site must be made aware of the new procedures. In addition, the safety meetings must reemphasize proper safety standards. The old philosophy that this accident will not happen to me and therefore I do not have to follow safety guidelines must be eliminated from the psyche of all the trades people working on site. Continuous safety training, site walking by the safety director, and numerous signs (see Exhibit 5-35 for a typical safety sign) are the only ways that accidents can be prevented.

Exhibit 5-35
Hard hat sign.

THE SAFETY POEM

All trades people should read the poem (Exhibit 5-36) that was written by an unknown worker from a Boston construction firm. This poem exemplifies the need

Exhibit 5-36
Safety poem.

Safety Poem

I Chose to Look the Other Way

*I could have saved a life that day,
But I chose to look the other way.
It wasn't that I didn't care,
I had the time, and I was there.
But I didn't want to seem a fool,
Or argue over a Safety Rule.
I knew he had done the job before,
If I called it wrong, he might get sore.
The chances didn't seem that bad,
I've done the same, he knew I had.
So I shook my head and walked on by,
He knew the risks as well as I.
He took the chance, I closed an eye,
And with that act, I let him die.
I could have saved a life that day,
But I chose to look the other way.
Now every time I see his wife,
I'll know I should have saved his life.
That guilt is something I must bear.
But it isn't something you need share.
If you see a risk that others take,
That puts their Health or Life at stake,
The question asked or thing you say,
Could help them live another day.
If you see a risk and walk away,
Then hope you never have to say,
I could have saved a Life that day,
But I chose to look the other way.*

Author Unknown

for adequate safety measures in the construction industry. If safety is not a primary factor on the job site, the viability of making money on the project becomes a question.

SUMMARY

- The construction industry has the worst safety record of any occupation in the United States.
- OSHA and local municipal building departments are trying to address the issue of safety by passing legislation and implementing new safety regulations.
- The construction industry now understands that good safety means good business. With escalation of premiums for liability and workers' compensation insurance, the CM/GC firms can now see that poor safety records affect the bottom line. In addition, poor safety on the job site creates slow down of projects and, in some cases, a complete stop until investigations have been completed.
- A valid safety program can only be effective when upper management of CM/GC firms take a positive interest in their safety initiative.
- When an accident does occur on the job site, the CM/GC must be prepared by having an emergency contact sheet and an on-site SWAT team that can react to the situation.
- The primary objective of all safety programs is to obtain a zero accident project.

6 Logistics
(How can you build on this minuscule site?)

LOGISTICS

Imagine a piece of vacant land that is surrounded on three sides by tall office buildings and a busy street located on the fourth side. Now, you, as the Construction Manager/General Contractor (CM/GC), have the responsibility of constructing a 60-story office building on the site. Due to the constraints of the site, the CM/GC must develop a plan that will achieve the objectives of getting material and labor to the site without affecting the surrounding neighbors, the pedestrians, and the local street traffic. In addition, the safety of the public and the trades people working at the site must be of paramount importance. The plan that will address these issues is called a logistic plan. Exhibit 6-1 is a logistics flowchart.

Exhibit 6-2 is an example of an access plan for the construction of a major office building in an urban setting. Exhibit 6-3 shows the location of hoist/elevator/loading dock for an urban project. Exhibit 6-4 is a plan showing the crane location on an urban site. Exhibits 6-2, 6-3, and 6-4 constitute a logistics plan. Exhibit 6-5 is a list of items that should be considered when preparing a logistics plan.

CRANES

Depending upon the height of the building constructed, a determination will have to be made on the type of crane that should be used for this particular project. For heights of up to 200 feet, a mobile crane can be utilized. Once this height is exceeded, then a climber tower crane will have to be considered.

Exhibit 6-1
Logistics flowchart.

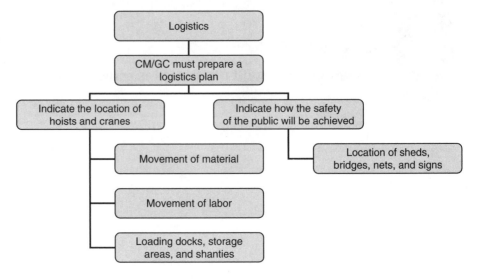

Exhibit 6-2
Site access plan.

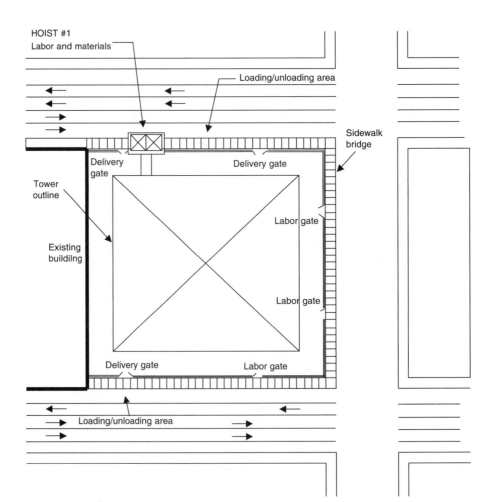

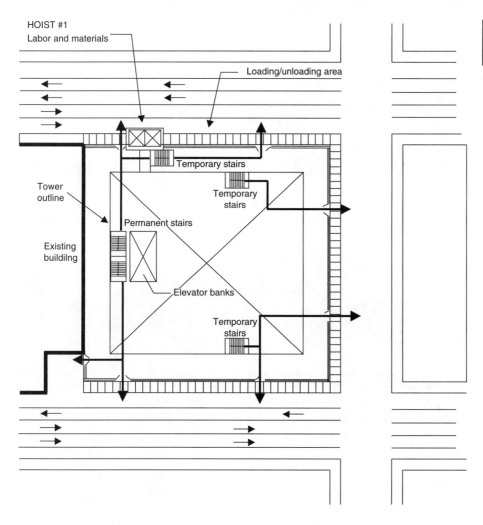

HOIST #1
Labor and materials

Loading/unloading area

Temporary stairs

Tower outline

Temporary stairs

Permanent stairs

Existing buildilng

Elevator banks

Temporary stairs

Exhibit 6-3
Hoists, elevators loading dock plan.

Large projects might have multiple cranes operating simultaneously. Both the prime contractor and subcontractor could have a role in selecting the number, type, and location of these cranes. Factors might include:

1. Project schedule and attendant requirements for crane picks
2. Available area for picks from the street and movement within the site
3. Cost of crane and operator versus productivity gain
4. Potential interference between cranes in operation or with other equipment operating on the site
5. Need for the hook to reach where needed with requisite lift capacity
6. Needs of trades to work independently of other trades with respect to lift scheduling
7. Potential interference between space requirements for cranes taken out of service and poised for storm winds
8. Potential benefit of having one crane erect or dismantle another crane
9. Load bearing of the ground to support the crane and its lifting load

Exhibit 6-4
Crane locations.

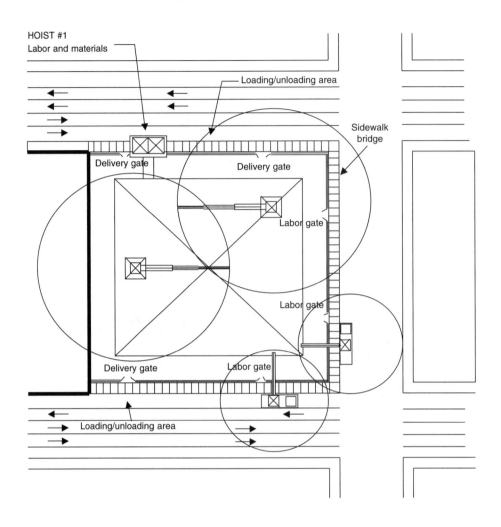

For a crane sitting on a city street, the load bearing capacity below its base will have to be analyzed by a structural engineer. The effects of the crane on underground utility vaults, transportation systems, and foundations in the area will have to be considered. In some instances, a crane is supported on the building structure and the building must be designed to sustain the loads in combination with other loads that occur during construction. As the self-raising crane starts "growing," then it has to be tied into the building structure for stability and the building must be designed to support its load. This usually occurs at every 60 to 100 feet. See Exhibit 6-6 for tower crane stages.

Exhibits 6-7 to 6-10 show various components of the crane including:

- Crane's cab (Exhibit 6-7)
- Climbing crane with structural support (Exhibit 6-8)
- Crane's structure (Exhibit 6-9)
- Crane's collar support to structure (Exhibit 6-10)

Item	Details	Requirements
Logistics		
Logistics plan	Plan of total area	
Hours of operation	Hours that the site will be open and any municipal restrictions	
Transportation stops	Bus, subway, and tram stops must be evaluated	
Moratorium periods	Restricted days when construction activitity has to be curtailed	
Field office and shanties	For CM/GC/Subcontractors/Owner/Consultants. Location of offices and shanties relative to site entrance	
Field office	Administrative offices	
Consultant's office	Offices for the architects and engineers	
Sidewalk protection	Safety for pedestrians	
Shanties	Subcontractors office and storage	
Storage areas	Location for storing large pieces of equipment	
Container Storage	Storage for containers for demo truck pick up	
Special yards	For the on-site fabrication of materials and staging of equipment	
Cylinder storage	Acetylene torch and concrete cylinders	
	Storage of flammable and hazardous materials	
Office power supply	Power for field office and shanties	
Loading dock	Off load of equipment and material	
Off-site storage	To be considered because of limitations on site space	
First aid office	Location for assisting injured workers	
Temporary toilets	Toilets for the workers	
	Location of temporary bathrooms	
	Heaters and A/C for construction shanties	
Utilities	Telephone, fax, Internet service for CM and trades	
	Utilizing permanent systems for use during construction	
Safety & Site Prep		
Safety nets	Nets required for safety of workers and to catch any falling objects	
Site security	Security of site and offices	

Exhibit 6-5
Logistics checklist.

(Continued)

Exhibit 6-5
(*Continued*)

Item	Details	Requirements
Safety & Site Prep		
Guards booth	To control site access	
Security fence	For securing the area from theft and access	
Security gates	Secure access for workers and trucks	
Bridges	Safety for pedestrians	
Security badges	Badges required to be worn by all persons working on the site	
Evacuation plan	Plan established for vacating the site in case of a problem	
Emergency system	Communication system for notification of fire, collapse, etc.	
	Emergency contact information	
	Communication system on site	
	OSHA protection, signage and logs	
Temporary Services and Facilities		
Temporary lighting	For security and safety	
Temporary power	For running the operation	
	For demolition and abatement	
	For excavation and foundation	
	For hoists	
	For loading dock	
	For hoist tower lights and outlets	
	For sidewalk bridge and sheds	
	Temporary light and power for construction	
	For specialty equipment	
	For elevators	
	Availability of permanent power	
	For saws	
	Welding equipment	
	Rigging equipment	
	Derricks	
	Project signage	
	Mixing equipment	
	For water pump	
Temporary water and pump	Temporary water for construction	
	Water barrels for water use and drainage	

Item	Details	Requirements	Exhibit 6-5 (Continued)
Water	Siamese connection to feed stand pipe		
Roofing	Temporary roof		
HVAC	Temporary heat		
	Interim cooling and dehumidification		
Security	Protection of facilities		

Maintenance

(General)	Maintenance of temporary systems		
	Work that may affect the Transit Authority work and zone of influence		
	Maintenance of temporaty facilities and standby operating personnel		
	Of permanent facilities used during the construction process		
	Refurbishment of equipment and extension of all guarantees and warranties		

Fire Protection

(General)	Fire resistant construction		
	Sprinkler protection of shanty areas		
Stand pipe	Fire department requirement		
	Fire extinguishers		
Siamese connection	Location for fire truck connection		
	Access to fire hydrants		

Signage & Traffic

(General)	Contractor's name and emergency numbers		
Authorities	Department of Buildings contact information		
Demolition	Demolition subcontractor's name and emergency numbers		
	Interim use of building loading dock		
Detours	Pedestrians directional		
	Sidewalk closing		
	Lane closing		
	Direction of the flow of traffic		
	Bus lanes and stops		
	Subway entrances and exits		
	Visibility of traffic signage or relocation of same		

(Continued)

Exhibit 6-5

(*Continued*)

Item	Details	Requirements
Permits		
Logistics	Permits required for implementing the logistics plan	
Construction	Permits from regent Building Department and Transit Authorities	
Labor	Overtime work permit	
Required permits for temporary facilites used during construction		
	Sidewalk closing	
	Lane closing	
	Relocation of hydrants	
	Relocation of parking meters	
	Relocation of affected transit signage	
	Installation of personnel hoists	
	Installation of material hoists	
	Installation of barricades in roadway and cross-walks	
	Relocation of bus stops	
	Relocation or closing of subway entrances	
	Sidewalk bridge and shed	
	Scaffolding	
Permits for storage	Dumpsters	
	Trailers	
	Construction materials	
	Portosans	
	Vehicles	
	Crane permit	
	Derrick permit	
	Rigging permit	
Vertical Transportation/ Sidewalk Bridge/Sheds		
Parking for pumper	Location of truck for pumping concrete	
Parking of trucks	Concrete, steel picks, equipment picks	
Sidewalk closings	Closing of pedestrian sidewalks for projects requirements	
Street closing	Closing of streets for project requirements	

Exhibit 6-5
(*Continued*)

Item	Details	Requirements
Road lane closing	Parking of equipment or storage area	
Location of hoist	Internal or external to building?	
	Construction personnel hoists	
	Construction material hoists	
	Combination personnel/material hoists	
	Lifting labor and materials	
	Loading dock for hoists	
	Barricades for hoists and loading dock	
	Signange	
	Flashing yellow caution lights	
	Concrete foundation pad for hoist	
	Tie backs of hoists to the structure	
	Runaways from hoist tower to building	
	Power for lighting and outlets at hoist landings and loading dock	
	Engineering	
	Sidewalk closing	
	Lane closing	
	Direction or flow of traffic	
	Zone of influence of the hoist for sidewalk bridge	
	Sidewalk bridge for pedestrian protection	
	Lighting under sidewalk bridge	
	Anchorage of sidewalk bridge	
	Weather-resistant enclosure at hoist entrance to building	
	Ramps from hoist entrance into building	
	Coordination of hoist location with permanent building facilities	
	Use of hoists for construction of tenant fit up program	
Elevators	Schedule for the availability of permanent service and passenger elevators	
	Interim use of service elevators	
	Interim use of passenger elevators	
Crane	Operating criteria	
Derricks	Operating criteria	

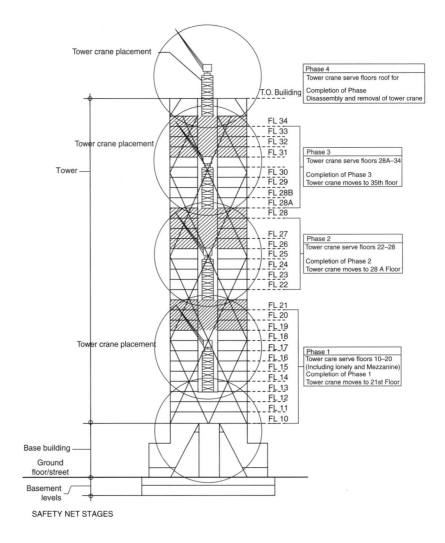

Exhibit 6-6
Tower crane phases.

Tower crane placement

Tower crane placement

Tower

Tower crane placement

T.O. Building

FL 34
FL 33
FL 32
FL 31
FL 30
FL 29
FL 28B
FL 28A
FL 28
FL 27
FL 26
FL 25
FL 24
FL 23
FL 22
FL 21
FL 20
FL 19
FL 18
FL 17
FL 16
FL 15
FL 14
FL 13
FL 12
FL 11
FL 10

Phase 4
Tower crane serve floors roof for
Completion of Phase
Disassembly and removal of tower crane

Phase 3
Tower crane serve floors 28A–34
Completion of Phase 3
Tower crane moves to 35th floor

Phase 2
Tower crane serve floors 22–28
Completion of Phase 2
Tower crane moves to 28 A Floor

Phase 1
Tower care serve floors 10–20
(Including lonely and Mezzanine)
Completion of Phase 1
Tower crane moves to 21st Floor

Base building
Ground floor/street
Basement levels

SAFETY NET STAGES

Exhibit 6-7
Climbing crane cab.

Exhibit 6-8
Climbing crane
with structural
supports.

Exhibit 6-9
Crane structure.

Exhibit 6-10

Climbing crane-support collars.

CRANE SAFETY

Crane safety is an especially poignant issue in the urban environment. Accidents in cities receive prominent news coverage that has prompted state legislatures to consider strengthening their regulations. A major crane accident occurred in New York City on March 15, 2008, killing seven people. On May 31, 2008, another crane accident occurred in New York City killing two construction workers. See Chapter 5 for more information on crane safety. California, New York, the State of Washington, and Dade County in Florida have enacted strong crane laws and other jurisdictions, including OSHA, are weighing the enactment of new rules. New York City and California have the most stringent crane requirements:

1. Only approved crane models are permitted to be used. Each crane must be registered and inspected annually.

2. Crane operators and riggers are licensed. Tower crane erection dismantling and climbing requires a licensed rigger.

3. Crane installation on new buildings must be engineered and the engineering is then reviewed by the building department.

4. Crane installations must be inspected by either a professional engineer or a building department inspector.

5. All "jumping" of cranes must be observed by a building department inspector and/or a licensed master rigger and independent crane consultant.

In 2008, 220 cranes were operating in New York City. Even with New York City's oversight, several accidents still occurred with cranes. After the May 30, 2008, incident, the New York City Department of Buildings closed down all construction sites that were using jumping cranes. These sites were not to continue work until a thorough inspection was made of every section and connection of each crane. This inspection caused major delays to the projects. Safety is the key to successful projects.

OSHA (Occupational Safety Health Administration) has jurisdiction for cranes on construction sites (Regulation Standards-29CFR Standard Number 1926.550). The ANSI B30 standards cover various types of cranes. These standards are not law but they are often treated as such by municipal building departments and they have high standings in the courts. A contractor has many incentives to keep safety of the cranes a primary objective and to heed these standards:

1. Problems with the crane can seriously affect the schedule.

2. The governing authorities can mandate heavy fines.

3. Insurance companies will increase premiums.

4. Bond sureties will increase bonding premiums or may decide not to bond the contractor.

5. Reputation can be tarnished.

6. The authorities may impose additional oversight.

7. The CM/GC could have a problem in obtaining "good" bids by subcontractors on future work.

8. With any injury or death, productivity of the workers on the job site decreases.

HOISTS

A construction hoist is a temporary elevator that transports workers and material to the various floors in a building. A hoist car that transports people is classified as an elevator and must conform to a more stringent set of standards then a mere material hoist. The hoist cars typically ride on rails that are tied to the structure every two to three floors. One or two cars ride on one mast. On a dual hoist, one car can be for workers and the other car can be for workers or material. Types of hoists and their car dimensions are indicated in Exhibit 6-11. Photographs of a hoist, hoist supports, and hoist protection are shown in Exhibits 6-12, 6-13, and 6-14. A hoist platform runway is shown in Exhibit 6-15. Hoist runways are required when the hoist is located outboard from the building structure. This would be necessary when there are set backs

Exhibit 6-11

Plan of hoist car.

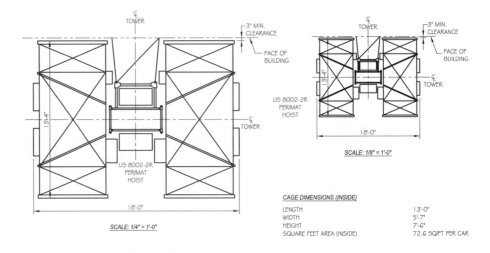

CAGE DIMENSIONS (INSIDE)

LENGTH	13'-0"
WIDTH	5'-7"
HEIGHT	7'-6"
SQUARE FEET AREA (INSIDE)	72.6 SQ/FT PER CAR

US-8002-2R PERSONNEL / MATERIAL HOIST

REGIONAL SCAFFOLDING & HOISTING CO., INC.
*3900 WEBSTER AVE
BRONX, N.Y. 10470
Tel: (718) 881-6200   Fax: (718) 324-4470*

Exhibit 6-12

Travelling hoist.

Exhibit 6-13
Hoist supports.

Exhibit 6-14
Hoist protection.

Exhibit 6-15

Plan of hoist platform runaway.

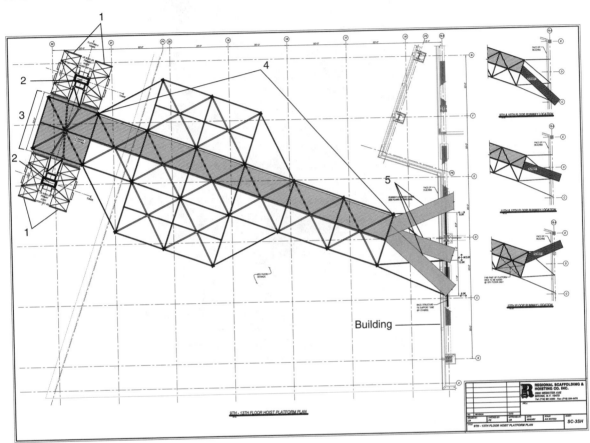

1. Rack and Pinion Hoists
2. Common Mast
3. Common Platform
4. Runaway to Building
5. Tie Backs and Supports

in the building or for ground logistics reasons. In addition, sometimes a common platform is utilized to allow several hoist assemblies to access one large platform structure adjacent to the building, thus allowing for more than two hoists in one location, if required.

The hoist noted can accommodate as many as 40 people and can carry loads of up to 8000 pounds. In a hoist designed for carrying people, the operator typically rides in the car, but in a machine that is not intended for personnel, this is not permitted. The car rides close to the face of the building so that the gap at each landing is small. If the geometry of the installation will not permit this, then a supplemental structure needs to be built to meet the car. A gate only operable from the outside and one on the structure side of the hoist are installed so that each opening is safely closed off. A typical personnel hoist for construction operates at a speed of about 300+ fpm and by certain local codes may not exceed 600 fpm.

Other safety considerations that must be observed include the following:

1. A 30-foot safety area should be maintained around the cars. This will include the installation of safety nets and a sidewalk bridge.
2. A loading dock shed should be constructed that will protect workers and trucks that will be unloading material at the loading dock.

The hoist location within the plan of the structure is determined by several factors:

1. Good access off the street for the unloading of material from trucks.
2. Close to the center of where material will be used on a floor with clear passage that will not interfere with critical finishes.
3. Close to the fire standpipe so that firefighters have ready access when they step off the hoist.
4. Minimal interference with the progress of construction by delaying closing or finishing critical parts of the building.
5. Avoiding potential conflicts with other activities such as crane picks.
6. Access to remove the hoist from the site after the work is completed.

The number of hoists on a project is determined by a cost benefit analysis considering the following:

1. Number of workers that need to go up and down the building at peak times. You want to avoid delays of bringing the workers up to the floors in the morning and then returning during the lunch break. These delays cause losses of productivity.
2. The amount of material that will have to be transported during the progress of the job. In most cases, hoists may be added when the interior work starts for the structure.
3. The cost of renting the hoist and the cost of the operators.
4. Availability of temporary power.
5. The cost benefit of schedule acceleration brought about by more carrying capacity versus the possible delaying effect of leave outs at the hoist locations.
6. A rule of thumb is one hoist car for every 250,000 square feet.

A checklist for hoists is shown in Exhibit 6-16.

Exhibit 6-16

Checklist for hoists.

Item	Included
1. Construction personnel hoists	
2. Construction material hoists	
3. Combination personnel/material hoists	
4. Location of hoist—internal or external to building	
5. Loading dock for hoists	
6. Barricades for hoists and loading dock	
7. Signage	
8. Flashing yellow caution lights	
9. Concrete foundation pad for hoist	
10. Tie backs of hoist to the structure	
11. Runaways from hoist tower to building	
12. Power for hoists	
13. Power for lighting and outlets at hoist landings and loading dock	
14. Permits	
15. Engineering	
16. Sidewalk closing	
17. Lane closing	
18. Direction of the flow of traffic	
19. Bus lanes and stops	
20. Subway entrances and exits	
21. Access to fire hydrants	
22. Visibility of traffic signage or relocation of same	
23. Zone of influence of the hoist for sidewalk bridge	
24. Sidewalk Bridge for pedestrian protection	
25. Lighting under sidewalk bridge	
26. Anchorage of sidewalk bridge	
27. Weather-resistant enclosure at hoist entrance to building	
28. Ramps from hoist entrance into building	
29. Coordination of hoist location with permanent building facilities	
30. Schedule for the availability of permanent service and passenger elevators	
31. Use of hoists for construction of tenant fit up program	
32. Structural support for the hoist pad, hoist tiebacks, sidewalk bridge, and sheds	
33. Structural reinforcing for the hoist area loading dock	

TEMPORARY SERVICES AND FACILITIES

It is extremely important during the planning of the construction project that all required temporary services and facilities are planned for to ensure that they are available in a timely and sufficient manner to support the construction process. In order for a construction site to function properly, it must have access to water, electricity, and a sewer connection. Exhibit 6-17 is a checklist for temporary services. Prior to starting construction, the CM/GC should contact One Call in the area in which you are working. (See Exhibit 6-18 for the One Call number in your area.) One Call will coordinate with the utilities, the marking out of all existing utility services coming in and out of the site, street and adjacent sites, and areas. This is imperative in an urban environment, where you have many utilities running below grade.

Once all existing services are identified, then the CM/GC needs to contact all of the respective utilities and agencies to arrange for the securing of the existing services, for those services that will be utilized during the construction process, and permanent services for the site. The following is a list of the utilities that should be contacted:

1. Electric utility company
2. Local gas company
3. Local steam provider (if available)
4. Department of water supply
5. Department of sewers
6. Local cable company
7. Local telephone company

An application to the appropriate utility along with an electric load letter (see Exhibit 6-19 for a typical load letter) is often required by the CM/GCs and owner to arrange for the temporary and/or permanent services required for the project. It often takes quite some time to arrange for services from the utilities depending on whether the service is readily available in the area, the work backlog of the utility, the moratorium period during the holidays, and emergencies that the utilities may be facing. The orders must be placed as soon as the requirements are known.

Once the utility services are provided to support the construction process, they need to be distributed within the site and building for use. The water service is secured with a valve at the foundation wall, with meters installed to measure consummation. Hydrostatic water pressure off the street mains is usually sufficient to allow water to rise to approximately the sixth floor of the building in most municipalities. To allow water to rise to higher levels, a pressure pump will be required. A hydraulic engineer should provide hydraulic calculations to ensure that an adequate quantity and pressure of water is available for construction purposes. Water will be needed for all temporary construction toilets on the site. These are normally located on the field office and shanty floors, along with approximately every five floors. Union jurisdiction and regulations of most construction trades working in urban areas will allow workers to walk up to five

Exhibit 6-17

Temporary
services checklist.

Item	Included
1. Power for demolition and abatement	
2. Power for excavation and foundation	
3. Power for hoists	
4. Power for loading dock	
5. Power for hoist tower lights and outlets	
6. Power for sidewalk bridge and sheds	
7. Power for field office and shanties	
8. Temporary light and power for construction	
9. Power for specialty equipment	
10. Power for elevators	
11. Availability of permanent power	
a) Saws	
b) Welding equipment	
c) Rigging equipment	
d) Derricks	
e) Project signage	
f) Mixing equipment	
g) Temporary water and pump	
12. Temporary toilets/portosans	
13. Temporary water for construction	
14. Water barrels for water use and drainage	
15. Standpipe for fire protection	
16. Siamese connection to feed standpipe	
17. Temporary roof	
18. Maintenance of temporary systems	
19. Utilizing permanent systems for use during construction	
20. Temporary heat	
21. Interim cooling and dehumidification	
22. Interim use of service elevators	
23. Interim use of passenger elevators	
24. Interim use of building loading dock	
25. Protection of facilities	
26. OSHA protection, signage, and logs	
27. Work that may affect the Transit Authority and zone of influence	
28. Maintenance of temporary facilities and standby operating personnel	
29. Maintenance of permanent facilities and systems used during the construction process, and the refurbishment of the equipment and extension of all guarantees and warranties	

UNITED STATES		
Region	**Website**	**Phone**
All United States	http://www.call811.com/state-specific.aspx	811
Arizona	http://www.azbluestake.com	1-800-782-5348
Northern California & Nevada	http://www.usanorth.org	1-800-227-2600
Southern California	http://www.digalert.org	1-800-422-4133
Colorado	http://www.uncc.org	1-800-922-1987
Delaware, District of Columbia & Maryland	http://www.missutility.net	1-800-257-7777
Florida	http://www.callsunshine.com	1-800-432-4770
Georgia	http://www.gaupc.com	1-800-282-7411
Illinois	http://www.illinois1call.com	1-800-892-0123
City of Chicago		1-312-744-7000
New England	http://www.digsafe.com	1-888-344-7233
New Jersey	http://www.nj1-call.org	1-800-272-1000
New York State	http://www.digsafelynewyork.com	1-800-962-7962
New York City & Long Island	http://www.nycli1calldsi.com	1-800-272-4480
North Carolina	http://www.ncocc.org	1-800-632-4949
Oregon	http://www.digsaelyoregon.com	1-800-332-2344
Pennsylvania	http://www.paonecall.org	1-800-242-1776
Texas	http://www.texasonecall.com	1-800-245-4545
Virginia	http://www.missutilityofvirginia.com	1-800-552-7001
Washington	http://www.callbeforeyoudig.org	1-800-424-5555

CANADA		
Region	**Website**	**Phone**
Ontario	http://www.on1call.com/	1-800-400-2255
Quebec	http://www.info-ex.com	1-800-663-9228, Ext. 2221
Montreal	http://www.info-ex.com	1-514-286-9228
British Columbia	http://www.bconecall.bc.ca	1-800-474-6886

Exhibit 6-18
One call service call numbers.

floors to arrive at the work area, shanty, and toilets. Temporary toilets are usually installed out in the open floor area outside of the building core, built out of temporary fire retardant wood construction, elevated off the floor to allow for drainage. Water for construction purposes is usually provided by a temporary water riser and controlled with a valve spigot on each floor, which will allow water to flow into a 55-gallon drum, with a drainpipe hooked up approximately $3/4$ of the height of the drum. These water risers need to be insulated and heat traced to prevent them from freezing during the winter.

Temporary Electrical Power

Temporary electrical power needs to be provided and distributed throughout the construction site. Temporary construction power is required for:

1. Temporary light and power stringers that are usually run on a grid of one light and outlet every 400 sq. ft (approximately one square column bay)
2. Personnel and material hoists

Exhibit 6-19

Utility load letter.

Utility Company
Urban City, USA

Re: Urban City Center
       150 Main Street – Electric Load Letter
       Urban City, USA

Dear Utility Company:

We have been retained to do the design of the electrical systems for the Urban City Center at 150 Main Street.

The owner is planning to construct a 255,000 s.f. office building:
- Basement – storage, mail room   -   5,000 sq. ft.
- 25 floors @ 10,000 s.f.   -   250,000 sq. ft.

We are herewith submitting the preliminary estimated connected load breakdown for the above-mentioned project:

Equipment	Connected Load	Remarks
Lighting	510 KW	2 watts/sq. ft. × 255,000
Miscellaneous Power	1020 KW	Included 4 w/sq. ft. office area floors × 255,000
Data Center Equipment	512 KW	
Kitchen Equipment	775 KW	Electrical kitchen
HVAC Load	1784 KW	Chillers, cooler tower, pumps
Total Connected Load	4601 KW	

3. Water pressure pumps

4. Lighting and outlets at the loading dock at the base of the hoist

5. Lighting and outlets at the floor hoist landings

6. Sidewalk bridge and shed lighting

7. Lighting to be installed on scaffolding

8. Welding, sawing, and special use equipment

9. Use of commissioning elevators

10. Stair lighting

11. Hanging scaffolds and rigs

12. Rigging equipment

13. Security lighting

The voltage, amperage, and power requirements of all construction equipment should be evaluated to ensure that sufficient power of the proper voltage is available. If high voltage power is available at the site, at 265/460 volts, 3-phase power, it should be utilized to the fullest extent possible. This higher voltage allows for more efficient distribution of electrical power, and minimizes voltage drop in the distribution of power. Transformers will need to be utilized to step down the power to 120/208 volts,

3-phase power for general lighting, small tools, etc. If only low voltage power is available, consideration should be given to stepping up the voltage to higher voltage for ease of distribution and compatibility with construction equipment. The electrical panels and circuit breakers need to have ground fault interrupting (GFI) protection to protect the workers and equipment properly, as well as comply with applicable codes and OSHA safety requirements.

STANDPIPE

A temporary fire standpipe needs to be installed in the building during the construction process to allow the fire department to fight a fire if required. One of the permanent standpipes is often used on an interim basis to provide this protection. The standpipe is dry, and fed from a Siamese connection that must be accessible from the street, outside of the sidewalk fence enclosing and securing the site. The height of the standpipe should be not less than two floors below the top floor. A sign must also be provided indicating the location of the Siamese connection at street level (see Exhibit 6-20). Portable fire extinguishers are also required on the site in addition to the standpipe. These fire extinguishers are often placed in the CM/GC field office, subcontractor shanties, at stairwells, and near welding, cutting, and use of flammable materials and fuels. The extinguisher should be of the proper size and type for the application.

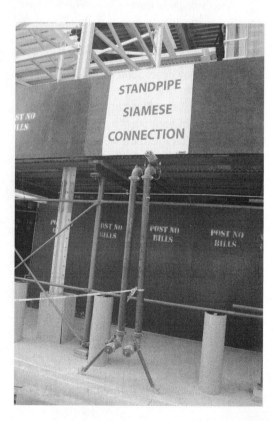

Exhibit 6-20
Siamese connection.

FIELD OFFICES AND SHANTIES

As part of the site logistics planning, it is important to identify the location for the CM/GC field office, and those of the subcontractors. See Exhibit 6-21 for the field office checklist. Initially during the demolition and then the excavation and foundation phase of the project, construction field offices are often housed in self-contained trailers that can be rented. The trailers can be located on the site, if open space is available. Often they are located adjacent to the site (see Exhibit 6-22 and Exhibit 6-23), or on top of the sidewalk bridge surrounding the construction site. If the trailers are to be located on top of the sidewalk bridge, then a heavy load bridge (300lbs/sq.ft.) to allow for the installation of the trailers is required.

The field offices and shanties are often then incorporated within the space inside the construction of the building, once the building structure is in place. Locations that should be considered are: (1) first floor retail space, (2) garage parking area, (3) below grade storage space, and (4) large contiguous space within the lower floors of the building. The

Exhibit 6-21

CM/GC's field office and shanties checklist.

Item	Included
1. CM's field office	
2. Subcontractor's field office and shanties	
3. Subcontractor storage of materials	
4. Fire resistant construction	
5. Sprinkler protection of shanty areas	
6. Emergency evacuation of construction personnel	
7. Power for field offices and shanties	
8. Location of temporary bathrooms	
9. Heaters and A/C for construction shanties	
10. Required shanties for Local 14, Teamster, and Master Mechanic	
11. Telephone, fax, Internet service for CM and trades	
12. Security of site and offices	
13. Storage of flammable and hazardous materials	
14. Location of office and shanties relative to site entrance	
15. Guard booth to control site access	
16. Fire extinguishers	
17. Types of interim heaters, air conditioners and fire protection	
18. Emergency contact information	
19. Communication system on the site	

Exhibit 6-22
Field office.

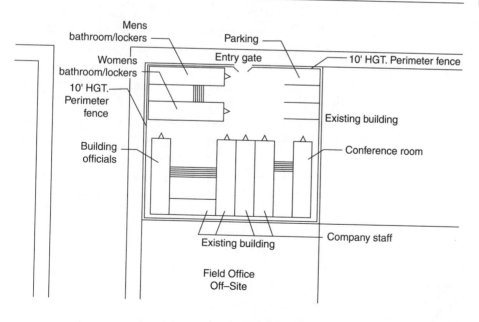

Exhibit 6-23
Office site field offices.

Mens bathroom/lockers

Parking

Womens bathroom/lockers

Entry gate

10' HGT. Perimeter fence

10' HGT. Perimeter fence

Existing building

Building officials

Conference room

Existing building

Company staff

Field Office Off–Site

location is ideally in walking distance from the personnel construction entrance to the site, and logistically in the middle of the action. The PM should try to avoid having to take an elevator or hoist to get to the CM/GC's field office if possible to maintain control of the site and overall security.

Field offices and shanties should be built out of fire retardant materials to minimize potential fire damage and comply with the local building codes, especially when the building is ready for a temporary certificate of occupancy (TCO). At this point in the project, all field offices and shanty spaces may have to be fully sprinkled, if required by code. The installation of a permanent sprinkler loop (if required by local codes) with temporary upright heads is often utilized to provide coverage of these temporary structures. As work progresses within the structure, the location of the shanties may have to change. Exhibit 6-24 shows a phasing plan for the relocation of shanties.

Exhibit 6-24

Staging for shanties.

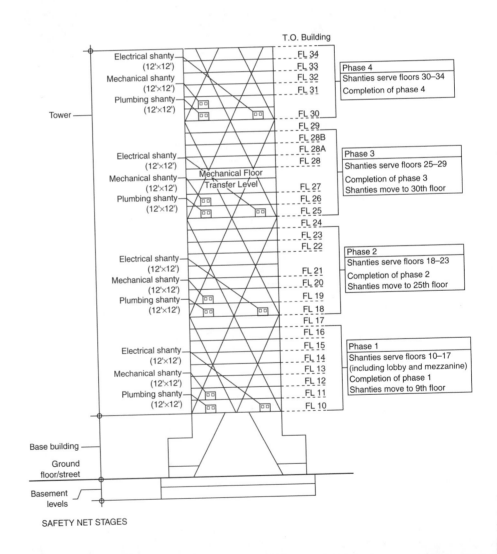

Construction field offices and shanties for the construction of the tenant fit up work is often accomplished by utilizing some of the permanent rooms that are fitted out on an interim basis as field offices and shanties by the CM/GC and subcontractors. This minimizes the costs for these types of facilities, as well as utilizing permanent systems, walls, doors, lighting, etc. to provide for the required spaces during the construction process.

If the project is large and is being built by union trades people, then jurisdictionally a special union representative may have to "sit" at the site for its entirety. A shanty for the conduction of union business may be required.

STORAGE

At most urban construction sites there is very limited space for storage of material, equipment, debris containers, and cylinders. Thus, creative means must be found for finding adequate storage areas for critical elements of the project. This may require the leasing of off-site storage areas, finding areas under stairs for temporary storage, or using off-hour areas such as loading docks for the temporary storage of debris containers.

Off-Site Storage

Often construction materials will be stored off site given the congestion found in the urban environment, with limited on-site storage. With virtually no available space at the site itself or in the immediate vicinity to utilize for on-site storage, the use of off-site storage facilities to store materials within reasonable proximity to the site will be necessary until they are ready to be incorporated into the project. See Exhibit 6-25 for a typical off-site storage facility. The CM/GC will determine what, where, and when various construction materials will need to be stored off site. Materials such as structural steel, pre-purchased equipment, generators, boilers, chillers, cooling towers, curtain wall, windows, electrical equipment, and light fixtures are examples of materials that may require off-site storage. If material is stored off site, and is to be requisitioned and paid for, ensure that the material is stored in a bonded warehouse, with an insurance certificate covering the material being paid for but not yet delivered to the site. An alternative approach would be for the manufacturers of the equipment to store the construction materials in their own warehouses, if they have sufficient space. However, the manufacturers' warehouse storage facility may be a long distance from the construction site, and may take a while to arrange for a delivery.

On-Site General Storage

Construction in the urban environment presents unique challenges for storage of materials at the construction site. Given the small size of the property, there is not much space surrounding the building on the site to allow for adequate storage. During the construction process, large quantities of construction materials will be delivered to the site for installation in the project. The CM/GC must make provisions for the interim storage of construction materials to ensure that they are available at the site when

Exhibit 6-25

Off-site storage.

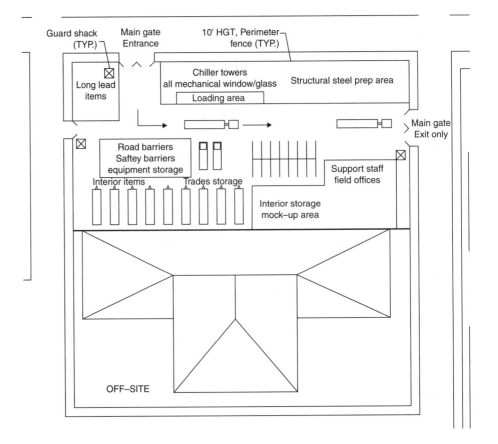

needed. Any materials being stored should be placed at a location where they do not pose a hazard to the workers, do not impede access to the construction areas, allow for the installation of other materials into the project, and are secured. The CM/GC will identify storage areas for each trade for their respective materials either on the floors where the materials are to be installed or at a central location, usually near the subcontractors' shanties. As an example, if the elevator doors and frames are stored in the floor space where sheet metal ductwork and piping is to be installed, there will be a logistical conflict. If possible, delivering and storage of the construction materials to the floor where they are to be installed would be preferable as to avoid double handling of the materials.

Dumpsters

Another challenge of construction in the urban environment is the storage of dumpsters at the construction site, given the limited amount of available space at the ground level. In the logistical planning of the project, the CM/GC will determine if a large dumpster can be accommodated on the site for the storage and removal of debris, or if mini trash containers, with a capacity of less than one cubic yard, will be utilized. Often mini trash containers are used given that they can be wheeled around the building to various floors to pick up debris, and then brought to the hoist or elevator loading dock area for a trash truck with a compactor to remove them. The CM/GC must coordinate the use of the hoist and elevators to allow delivery of materials and personnel and the removal of debris, without causing congestion at the loading dock area, and restricting the access and availability of the hoists and elevators. This often requires deliveries and removal of debris before or after hours.

Storage of Gas Cylinders and Other Hazardous Materials

During the construction process, cylinders of flammable gases will often be utilized for demolition, welding, and brazing. These cylinders of gas tend to be disbursed around the site to perform construction work in various areas as required. They need to be kept track of, and locked up in a safe location each evening at the completion of the work. The lock up is either a locked steel cage secured to the structure or a chain and lock with the cylinders secured to the structure, such as a column. The cylinders should be identified as to their contents, have caps placed over them to secure the tops and valve areas, and be in good condition. A cylinder of flammable gas has a tremendous amount of explosive power. If the gas were to escape and ignite, this could result in a significant amount of personnel and property damage. It is important for the CM/GC to check the local codes and with the fire department to ensure adequate and safe storage of all materials at the construction site. Upon completion of the use of the flammable gases, they should be removed from the site as soon as possible to avoid any unnecessary exposure.

REQUIRED PERMITS

Prior to the start of the construction process, it is important that the CM/GC identify with the local building department, fire department, expeditor, and transportation department those filings and permits that are required to support the required logistics of the construction project, aside from the construction process and permits to perform the work itself. The owner often retains these expeditors for the filings, permits, inspections, sign offs, and approvals of the project. The CM/GC should utilize the services of these specialized professionals to help facilitate the process. See Chapter 4 for additional information.

Once the logistical plans are developed for the project, the required filings and permits must be identified and obtained in a timely manner to support the construction process. The filings and permits that are sometimes required from the local building department are:

1. Hoist
2. Sidewalk bridge and shed
3. Scaffolding
4. Fence
5. Crane
6. Derrick
7. Water supply and services
8. Gas supply and services

9. Steam supply and services

10. Electrical supply and services

11. Storm sewer connections

12. Sanitary sewer connections

13. Overtime work permits

Additional filings and permits that are sometimes required from the local transportation department may include:

1. Lane closing

2. Sidewalk closing

3. Pedestrian crossing relocation and/or closing

4. Bus stop relocation

5. Closing of the local transportation system entrances

6. Removal and/or relocation of parking meters

7. Removal and/or relocation of fire hydrants

8. Removal and/or relocation of traffic signage

9. Removal and/or relocation of traffic signals

Filings and permits that are sometimes required from the local fire department are:

1. Storage of flammable materials

2. Storage of metal cylinders containing flammable gases

3. Permit to burn fuels

4. Permit to store fuels

These filings and permits should be initiated early in the construction-planning phase of the project to ensure that all required permits are obtained in a timely manner, when the construction of the project is ready to proceed. Given that the logistics of the project will change over time, it is important to identify when the above referenced items are required, and filed and permitted accordingly. Permits once obtained are valid for a specified period of time, which varies with the type of permit and particular agency involved. A permit log should be maintained to properly administer and control the application and renewal of all required permits, to ensure that none of the required permits expires (see Chapter 4). In certain locales, permits will expire when the CM/GC's insurance policy is up for annual renewal, regardless of the timing of the normal duration of the permit itself.

MORATORIUMS

Some of the major cities in which the CM/GC will be working have a moratorium period for things such as major parades or visits by foreign and domestic dignitaries. During the moratorium period, the local government might not issue a permit for new work or logistics that would potentially disrupt the flow of pedestrian and vehicular traffic. It is prudent to file and obtain all required permits prior to the moratorium period to avoid this potential problem. At times, alternate logistical plans need to be put in place during this period to allow for the proper logistical access and support of the construction project.

SIGNAGE

In order for the site to comply with traffic, safety, and notifications regulations, numerous signs must be installed in and around the site. A list of proposed construction site signs is noted in Exhibit 6-26, depending on the municipality. Typical site signs are shown in Exhibits 6-27, 6-28, 6-28A, and 6-28B.

Item	Included
1. Contractor's name	
• Address	
• Emergency telephone number	
2. Building Department	
• To contact in case of an emergency with telephone number	
• To contact in case anyone notices a violation	
3. Demolition subcontractor	
• Name and emergency telephone number is listed	
4. Excavation subcontractor	
• Name and emergency telephone number is listed	
5. Directional sign for pedestrians	
• No Entry	
• Sidewalk closed	
• Follow the sign for the temporary walkway	
6. Security signs	
• Authorized personal with proper badges	
• Gate locations	
• Personal gate locations	
7. Height signs	
• At loading dock for truck information	
8. Hydrant sign	
• For fire department identification	
9. Street closing	
• Lane closing as per DOT permits	
• Street closing as per DOT permits	

Exhibit 6-26

Proposed costruction site signs.

(Continued)

Exhibit 6-26

(*Continued*)

Item	Included

10. No smoking signs
  • Hang around the site so that workers refrain from smoking
11. Hard hat work area
  • Everyone on the site must wear a hard hat
12. Eye protection
  • Workers who are cutting, sawing, or burning material should all wear eye protection
13. General safety signs
  • Remind all workers that safety is a primary concern
14. No parking sign
  • Authorized parking only
  • Parking required for trucks, pumpers, concrete, steel delivery, etc.
15. Flag personal
  • On occasion, a flag person is required to: (1) divert traffic, (2) hold pedestrian traffic while picks are occurring, and (3) when buckets are being raised or lowered
16. Transportation signs
  • Bus stop location
  • Underground transportation location
  • Tram location	

Exhibit 6-27

Typical
construction sign.

Exhibit 6-28
Safety zone sign.

Exhibit 6-28A
Follow the
flagman.

SIDEWALK BRIDGES, SHEDS, AND FENCING

In order to protect the public, the workers, and the equipment, a form of protective "shell" has to be created. The form of shell that is commonly used on construction sites is the sidewalk bridge. A checklist for bridges and sheds is shown in Exhibit 6-29. This bridge consists of lolly columns spaced approximately 5 feet on center. Small steel sections are connected at the top to the lolly columns. Then wood planks span the steel beams. Typical pedestrian sections are indicated in Exhibit 6-30 and Exhibit 6-31. The

Exhibit 6-28B

No smoking sign.

NO SMOKING ALLOWED IN THIS BUILDING AT ANY TIME UNDER PENALTY OF LAW CARRYING LIGHTED CIGARS, PIPES OR CIGARETTES STRICTLY FORBIDDEN

roof of this shell should support a load of 300 pounds per square foot. Safety lights are hung under the planking so pedestrians can see where they are walking. At truck gates (that intersect the pedestrian walkway) strobe lights are placed at the intersections so that pedestrians will know when a truck is pulling in or out of the loading dock. In addition, a flag person is assigned to stop pedestrian traffic. The bridge should extend 50 feet beyond the limits of the site under construction.

At areas where workers will be loading or unloading material at the hoist, a bridge has to be provided to protect the workers. The bridge should support a load of 300 pounds per square foot and should extend at least 30 feet out from the hoist structure. It is best to check with the local jurisdiction to determine their requirements.

In some areas, on the first floor it may be necessary to enclose a linear portion of the space. In this instance, a shed is provided. The shed is used to help keep out the public and minimize the infiltration of adverse weather elements. The shed is usually constructed of plywood.

SECURITY

Security of the total site is of extreme importance to the owner and workers. The owner does not want unauthorized personnel walking around the site. This could create a liability problem if someone is injured on the site and is not authorized to be there in the first place. In addition, expensive tools, equipment, and components have to be protected from theft. Therefore, a protective fence should be installed to protect the site from unauthorized entry and to minimize potential theft. The fence is usually made of

Item	Included
1. Construction personnel hoists	
2. Construction material hoists	
3. Combination personnel/material hoists	
4. Loading dock for hoists	
5. Barricades for hoists and loading dock	
6. Signage	
7. Flashing yellow caution lights	
8. Concrete foundation pad for hoist	
9. Tie backs of hoist to the structure	
10. Runaways from hoist tower to building	
11. Power for hoists	
12. Power for lighting and outlets at hoist landings and loading dock	
13. Permits	
14. Engineering	
15. Sidewalk closing	
16. Lane closing	
17. Direction of the flow of traffic	
18. Bus lanes and stops	
19. Subway entrances and exits	
20. Access to fire hydrants	
21. Visibility of traffic signage or relocation of it	

Exhibit 6-29
Sidewalk bridge/
sheds checklist.

Exhibit 6-30
Typical bridge
section.

Exhibit 6-31

Fancy urban bridge section.

material that is difficult to penetrate. Gates are provided for access of authorized trucks and personnel. Guards are posted at the gates to check everyone's identification. See Chapter 11 for additional security information.

UNDERPINNING AND SHORING

In the urban environment, there is a lack of available land on which to construct new structures. Thus, older buildings are torn down to make way for new structures. When this occurs, older buildings surround the new building's site. The new building's footings and basement excavation may extend below the footings and perimeter walls of the older buildings. When this occurs, the CM/GC must analyze the conditions that exist and make sure that the older surrounding buildings will remain stable during the excavation of the new building. This may require the temporary support of the older buildings' footings or the placement of a wall around the new site that will support any load emanating from the older buildings' footings. Whichever stabilization process is selected, the CM/GC needs to review the foundation plans of the surrounding buildings. If the plans are not available, then other sources must be sought out. The public library, local building department, and historical societies are good sources of information.

If no documents are available, then bore scopes or test pits may have to be dug around where the existing footings are located. This must be done with care so that the existing footings are not disturbed. The owner's design team and the CM/GC and a specialist subcontractor should be brought in to review the conditions. Recommendations will then be made as to which method should be employed with associated cost and schedule. In some cases, a balance may have to be made between the cost to stabilize the existing structures and the proposed schedule. Several underpinning and sheeting methods can be used to stabilize adjacent buildings and the surrounding soil of the new building's excavation. In addition, a surveyor should be retained to monitor the condition of the existing building to make sure no movement is occurring.

Underpinning

Jet grouting—A series of pipes are run in the ground under the footings of the existing buildings. Grout under high velocity is then sent through the pipes. The grout breaks up the soil structure and creates a solid grout–soil mass under the footings. Work could then continue with the excavation of the new building.

Freezing of the soil—A closed pipe system surrounds the footings and refrigerant then runs through the pipes. The soil then becomes frozen under the existing footings and the new excavation work could continue. In addition, this may be a temporary method until a more permanent underpinning method is employed as will be described further. This method can also be used to temporarily stop water flow that may be affecting the site construction.

Minipiles—Small diameter (4 to 10 in.) piles are driven through the foundation of the existing building (and the new building's partly excavated area). These piles can be made of steel, steel pipe (filled with concrete), or concrete and would then be driven into the soil surrounding the existing footings. If vibration is a problem, then pre-bored concrete or pipe piles would be placed in the ground and then grouted in place. The minipiles could be used as end bearing or friction piles (if sufficient pile surface area can develop the required frictional stresses). See Exhibit 6-32 for a typical minipile installation.

Straddling footings with needle beams—New footings can be installed on either side of the existing footings. Then steel beams (called needle beams) would be pushed through the soil to be supported by the new footings. The needle beams would then support the old footing.

Wall replacement—Walls can be replaced or extended by digging under the existing wall and placing a new wall footing foundation. This would be accomplished by removing small sections of the wall (to be confirmed by the engineer) and then placing new walls on top of the newly created wall footing. See Exhibit 6-34A for details on wall replacement methods.

Prior to starting any work, the checklist shown in Exhibit 6-33 should be reviewed.

Shoring and Sheeting

Soldier beams and lagging—Steel H piles are driven into the soil to a depth below the excavation level. As the earth is removed for excavation, wooden slots (lagging) are placed between the H piles (see Exhibit 6-34). Depending on the stability of the soil, it may also be necessary to anchor the H piles. This is accomplished by placing rods (or

Exhibit 6-32

Minipiles detail.

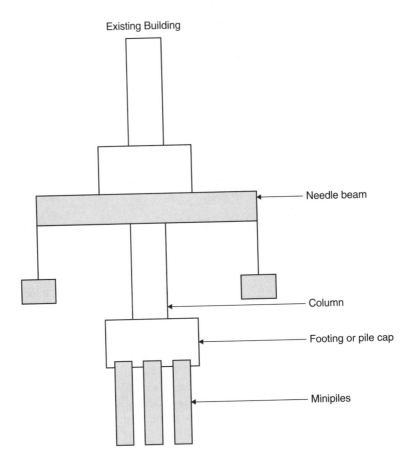

Existing Building

Needle beam

Column

Footing or pile cap

Minipiles

Exhibit 6-33

Underpinning and shoring checklist.

1. Obtain drawings of the existing building's foundation plan.
2. If required, dig small test pits around the foundation without disturbing the structure.
3. Prepare a plan of the new excavation and foundation in relationship to the existing building's foundation.
4. Show a section at the elevation of the new foundations in relation to the old foundation and the distance between the two. This has to be accomplished at all impacted foundation locations.
5. Indicate the type of existing foundation (spread footing, piles, mat, etc.).
6. Indicate the material of the existing foundation (stone, concrete, steel).
7. Indicate type of soil or rock encountered under the old foundation.
8. Determine estimated load on the existing foundation.
9. Locate any water and determine the elevation.
10. Review all information with the soils engineer and underpinning specialty contractor.

Exhibit 6-34
Soldier beams details.

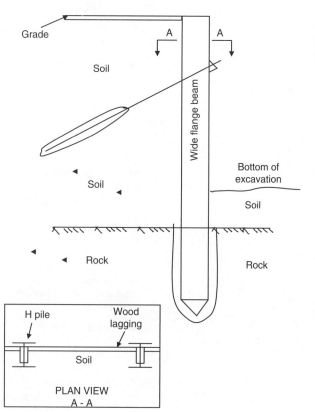

cables) into the soil or rock behind the H piles and lagging and then anchoring them in place by grouting the rods or cables. If the soil is not capable of taking the load of the anchors, then rackers are placed to brace the soldier beams and lagging from inside the excavation. This method is problematic because it restricts working around the rackers within the excavated area.

Steel sheeting—Four feet length (or longer) of steel sheeting is "piled" into place to the depth of the excavation. The sheeting sections are attached to one another by interlocking with the next section. As the earth is excavated, the sheeting is anchored back to the soil or rock as described in the previous paragraph.

Slurry wall—A clam bucket excavates a trench around the site. As soon as the trench reaches a depth where the earthen walls are no longer stable (and will collapse) a slurry mixture (bentonite) is placed in the trench to stabilize the walls. Once the trench and slurry reaches a depth below the excavation elevation, then a reinforcing steel cage is placed in the trench. Once this process is completed, concrete is poured into the trench, replacing the slurry mixture (which is reused). Once the concrete hardens, the concrete wall is then anchored to the soil or rock as stated previously. In some cases, precast prestressed concrete panels are used in lieu of pouring the concrete (with the reinforcing steel cage). See Exhibit 6-35 for a slurry wall section and Exhibit 6-36 for a photograph of a slurry wall with tiebacks.

Exhibit 6-34A
Alternative
supporting walls.

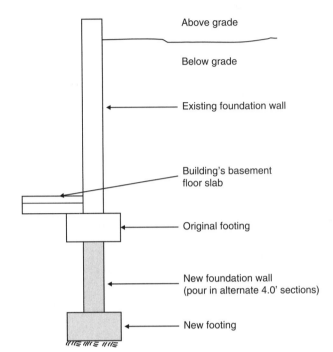

- Above grade
- Below grade
- Existing foundation wall
- Building's basement floor slab
- Original footing
- New foundation wall (pour in alternate 4.0' sections)
- New footing

Exhibit 6-35
Slurry wall detail.

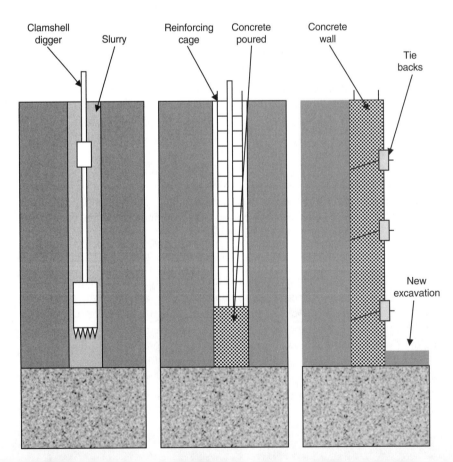

Clamshell digger   Slurry   Reinforcing cage   Concrete poured   Concrete wall   Tie backs   New excavation

Exhibit 6-36
Slurry wall with anchors.

Soil mixing—This procedure mixes grout with the soil (by rotating paddles) to create small long cylinders (that reach the bottom of the excavation). The continuous linking of these cylinders would provide a wall around the site.

Soil nailing—Reinforcing steel bars of approximately $1/2$ in. to 2 in. are placed horizontally into the soil with plates on the outside of the soil. Then grout is injected along the length of the reinforcing bars. This nailing creates a solid block of soil that can sustain the earth's pressures. This method can increase the shear strength of the soil.

EXCAVATION

Excavation is a method for removing earth and rock from a site in order to get to a certain elevation, usually below street level. The reasons for getting to a lower elevation could be for:

1. Basement storage area
2. Mechanical equipment room
3. Electrical rooms, especially for entering electric service that would emanate from a transformer that is installed below the street level
4. Utilities coming into the site from those that are buried in the street
5. Spread footings for the foundation
6. Parking garage
7. Special functions for the building (meeting rooms, gym, etc.)

Initial Layout

The site engineer indicates on the site plan the boundary of the area to be excavated. The subcontractor responsible for the excavation checks the layout that was initially created by the CM/GC's surveyor. Once the layout is confirmed, a fence is established around the site (with access gates for equipment), so the public is protected from the excavation. In addition, the CM/GC is protecting the site for security reasons and for possible intrusion from unauthorized personnel. The CM/GC has to be concerned about injury to unauthorized people who may enter the excavation for whatever reason (vandalism, burglary, or just for fun).

The Equipment

Heavy equipment is brought in to start the removal work. This can consist of bulldozers, rock crushers, trucks, backhoes, compressors, tunnel drills (for any required tiebacks), or even a crane with a scope bucket attached. Trucks will need access to the site for removal of the dirt. In order to protect the public, when equipment enters or leaves the site a flag person is stationed at the access gates stopping pedestrians from crossing in front of the equipment that may be arriving or leaving the site. When the excavation is completed, the equipment is removed from the bottom of the excavation by the use of a crane.

Excavation Stability

When the excavation reaches a certain elevation below the initial grade level and the point at which the sides of the excavation may become unstable, then some retaining system is required. The side of the excavation may become unstable because the soil cannot maintain a 90° cut or the pressure being exerted by the soil from street traffic loads or other conditions becomes too great. When this condition exists then the methods outlined in the section on sheeting will have to be implemented. If an existing building is close to the excavation, then shoring and underpinning may be required. Exhibit 6-37 is a photograph of a building being braced due to adjacent excavation. Exhibit 6-38 is a photograph of a typical urban site where contiguous buildings are very close together to the proposed building site. See the section on shoring and underpinning for the methods used to stabilize the existing structure. In addition, a ramp has to be created so that the trucks are able to climb out of the excavation. If water is found when excavation continues, then some form of dewatering process will be required. See the section on dewatering for the methods used to contain or divert water.

Rock Excavation

At one time, rock was removed by blasting with dynamite. In order to protect the public, metal mats were placed so that the rock pieces would not fly into the air and injure someone. Now, in most urban situations, blasting is not allowed due to the potential dangers, dust, vibration, and noise. Rock crushing and drilling machines are now used to break up the rock. Exhibits 6-39 and 6-39A are photographs of a drilling machine and rock crusher. After the rock is crushed, the heavy equipment is brought in for placing the rock onto the trucks. This excess excavation material is sold to a third party for use in another project or process (sea walls, road and railway beds, etc.).

Exhibit 6-37
Bracing of wall.

Exhibit 6-38
Buildings surrounding proposed site.

Exhibit 6-39
Rock driller.

Exhibit 6-39A
Rock crusher.

DEWATERING

Dewatering is the process of removing water from the construction site, utilizing berms to keep surface water away, and pumps to drain the subsurface of the site and lower the water table so that the work can be performed in a dry site. When building in an urban environment, the CM/GC often encounters subsurface water conditions given that the major cities around the world were originally founded near bodies of water to facilitate access and transportation. The specifications, local building codes, and project safety usually require that excavation and foundation work be performed in a dry site. In addition, given the limited amount of space to build on in an urban environment, buildings tend to be taller and have several sub grade levels, thus having a greater tendency to encounter subsurface water.

Water tables are usually found below grade level. Dewatering is usually accomplished by pumping the water from the site, from sump pits at the bottom elevation of the site. In order to ascertain the level of the water table and the type of sub-grade soil and rock conditions that exist at the site, borings need to be done at the beginning of the project. This will indicate the existing water table elevation. The PM needs to be aware that if the construction site is near a body of water that is affected by the tides, then the water table may vary depending on the tidal conditions and whether the site is in the zone of influence of the tidal body of water. Exhibit 6-40 is a diagram of a typical well point installation.

Methods for Dewatering

Depending on the type of soil encountered, two different types of approaches are generally utilized. The first approach is utilized when sand and silt are being excavated. Sand and silt have a tendency to seep into the sump where the water is being pumped,

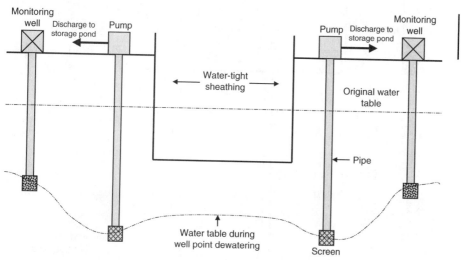

Exhibit 6-40
Well point dewatering.

and possibly soften and shift the soil in the surrounding area on which the foundations of the building may bear. In order to stop groundwater from entering the site, draw down sumps can be created around the site to depress the entire water table in the surrounding area. This process utilizes vertical pipes with screened openings at the bottom to keep soil in place and allow water to enter. The water has to be discharged away from the site to prevent it from potentially re-entering. Discharge from such wells is often regulated by various jurisdictions. The PM may be required to obtain permits for the project to allow for the legal discharge of the water. Piezometers, which measure the water elevation in the surrounding area, are also utilized to monitor the draw down curve of the surrounding area and its water table. Sometimes, when excavating to a lower elevation at the site, a series of rings of well points may be required to properly draw down the water table of the site and the adjacent areas. The lowering of the water table by well points in the surrounding area can have an adverse effect on neighboring buildings. It can cause consolidation or settlement of the surrounding soil under the adjacent building foundations. It can also expose untreated wood pilings that were previously immersed in subsurface water to air and possible decay. It is recommended that a geotechnical engineering consultant be retained to determine the subsurface conditions and the appropriate method of dewatering. Exhibit 6-41 indicates water being pumped utilizing well points monitoring wells in the surrounding area.

Another approach is to erect a watertight barrier creating a bathtub effect around the entire site. See the section on shoring and sheeting. This only works if the watertight barrier walls go down to the bottom edge of an impermeable layer of subsoil material to prevent water from seeping under the walls. A slurry wall is often utilized in this situation, and makes a good watertight barrier. Sheet piling is sometimes used; however, it has a tendency to leak around the joints in the piling. The hydrostatic pressure

Exhibit 6-41
Dewatering draw down curve.

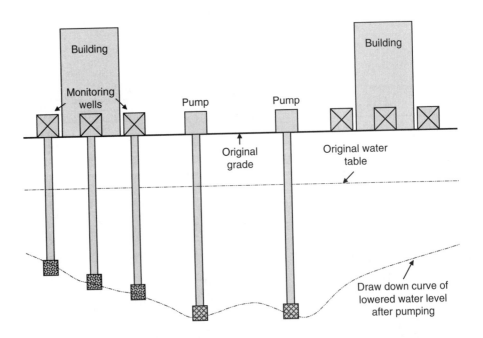

of the water surrounding the site increases as the water table is lowered at the construction site, and therefore the watertight barrier system must withstand the additional lateral pressure with a strong system of bracing and tiebacks.

Once the water table is lowered and the foundation is put in place at the site, the CM/GC must be careful as to when and at what rate the water table can be raised to its original elevation. The new foundation is like a watertight bathtub, which will experience hydrostatic uplifting pressure once the water table is raised. The building has to have a sufficient weight to counteract this uplifting force, which has to be properly engineered and sequenced.

SUMMARY

- A good logistics plan is one of the key ingredients for a successful urban project.
- Protection of the public by fences, bridges, and other barriers is a paramount consideration for an urban construction project.
- In order to be efficient and to meet critical schedules, the movement of trades people and material must be analyzed. Thus, the location and number of hoists and cranes must be determined by the CM/GC and related subcontractors.
- Signs must be placed around the site for directing the public around the site, notification in case of emergency, traffic flow, and safety considerations on the site.
- In urban settings, storage of material, cylinders, and dumpsters is extremely difficult due to the limited amount of space at the site. Therefore, off-site storage must be considered and space must be set aside on site for dumpster and cylinder storage.
- Urban sites require the evaluation of structures adjacent to the proposed site. Underpinning of the adjacent structures may be required for stability. Several methods could be used, such as minipiles, shoring, and needling.
- When excavation takes place, the side walls have to be stabilized. This could entail soldier beams, sheeting, or slurry walls.
- Since urban centers are located near major waterways, the groundwater table may be close to where the site will be excavated. Thus, several methods are used to remove the water from the site, such as sump pumps and well points.

7 Layouts and Surveying
(We are only off by 1 foot. Big deal!)

SURVEYING

Building surveyors have been around before the great pyramids of Egypt were built. For the great pyramid at Giza, the surveyors understood how to calculate angles (52° on all sides), level the base within $1/4$ in. and measure lengths (750 feet on all four sides with an error of only 8 in.). This is quite an accomplishment considering the fact that they had no sophisticated surveying equipment or the use of GPS. There are no indications that corrections were made for heat, tension of the measuring rope, or for keeping the measuring rope from sagging. The need for surveyors became more prominent in history as land was divided by the various monarchs of the time. The need to measure property areas and boundaries became the requirement for not only indicating land ownership but for tax purposes as well. During the Middle Ages, the surveyor had to lay out fortifications, castles, religious buildings, municipal buildings, and other structures. This all gave rise to the modern building urban surveyor who is responsible for the layout of the building site, location of the footings, determination of the floor elevations, and the complete alignment of the building being constructed. Due to limited access to sites in the urban environment, the surveyor has to deal with vibration from traffic and mass transit, the set up of equipment with pedestrian traffic (potential for knocking over the equipment), and trying to obtain line of sight with numerous impediments in the way (buildings, trucks, cars, cranes, construction equipment, etc.). See Exhibit 7-1 for the surveyor setting up within an urban environment. The surveyor uses several bench marks for setting up the instrument and for determining elevations. Exhibits 7-2, 7-3, and 7-4 show markers in the sidewalk that are used by the surveyor. However, the surveyor prevails, and the work is accomplished with the utmost accuracy. See Exhibit 7-5 for the site survey flowchart.

Exhibit 7-1
Surveyor working in an urban environment.

Exhibit 7-2
Setup marker in a sidewalk.

Exhibit 7-3
Control point
sidewalk-2 marker.

Exhibit 7-4
Benchmark in
concrete base.

BASIC INFORMATION

All surveyors should be registered in the state where the Construction Manager/General Contractor (CM/GC) is performing the work. In most states, the surveyors must pass a state exam in order to qualify as a registered surveyor.

Exhibit 7-5

Site layout
flowchart.

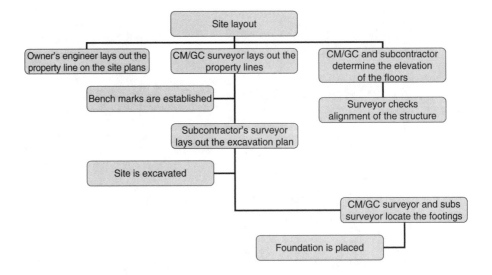

The owner's civil engineers prepare the site plan for the project, indicating the boundary of the property either by meets and bounds or by bearing and distance. Meets and bounds tend to be used in the urban environment because the reference points are the major buildings and streets in the area. So the boundary descriptions are tied into the address of a building and the associated streets. This would not be true if a project were being constructed in the farm land of Kansas. In this particular case, the surveyor would use compass points and distance to determine the property lines for a potential project. The two types of methods for indicating property boundaries are noted on Exhibit 7-6 and Exhibit 7-7.

The owner's design team will indicate the location and elevation of the footings and will note the elevation of each floor. The CM/GC's surveyor takes the basic information from the construction documents and lays out the property lines. See Exhibit 7-8 and Exhibit 7-8A for the photograph of a property line indicated on a sidewalk. When excavation starts, the surveyor determines the elevation of the excavation and the location and elevation of the footings.

Exhibit 7-6

Meets and bounds.

Description of Land and Premises

The zoning lot on which the premises are located is bounded as follows:
BEGINNING at the point on the *South* side of **Main Street** distant *150.10 (East)* feet
of the corner formed by the intersection of **Main Street** and *57th Street*

running thence	*110.50 (east)* feet; thence *100 (south)*	feet
thence	*110.50 (west)* feet; thence *100 (north)*	feet
thence	feet; thence	feet
thence	feet; thence	feet
to the point of beginning.		

Exhibit 7-7

Bearing and distance.

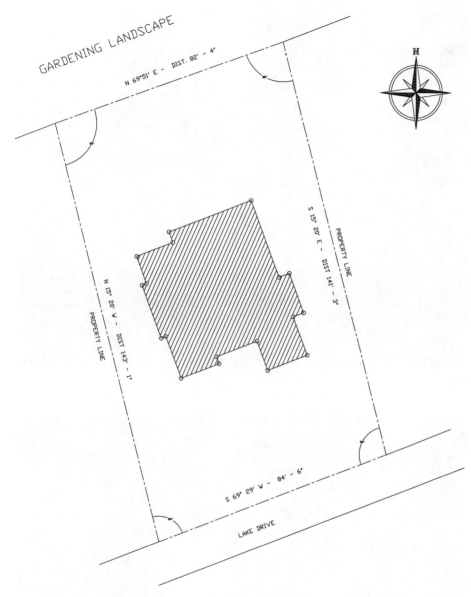

After the CM/GC's surveyor lays out the basic parameters of the site, the subcontractor's surveyor takes over and establishes benchmarks (elevations) for the components that have to be constructed. Thus, the concrete subcontractor will lay out the location of the footings in relation to the property line of the building. In addition, stakes (wooden slates) will be driven into the ground at the lowest point of the excavation. These stakes will note location of the footing and the bottom elevation of the footing. As the excavation of the footing continues, the subcontractor's surveyor will constantly check the location and elevation of the footings.

Exhibit 7-8
Property line.

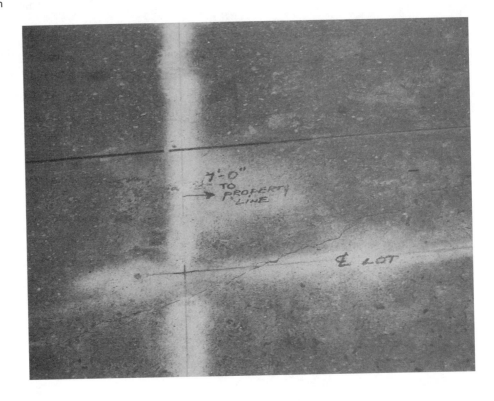

Exhibit 7-8A
Property line to
face of steel.

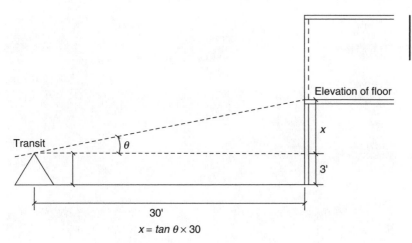

Exhibit 7-9
Surveying by
trigonometry.

Elevation of floor

x

3'

30'

$x = \tan \theta \times 30$

The subcontractor's surveyor again establishes the elevations of each floor of the building. Thus, for a steel structure the steel subcontractor will have their surveyor determine the elevation of the top of steel at each floor. This is usually done by trigonometry as noted in Exhibit 7-9. When the concrete has to be poured over the metal deck of a steel structure, the concrete subcontractor's surveyor establishes the top slab elevation. In most cases, the elevator sill height is maintained as the benchmark for the floor.

The use of trigonometry for determining the elevation can be used for all floors. As the surveyor starts determining the elevations of higher floors, the calculations may be affected by temperature differential and the curvature of the Earth. In addition, the compression of the structure has to be taken into account. The owner's design team should determine potential compression for a steel or concrete structure.

Elevations at a site are measured from established datum. Most datum use the average (mean) sea level as "0" datum. The mean sea level (0 datum) is determined by measuring tide gauges (high and low tides) at the seacoast city you are using for a period of 19 years. For inland areas, the mean sea level is "brought" to the interior land area that is being considered. Most cities have monuments established by the National Geodetic Survey (NGS), which is part of the U.S. Department of Commerce, National Oceanic and Atmospheric Administration. Information on NGS can be obtained from their Website www.ngs.noaa.gov.

STEEL STRUCTURES

The structural frames for steel buildings are erected at a fast pace. Thus, the alignment of the structure must be "true" if the building is not to lean like the tower of Pisa. To avoid these problems, the surveyor checks the alignment of the columns with

Exhibit 7-10
Turnbuckle.

a transit, and then the iron worker checks each column with a plumb bob. If the surveyor or the iron worker find out that the columns are out of alignment, then cables are placed at the top and bottom of the columns and turn buckles are installed (see Exhibit 7-10) to straighten the structure. The permissible alignment of interior columns for the first 20 floors is 1 in., with an increase of 1/32 in. for each additional floor with a maximum for the full building of 2 in. For exterior columns, the displacement for the first 20 floors is no more then 1 in. inside the building line and no more then 2 in. outside the building line. Above the 20th floor, the limits may increase by 1/16 in. for each additional story but no more then 2 in. toward the building line or 3 in. away from the building line. However, all alignment criteria should be obtained from the owner's design team.

FACTORS AFFECTING THE SURVEYOR

Due to the geometry of the building, its height, construction loads, and other factors, the surveyor has to make many corrections in order to develop reliable information for setting benchmarks and elevations. These corrections are noted in Exhibit 7-11.

In order to avoid any problems with the setting of critical points, the surveyor must do the following:

1. Set elevations in a "monument" (fixed object set in concrete) that will be permanent and cannot move near the building.

2. Set benchmarks at every floor of the building.

1. Wind loads
2. Thermal expansion from the sun
3. Contraction from cold weather
4. Shortening from the force of upper floor loading and façade
5. Possible minor settlement of the foundation
6. Any large construction equipment that is temporarily supported by the building structure
7. Vibration caused by traffic and underground mass transit systems
8. Vibration caused by construction equipment
9. The design of the structure with potential offsets and setbacks
10. Height of the building and the curvature of the earth
11. Adjacent buildings that may be out of alignment

Exhibit 7-11

Surveyors corrections.

3. Critical operations should be accomplished when the sun is not shining, and the temperatures are in the 50 to 60°F range (where possible). If these conditions do not exist, then corrections will have to be made to the calculations.

4. The owner's structural engineer should determine settlement calculations at various stages of the project.

5. The owner's structural engineer should determine potential shortening of the structure due to the loads from other floors and construction equipment.

Surveyors have used many methods to achieve their objective of constructing a straight structure with the proper floor elevations. These have included:

1. Cutting four small holes on each floor around the core and then checking floor alignment and elevation at each floor via a laser beam and prism targets.

2. The use of GPS on each floor where coordinates and elevations can be determined.

3. The old standard of using a transit and measuring the angles and distance from known benchmarks.

4. In major cities, buildings are being constructed contiguous to other buildings. Then the surveyor places benchmarks on the existing structure. The benchmarks are then used as reference points as the structure climbs from the first floor to the upper floors.

Whatever method is used, it is critical that the alignment and elevations of the structure be set to the parameters established by the design team. Any deviations can create major problems for the project. Thus, in most cases not only is the CM/GC's surveyor involved in the surveying process but the subcontractors who are performing the actual work use their surveyors for confirming all calculations. See Exhibit 7-12 for the construction survey checklist. This list should be reviewed prior to and after the surveying process has begun.

Exhibit 7-12

Survey checklist.

Project_____ Date _____
Owner_____ Job No._____
Contractor_____ Contract No._____
Project Manager_____

Item	Name	Date	Remarks (field book, page)
Preliminary Bench marks			
Base line			
Existing grades			
Property lines			
Buildings Layout stakes			
Footing grades			
Finish floor elev.			
Compaction tests (slab on grade)			
Utilities Layout stakes			
In place grades: sewer			
Steam			
ELectric			
Gas			
Storm drain			
Comp. tests in backfill			
SITE WORK Compaction (sub-grade)			
Curb & walk layout stakes			
'In place' grades (manhole covers, catch basins, etc.			
FINISH GRADES (1) Large landscapes			
(2) Parking, etc.			
SPECIAL			

SUMMARY

- The proper layout of the building including footings, building alignment, and floor elevations is a key component of the project. Thus, without a surveyor establishing the critical benchmarks and elevations, a project cannot proceed.
- The surveyor must take into account external elements that may impact the accuracy of the building's alignment and floor elevations. This includes:

- Wind loads
- Solar loads and high temperatures
- Cold temperatures
- Building compression
- Settlement
- The surveyor uses numerous procedures to obtain the accuracy required for a major building.
- Due to the critical nature of the required measurements, more then one surveying team will check all alignment and elevation calculations.

8 Drawings and Specifications
(The 100% documents. Not!)

DRAWINGS AND SPECIFICATIONS

Drawings and specifications represent the scope of the project and the details required for its construction. These documents will include plans, elevations, sections, reflected ceiling plans, blow-ups, definitive information (such as the type of pipe to use), and testing and quality control (QC) that will be used for the project. See Exhibit 8-1 for a drawings and specifications flowchart.

The architects and associated consultants (design team) prepare the construction documents. The construction documents are supposed to be coordinated by the design team that is involved with their preparation. The coordination process entails the review of the architectural, structural, mechanical, electrical, plumbing, sprinkler, and other plans to make sure conflicts do not occur between the various design team discipline drawings. Unfortunately, the reality of the consultant's world is that the consultants usually do not have the proper time to coordinate and check the construction documents. Sometimes the owner who wants to begin construction immediately dictates this. Thus, the reality of the preparation of construction documents is that it is very rare that the construction manager/general contractor (CM/GC) will receive 100% completed documents.

The CM/GC must review the documents for constructability. This is not to say that all discrepancies will be picked up. However, the CM/GC is trying to minimize potential problems.

One of the first things the CM/GC wants to do is to make sure the latest construction documents have been submitted. This is achieved by requesting from the design team a list of all the construction documents indicating the latest revision dates. The design team's list should also include all addendums, sketches, bulletins, specifications, and all other documents relevant to the construction of the project. Exhibit 8-2 is a design document log that should be kept by the project manager (PM).

Exhibit 8-1

Review of design documents flowchart.

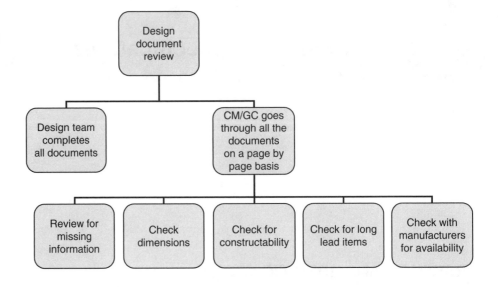

Exhibit 8-2

Design documents log.

Document	Type	Designation	Submittal Date	Revision Date	Comments
Drawings	Site plan				
	Boring log				
	Landscaping				
	Architectural				
	Structural				
	Mechanical				
	Electrical				
	Plumbing				
	Sprinkler				
Specifications	1				
	2				
	3				
Sketches	1				
	2				
	3				
Addendums	1				
	2				
	3				
Bulletins	1				
	2				
	3				
E-Mail	1				
	2				
	3				
Letter	1				
	2				
	3				
Field authorization	1				
	2				
	3				

REVIEW MEETING

Prior to the start of the project, a meeting should be set up with the design team so that the drawings and specifications can be reviewed on a page-by-page basis. The CM/GC's team should include the PM, superintendent, estimator, scheduler, and purchasing agent. The factor that commonly prevents this meeting is time. The excuse has always been that the construction business does not have time to perform this exercise. Guess what? Either you find the problem early, or you wait until the problem exposes itself later at a critical stage of the project. You make the choice.

The construction documents need to be checked to make sure the equipment components as specified can be delivered in a timely manner. This is achieved by constantly checking the status of the specified component. Chapter 15 provides definitive information on how to track critical components. In addition, check to see if all the equipment specified is still being manufactured.

The construction documents should also be reviewed in the context of cost savings. This would include reviewing certain components of the design where potential value engineering analysis may be considered. Value engineering is the process where less costly alternatives are reviewed for components that have been initially specified (usually specified by the architect or the consultants). This may include types of curtain walls, complex structural systems, and other costly systems.

After the construction document review session, the owner should be made aware of any problems found so that they can be addressed immediately. The owner may also want to evaluate alternatives that may have been suggested by the CM/GC.

CONSTRUCTION DOCUMENTS CHECKLIST

The following is a checklist to help navigate through the abundant construction documents that have to be analyzed. The intent is to give the PM a guide when reviewing and analyzing the construction documents. Every project is different; thus, this checklist must be viewed in the context of the type of project being evaluated.

1. Check that all of the drawings and specification sections are included by reviewing the index on the first sheet of drawings and specifications. The following should be included:

   - Site plan
   - Geotechnical
   - Foundation
   - Underpinning (may be included with the under pinning subcontractor)
   - Architectural
     - Façade
     - Roof

- Finishes
- Millwork
- Special areas
- Structural
- Plumbing
- Mechanical
- Electrical
- Generators
- Life safety
- Sprinkler
- Elevators (and other conveying systems)
- Landscaping

2. Contact the design team for a list of the latest documents. Make sure the latest drawings, specifications, and other construction documents have been provided.
   - Addendums
   - Bulletins
   - Sketches
   - Clarifications
   - Contact the architect and the consultants for a list of the latest documents

3. Review all the notes on the drawings.
   - Look for special requirements
   - Look for special testing
   - Be aware of municipal and state special requirements
   - Be familiar with the local codes
   - Be familiar with the state's energy code

4. Read the specifications.
   - Cross-check against the drawings for more complete information
   - Make sure there is consistency with the drawings
   - Look for special requirements (i.e., testing, factory visits, etc.)
   - Look for attic stock requirements (to provide extra material at the end of the project)
   - List all warranties and guarantees spelled out in the construction documents
   - Determine if maintenance is required for any equipment and if it is part of the requirements of the project

5. Preliminary coordination review
   - Look for "obvious hits" (conflicts with other construction items)
   - Determine if the equipment shown can fit into the areas provided

- Ensure consistency between the architectural drawings and the consultant's drawings as far as dimensions and locations
- Ensure mechanical and electrical coordination (i.e., power shown for hot water heater)

6. Review shaft openings for proper fit of all devices.
   - Elevators (and other conveying systems)
   - Pipes (drainage, sanitary, water, etc)
   - Duct work
   - Conduits
   - Chilled and condenser water lines
   - Vents
   - Dampers (control, smoke, and fire)
   - Clearances
   - Conflicts

7. Make a list of major components for potential long lead items.
   - Elevators (and other conveying systems)
   - Mechanical equipment
   - Curtain wall
   - Special steel sections
   - Utility equipment
   - Electrical switch gear
   - Bus ducts
   - Transformers
   - Generators
   - Pumps
   - Millwork
   - Stone
   - Light fixtures
   - Hardware
   - Special doors

8. Review utilities
   - Are all of the following shown?
     - Water
     - Sanitary
     - Electrical
     - Gas
     - Steam
     - Cable
     - Communications

- How do the utilities come into the site and who will be responsible (owner or the CM/GC)?
- List all utilities and prepare contact list
- Bridging that may be required to support the temporary utilities
- Temporary utilities required during construction:
  - Electrical
  - Water
  - Fire protection (standpipes)
  - Bathrooms

9. Review site plan
  - Utility locations noted?
  - Topographic survey included?
  - Develop a logistics plan (which may have to be presented to the local building department)
  - Any partial street closings required?
  - Benchmarks for surveyors noted?
  - Any problems with bringing trucks or equipment into the site?
  - Protection of neighbors required?
  - Any potential noise problems?
  - Will turning radius for trucks with large equipment be a problem?
  - Will flag people be required for traffic control?
  - How many guard stations will be required?

10. Review major dimensions
  - Make sure the dimensions add up
  - Look at floor-to-floor elevations
  - Look for varying floor elevations on the same floor (mechanical rooms and elevator lobbies)
  - Note any missing dimensions, such as for the construction of a wall; do not scale the drawings but obtain the dimensions from the design team
  - Check the "blow up" details shown verses the detail on the plans

11. Constructability
  - Is structure unique—does the building consist of angles with no vertical sides?
  - Are details sufficient?
  - Will 3D drawings be required?
  - Wills subs have a problem with understanding the drawings?

12. Look at heavy equipment
  - Is structural reinforcement provided for heavy equipment?
  - Potable and sprinkler water tanks
  - Elevator equipment

- Escalators
- Generators and fuel tanks
- Switch gear
- Batteries
- Pumps
- Vaults
- Cooling towers
- Chillers
- Chilled and condenser water pipes
- Water tanks
- Thrust loads on pipes
- Ice storage for mechanical systems
- Roof gardens
- Curtain wall supports

13. Review roof plan
- Proper penetrations and protection?
- Adequate slope provided for drainage?
- Expansion joints required?
- Overflow scuppers indicated?
- Drains indicated in the proper location and are there sufficient number of drains?
- Lightning protection provided and grounding shown?
- Safety lights located and electrical circuits provided?
- Mechanical equipment penetrations and flashing indicated?
- Walkways in the proper locations?
- Hatches for roof access shown?
- Proper flashing details?
- Sufficient clearances as required for maintenance?
- Communication dishes shown with proper penetration details?

14. Consistency
- Toilets line up?
- Bus ducts (electrical risers) in the proper risers?
- Equipment rooms indicated?
- Communications closet indicated?
- Any offsets?
- Core plans consistent?
- All pipes (storm, chilled water, condenser water, sanitary, domestic water, vents) in their proper chases?
- Ducts in proper chases?

- Access provided for all electrical and mechanical equipment?
- Wet column locations?

15. Foundations
    - Sufficient soil borings indicated?
    - Water anticipated in the site and what provisions will be required?
    - Bottom of footing elevations shown?
    - Special types of foundations?
    - Special bracings for foundation walls (i.e., slurry walls, soldier beams, etc.) required?
    - Underpinning required?
    - Entrance of utilities shown?
    - Any underground tunnels to protect?
    - Monitoring of adjacent structures required?
    - Excavation and access to the site problematic?

16. Have all the riser and associated schedules been included, such as:
    - Bus duct
    - Sprinkler
    - Chilled and condenser water lines
    - Domestic water
    - Roof drains
    - Sanitary lines
    - Communications
    - Column schedules
    - Mechanical equipment schedules
    - Electrical equipment schedules
    - Pump schedules
    - Electrical panel schedules
    - Curtain wall wind pressures

17. Layouts
    - Equipment rooms and associated equipment
    - Code requirements (electrical panel clearances)
    - Maintenance considerations (i.e., can tubes be pulled for the chillers?)
    - Access for equipment
    - Core toilets including American for Disabilities Act (ADA) requirements
    - Locations of valves and switches
    - Electrical equipment clearances for Nation Electric Code (NEC)

18. Testing and QC
    - Prepare a list of tests required by the construction documents (see Chapter 3)
    - Add special manufacturer's QC requirements to the CM/GC's list
    - Include municipal building department requirements

19. Mock-ups that may be required as part of the construction documents include the following:

- Curtain wall
- Offices
- Exterior stone walls
- Interior floors and walls
- Special design considerations
- Lobby finishes
- Toilets
- Elevator cabs
- Special ceilings
- Special floor systems (e.g., under floor air conditioning delivery system)

COORDINATION OF DRAWINGS BY THE DESIGN TEAM

Coordination of drawings is a process where all the various elements of the project are checked to make sure no conflicts exist. In most cases, the spaces where the building structure shares space with the mechanical, electrical, sprinkler, plumbing, and cables are the most vulnerable for conflicts. With the design team using the various levels of CAD (computer aided design), it is routine to check for conflicts. However, just like checking the other aspects of the drawings, only if the design team has the time will conflicts be evaluated.

COORDINATION OF DRAWINGS BY THE CM/GC

Just as the design team checks for conflicts, the CM/GC must also check for potential conflicts. This must be accomplished as early as possible in the construction process, so that problems can be resolved prior to installing all the equipment and building elements. If conflicts are found during construction, then this will only create delays. These delays can be avoided with proper coordination. The same areas are evaluated as with the design team—the space that is occupied by the building structure in combination with the mechanical, electrical, plumbing, sprinkler, and cable systems. The coordination process usually starts after the building structure has been erected. Since the ductwork is the largest component after the installation of the building structure, the ductwork shop drawings then become the basis for conflict investigation. The other trades people then use the ductwork shop drawings to overlay their components on the drawings and determine where conflicts may exist. Areas of a building that should be checked for potential conflicts include, but are not limited to, the following:

1. The plenum space above ceilings
2. Mechanical and electrical rooms

3. All chases (openings) for duct work, electrical conduits/bus ducts, water lines, drainage pipes, sanitary pipes, telephone conduits, cables, sprinkler lines

4. All elevator and other vertical transportation openings or shaft ways

5. Toilet chases

6. Transformer vaults

7. Water/sprinkler tank rooms

8. Underneath raised floor areas

9. Special areas such as cafeterias, auditoriums, studios, trading areas, training rooms, and large conference rooms

10. Special mezzanine areas

ACCESS FOR CRITICAL ELEMENTS

When evaluating the coordination of all the elements of a project, the PM has to be cognizant of the fact that all systems and lines have to be maintained. Thus, access to these critical elements has to be provided. The design team has the initial responsibility for providing the required access. However, often this is not the case and it becomes the responsibility of the CM/GC to notify the design team of where access should be provided. Even though it is not the responsibility of the CM/GC to locate these access openings, it makes sense to assist the design team and the owner to provide this service. Also, it may be a requirement of the contract. Access should be provided for at least the following critical elements:

1. All mechanical equipment

2. Other type of equipment that has to be maintained or checked

3. Fire/smoke dampers

4. Balancing dampers, variable air volume (VAV) boxes

5. Lights that have ballasts

6. Lights with bulbs that are inaccessible from the bottom of the fixture

7. Splice boxes

8. Electrical cut-off switches

9. Valve areas

10. Pipe clean-outs

11. Control point connections

12. Elevator shafts

13. Roof mounted equipment

14. Bus duct risers

15. Sprinkler rig (valve, water flow switch, pressure gauge)

16. Sprinkler drain down pipe

17. Cable ladder racks

SUMMARY

- When the drawings and specifications are submitted to the PM, he or she should make sure all construction documents have been provided. A log of all the documents must be requested from the design team. This will show the list of drawings, specifications, addendums, bulletins, sketches, all with the latest revision dates.

- Time must be set aside by the PM so that sufficient effort is devoted to this process of document review.

- The PM must check all the construction documents on a page-by-page basis to evaluate the constructability of the project.

- A list of long lead items has to be prepared so those items that may fall outside the initial schedule can be ordered early.

- The construction document review will also assist in a value engineering analysis that may be required during the document preparation phase of the project.

- The owner must be made immediately aware of any problems found with the construction documents review.

- Coordination of all the design team drawings must be accomplished by the PM to avoid conflicts.

- Access for maintaining equipment, pipes, etc. must be established by the PM.

9 Contracts
(Tough negotiations.)

WHAT IS A CONTRACT?

A contract binds two parties together with each party having certain obligations to the other. The construction manager/general contractor (CM/GC) has an obligation to complete a project within a certain period, with good quality, and at a definitive cost. The owner has the obligation to provide technical documents, the site, permits (in certain cases), adjacent property approvals, and to pay the contractor in an expeditious manner when invoices are submitted (on a regular basis). The objective of the owner when constructing a facility is to obtain quality workmanship at the lowest possible price. The secondary objective is to have the facility completed within a satisfactory period.

Many forms of contracts are used to make sure that the owner's objectives are achieved. The American Institute of Architects' (AIA) documents are used extensively throughout the construction industry. Other types of contract forms are available from trade organizations such as the Associated General Contractors of America and the Contractors Association of America. Whatever forms are used, it is important to understand that they are only a starting point and the CM/GC's lawyers must review them for potential risks and applicability to the project at hand. In addition to the base contract, two other sections of the contract form the total contract. These are the general conditions and supplemental conditions. They are the real "teeth" of the contract.

The project manager (PM) must also review and understand all construction contracts and the associated general conditions and supplemental conditions. Many sections of the contract have a profound impact on the direction and completion of a project. In most cases, the contract is not looked at again. When a contract is initially read, not all of the elements are understood. Therefore, it behooves all PMs to read the contract several times. However, when a problem arises, the first document to come out is the contract (see Exhibit 9-1, contract flowchart).

Exhibit 9-1

Contracts flowchart.

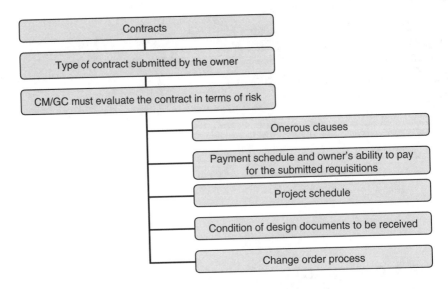

The most prevalent forms of contracts include stipulated sum (lump sum), cost plus, guaranteed maximum price (GMP), construction management, construction management with risk, construction management with no risk, unit price, and time and material (T&M). Hybrid forms of these contracts are usually the rule, not the exception, as one might surmise. In the theoretical world, the cost of constructing a particular facility should generally be the same, no matter what form of contract you use. However, in the real world, you only have the opportunity to choose one form of contract (or hybrid). Thus, it is almost impossible to state that one type of construction contract would be more advantageous than another type. The only way this could be done is if two identical projects were built at the same time on the same site, using different contracts with the same contractor. Each contract form has its place, depending upon the circumstances at the time of consideration. CM/GCs are usually given a contract by the owner. The owner will decide upon the contract to use based on the items listed in Exhibit 9-2.

When dealing with work in the urban environment, private owners prefer two major forms of contract. They are the GMP (as defined later in Types of Contracts section) and the construction management (as defined later in Types of Contracts section) contracts. The GMP provides the owner with a definitive price for the project. The actual construction can start early because of fast tracking. These two advantages give the owner the ability to assist in the financing of the project and to complete the project in an expeditious manner so that cash flow from the occupancy of the project can start earlier than under a normal construction contract.

The construction management contract provides the owner with the ability to analyze the cost and complexity of a project from conception to completion. This offers the owner the ability to construct the designed project at presumably the lowest cost and with a minimum of cost changes (excluding scope changes). It is also possible to start the project early by using fast tracking methods as indicated in Chapter 15. See Chapter 1 for the preconstruction services that are offered by construction management contracts.

1. Quality of construction
2. Anticipated cost of the project
3. Schedule
4. Company policy
5. Construction market conditions
6. Availability of good contractors and subcontractors
7. Size of the project
8. Inflation rate
9. Insurance consequences
10. Critical nature of the work
11. Union considerations
12. Renovation vs. new construction
13. Availability of owner's staff
14. Control of the project
15. Fast tracking (for better cash flow at the completion of the project)
16. Dictated by the legal department
17. Risk factors
18. Familiar with a form of contract
19. Lender's requirements
20. Local municipality requirements

Exhibit 9-2

Selection on the type of contract.

TYPES OF CONTRACTS

The following is a brief description of each type of contract from the owner's prospective:

1. Stipulated sum (lump sum) (traditional method)
   a. Procedures
      - Drawings and specifications are complete (however, no construction documents are 100% complete).
      - Documents are submitted to several contractors.
      - GCs receive proposals from the various trade subcontractors.
      - One price bid is received from each contractor.
      - Contract is awarded for a lump sum (to the lowest responsible bidder) and a definitive completion date is agreed upon.
   b. Advantages
      - GC is responsible for all related construction tasks of the project.
      - Total cost of the project (without changes) is determined initially.
      - Constant auditing of the project is not required.
      - Schedule is defined based on the completed documents.
      - Owner has the option of selecting the lowest price.

    c. Disadvantages
- GC usually bids what is shown on the documents (even if the drawings are not consistent with intent).
- Selection of the subcontractors is usually based on the lowest bid and not necessarily on the quality of work.
- Interpretation of the documents may have future cost implications for the owner and the GC.
- The GC's staff may not be up to organizational standards required for the project (unless spelled out in the documents). This is done in order to submit a competitive price.
- It is a closed-book relationship. The owner does not see any of the subcontractor's pricing.

    d. Suggested use
- If dictated by company policy.
- When construction market is weak.
- When several good GCs are bidding on the project.
- When owner's staff is limited.
- Government/municipal-type projects.
- May be required under the Sarbanes–Oxley Act, which is a government mandate to enforce ethical standards for public corporations.

2. Cost plus

    a. Procedures
- CM/GC will be paid for all trade costs and expenses plus a fee to complete the work.
- This method could be used when all the construction documents are completed, but in most cases, it is used when the construction documents are not complete.
- Can be used to negotiate with only one CM/GC.

    b. Advantages
- Can expedite construction when the scope is not clearly defined.
- Long lead items can be purchased early.
- Is usually owner/CM/GC "friendly" because the scope is reviewed when bids are obtained.
- Can be open book when spelled out in the contract. (Open book is the process whereby the owner reviews the bids of the subcontractors as submitted to the CM/GC.)
- The owner will know the subcontractors bidding the work and thus will give approval for the subcontractors being selected. A full leveling process has to be accomplished by the CM/GC (see Chapter 16).
- Clearly defines the CM/GC's team and time to be spent on the project.
- It is easier for the CM/GC to perform value engineering.

- The value engineering method is used to reduce cost by evaluating alternative materials, methods, and scope of the project.
- CM/GC has full responsibility for the project.

c. Disadvantages

- Auditing of the project by the owner is usually required.
- Total cost of the project may not be known.

d. Suggested use

- When documents are not 100% complete.
- To expedite a project (i.e., fast tracking—starting a construction project without 100% completed documents, such as the foundation work starts without the curtain wall documents completed).
- When the owner requires a CM/GC early on in the project to evaluate cost and schedule.
- When the owner wants to negotiate the contracts with only one CM/GC.
- When critical long lead items have to be purchased early.
- Can be converted to a GMP contract.

3. Guaranteed maximum price (GMP)

a. Procedures

- CM/GC's cost is fixed in advance and guarantees a limit on the cost of construction. A contingency is also included in the cost base. Any money left over at the end of the project is usually shared between the owner and the CM/GC.
- Usually done on a negotiated basis between the owner and the CM/GC.
- The documents are not complete (usually 80% completed documents).
- CM/GC will not establish a GMP until approximately 80% of the trades have been bought out.
- Costs are based on pricing received from the subcontractors and where items are unknown, an allowance is added. An allowance is the estimated cost usually based on preliminary construction documents.

b. Advantages

- Can expedite construction when the scope is not clearly known.
- Maximum cost of the project is known even though the documents are not complete.
- Work is usually accomplished in an atmosphere of cooperation.
- Savings can be achieved and shared between owner and CM/GC through the use of alternatives or construction method efficiencies (value engineering).
- It is easier to resolve problems when working within a cooperative environment.
- Material can be purchased early, thus avoiding potential inflationary factors.

c. Disadvantages

- Every minor change has to be negotiated with the owner.
- The contract has to be very specific on what is to be included in the cost and what is to be excluded.

- All allowances must be clearly defined.
- Design concepts must be "frozen" early, thus reducing design flexibility.
- Large contingency will be required by the CM/GC to protect the GMP.
- Any contingency money left over after the job is completed must be negotiated between the owner and the CM/GC.
- The owner may require an audit.
- Changes to the project have to be defined as to where the associated cost will be taken from (either from the contingency or additional change order cost).

  d. Suggested use

- When construction has to be expedited and drawings are not complete.
- In an inflationary market.
- When the owner wants to know the cost exposure early in the project.
- Large base building projects in an urban environment.
- Cost plus contracts is sometimes converted to a GMP.

4. Construction management (no risk)

  a. Procedures

- Owner retains the services of a second party to act as the owner's agent.
- Usually works on a CM's out-of-pocket expenses plus a fee basis.
- CM is the "owner" of the project.

  b. Advantages

- Construction documents are reviewed and evaluated for constructability, cost, and schedule.
- Subcontractors for bidding are selected on a pre-qualified basis.
- Construction cost is based on subcontractor's cost without a CM's mark-up.
- Professional CM organization is used for the management of the project.
- Value engineering and cost control can be maintained throughout the project.
- Fast track method of construction can be utilized.
- Pre-ordering of long lead items can take place early.
- Coordination of all drawings is accomplished throughout all phases of the project.
- Change orders should be kept to a minimum.
- Quality control (QC) should be higher.

  c. Disadvantages

- No cost exposure except reputation.

  d. Suggested use

- Expediting of project is required.
- Owner does not have the resources to manage the project.
- Large projects.
- Complicated projects.
- When you have critical long lead items.

5. Construction management (at risk)
   a. Procedures
      - Usually works on a CM's out-of-pocket expenses plus a fee basis.
      - Fee is higher than at no risk.
      - CM is the contractor for the project.
   b. Advantages
      - Construction documents are reviewed and evaluated for constructability, cost, and schedule.
      - Subcontractors for bidding are selected on a pre-qualified basis.
      - Professional CM organization is used for the management of the project.
      - Value engineering and cost control can be maintained throughout the project.
      - Fast track method of construction can be utilized.
      - Pre-ordering of long lead items can take place early.
      - Coordination of all drawings is accomplished throughout all phases of the project.
      - Change orders should be kept to a minimum.
      - QC should be higher.
      - Cost of the project is guaranteed by the CM.
      - Open book relationship between the owner and the CM.
   c. Disadvantages
      - Risk could be high for the CM.
   d. Suggested use
      - When expediting of project is required.
      - Owner does not have the resources to manage the project.
      - Large projects.
      - Complicated projects.
      - Fast tracking required.
      - When there are critical long lead items.
6. Unit price or time and materials (T&M)
   a. Procedures
      - CM/GC submits cost based on unit prices or labor time and materials used.
      - In some cases, the units are taken off by a quantity surveyor (British method), and then the CM/GC fills in the unit prices.
      - Total price is based on the addition of all the quantities plus the cost for the general conditions.
   b. Advantages
      - The owner only pays for the materials that are actually installed and the labor that is expended.
      - Easy to calculate any changes.
      - Easy to evaluate bids.

    c. Disadvantages

- This type of contract is usually not used in the United States except for change order pricing and small projects. It has not been accepted because the unit prices may change as the job progresses. The unit price may not reflect the actual condition in the field.
- Need auditing staff to keep track of all quantities (labor and material).
- Need good outside estimating company to complete the "take-offs."

    d. Suggested use

- For change orders.
- When simple elements (that are repetitive) are used in a project.
- T&M to be used where change order needs to be expedited and change is not clearly defined.
- Small or maintenance-type projects.
- If a CM/GC has to be replaced.

CONTRACT SUMMARY

Whichever contract is selected, it behooves the PM to summarize the various contract provisions. Exhibit 9-3 is a summary sheet for a GMP contract. Having this information available gives the CM/GC a better understanding of the financial aspects of the contract, the type of information and details that may be needed, insurance and bond limits that are required by the subcontractors, and a list of any potential risk areas.

CONTRACT PROVISIONS THAT MUST BE REVIEWED FOR CLARIFICATION

1. Owner's responsibility

- Drawings and specifications (construction documents)
- Site layout
- Soil borings and site conditions
- Environmental reports and approvals
- Abatement of any hazardous material
- Special studies (traffic, water, utilities)
- Permits clarification
- Utilities
- Disputes with adjacent owners or municipality
- Testing responsibility
- Other areas where the owner has control and must submit the information

	Comments
Owner	
Location	
State Law	
Architect	
Owner's Representative	
Start Date	
Completion Date	
Substantial Completion Date	
Amount (1)	
Preconstruction Fee	
Construction Fee	
Preconstruction Fee	
General Conditions	
Contingency (1)	
Contingency % Split	
Performance Bond Amount	
Completion Bond Amount	
Lien Bond Amount	
Liability Amount	
Comprehensive Amount	
Automobile Amount	
Builder's All Risk Amount	
Additional Insured	
Requisition Submittal	
Payment of Requisition	
Retainage % Amount	
Liens Required	
Permits Required	
Claims and Disputes Resolution	
Special Administrative Programs	
Documents for Changes	
Notice Provisions	
Type of Damages (if late)	
Owner's Work to Be Coordinated	
Potential Risk Provision-1	
Potential Risk Provision-2	
Potential Risk Provision-3	
Potential Risk Provision-4	
Potential Risk Provision-5	
(1) Determined at 80% trade buy-out	

Exhibit 9-3

GMP contract summary.

2. Schedule
   - Type that has to be prepared
   - Updates required
   - Penalty clauses
   - Substantial completion provisions
   - Delays (not covered by force majeure)
   - Extensions
   - Delays caused by owner or owner's subcontractors or vendors

3. Costs
   - As defined in the contract
   - Allowances and how to be used
   - Definition of allowance items
   - How will GMP be impacted by actual cost vs. allowances allocated?
   - Contingency and how is it to be used and who controls it?
   - General condition's clarifications (what is included and what is excluded?)
   - Inflation clause
   - Costs covered by the owner
   - Value engineering
   - New regulations by municipal and government agencies
   - Auditing of project by the owner

4. Change orders
   - How defined
   - Method for developing costs
   - Approval process
   - Who signs off on any change orders?
   - Will labor rate change if new contracts are agreed upon?

5. Dispute resolution
   - Records that have to be kept
   - Type of daily reports that have to be maintained
   - Owner and CM/GC executive review
   - Independent review board
   - Mini trials
   - Mediation
   - Arbitration
   - Other legal means, including litigation
   - Notice provisions

6. Requisitions and retainer
   - Schedule of values required? When is it paid and how often?
   - What is required with submittal?

- Waiver of liens and from whom?
- Subcontractors invoices
- Percentage of retainer
- When will the retainer be released?
- Will retainer be held on fee, bonds, or insurance?
- Final payment (In what manner?)

7. Insurance
   - Coverage
   - Is builder's all risk policy required?
   - Will wrap-up insurance be considered?
   - Understanding the requirements
   - Who is to be covered?

8. Bonds
   - Types (i.e., bid, performance, payment, lien)
   - Enforcement
   - Circumstances in which they will be evoked

9. Unforeseen conditions
   - Foundations and bearing capacity
   - Property lines
   - Utilities

10. Owner's representative, architect, and consultants
    - Definition of role
    - Changes of scope
    - Means and methods
    - Review shop drawings and RFIs

11. Coordinated drawings
    - CM/GC vs. the design team
    - What is to be coordinated?

12. Close-out
    - Punch lists
    - Final waiver of liens and claims
    - Department of Buildings sign off
    - Close-out documents
    - Warranties (what is covered and when does it start?)
    - Maintenance agreements (if any)
    - Attic stock
    - Specific equipment use and maintenance classes
    - Maintenance costs required?

13. Termination
    - Under what circumstances?
    - Define payment

POTENTIAL ONEROUS CONTRACT CLAUSES

The CM/GCs should review certain clauses in the contract to make sure that any potential risk involved is within acceptable levels, including, but not limited to the following:

1. Payment
   - Owner should make payments within a 30-day period.
   - If payments will be made outside of 30 days, then the contractor and the subcontractors must be aware of this delay, and bid accordingly.
   - Be careful of states that refuse to enforce "pay when paid" clauses, thereby requiring payments to subcontractors even when the contractor is not paid by the owner in a timely manner.
   - Check to see if the state in which you are working gives certificates of capital improvements (usually means that you do not have to pay sales tax on the labor portion of a capital improvement).

2. Retention
   - Most owners require 10% retention. This means that when a subcontractor submits a monthly invoice, only 90% of the submitted costs will be paid. The remaining 10% is held by the owner (or financial institution paying for the project).
   - The contract should define at what point in the project the retention will be released or reduced.

3. Change orders
   - May require the contractor to proceed with the work even when the cost is in dispute. In this particular case, all involved parties must retain accurate cost, scheduling, and photographic information.
   - Negotiating skills must be exercised so that the cost issue can be resolved as expeditiously as possible. Many disputes and claims are based on change order work that has progressed without resolving the monetary issues.
   - Method for providing data for a change order should be spelled out in the documents.
   - Make sure that change orders will only precede after the owner's authorized representative signs a proposal.

4. Authorization
   - The owner must assign one person who has the authority to make changes and to approve change orders.
   - Be careful of multi-approval process (i.e., architect and owner representation).

5. Liquidated damages or consequential damages

- The contractor is usually better off with a specified sum for each day lost than having to determine what the cost expenses might be at some future date.
- The schedule has to be clearly defined.
- A change initiated by the owner must be detailed as to the possible extension of time.
- Never agree to any consequential damages. They may include items by the owner that would be completely out of the scope of the project. If it cannot be avoided, then a list of all the consequential damages (with the costs) must be enumerated. The contractor should review this and unreasonable items should be eliminated.

6. Force majeure

- Events or circumstances beyond the control of the parties. Examples could be war, flooding, earthquakes, tornados, hurricanes, and volcano eruption.
- Force majeure clause must be part of the contract. Thus, weather information history for the area in question must be analyzed.

7. No damages for delay (by the owner)

- This is an onerous contract provision and should be deleted. It typically states that the contractor's sole and exclusive remedy, if delayed, is a time extension. The CM/GC should be paid for any event precipitated by the owner that will cause the project schedule to extend beyond the original agreed upon date.
- Any delay caused by the owner will cause extensive costs to the CM/GC. This could be additional general conditions, subcontractor costs, and potential inflationary costs for material and labor.

8. Warranty

- Warranty for construction is usually for 1 year (except for certain work and equipment that may have an extended warranty).
- When does the warranty period start? AIA documents state that the warranty period will start from substantial completion of the project.
- Substantial completion as defined by the AIA is the period when the facility can be occupied for the owner's beneficial use.
- If the owner should occupy the space prior to substantial completion, then the responsibility of the owner must be detailed in writing as to the owner's obligation for non-completed areas. This could include protection of the walls, ceiling, and flooring; rebalancing of the mechanical systems; and early start-up costs for all systems.
- Make sure that any warranty does not include maintenance of the equipment during the period unless expressly requested in the design documents.
- If the owner wants the CM/GC to attend any meetings after the warranty period, then define the meetings and how the CM/GC will be paid.

9. Allowances

- When documents are presented that are not complete, an allowance must be indicated.
- Make sure that any allowance is based on as much information as possible provided by the design team.

- In order to protect the CM/GC, the allowance cost item must be clearly defined (i.e., curtain wall is standard $1/4$ in. insulated glass in aluminum mullions that will be spaced 5 ft apart and can withstand a wind load of 60 lb/ft^2 with no air or water leakage. The total area of the curtain wall will be 80,000 ft^2).
- When the drawings are complete (no drawings are 100% complete) and you receive a bid from a subcontractor, the cost is going to be compared to the allowance you developed. If the bid is lower, there is no problem. If the bid is higher than the allowance, then the problems start. The CM/GC may be required to pay the difference if the allowance was not clearly defined.

10. Construction documents
    - The contract documents (drawings, specifications, addendums, bulletins, sketches) must be listed with the latest revision dates.
    - These documents will act as your base for the construction of the project.

11. Dispute resolution
    - Some form of dispute resolution should be included in the contract.
    - This would take the form of collaborating, arbitration, mediation, or a dispute resolution board.
    - If the parties do not agree to arbitration, they can end up in court.
    - See Chapter 22 for more information on dispute resolution.

12. Construction acceleration
    - If the owner has this provision in the contract, then the CM/GC must prepare a cost (with inflation) that would account for the economic effect of finishing the work earlier then anticipated.
    - It is of utmost importance that the CM/GC does not proceed until a cost has been negotiated with the owner.
    - It must also be stipulated that at least 1 month would be required to obtain the costs for acceleration.

13. Schedule
    - The contract should be specific as to the time to complete the work (i.e., 100 workdays, no weekends or holidays).
    - The approved schedule should be based on one that is submitted by the CM/GC and approved by the owner.
    - All holidays, weekends, and restrictive days (no work mandated by a municipality) must be noted on the schedule.
    - Owner-provided equipment and finishes must also be shown on the schedule.

14. Approval process
    - Be careful of clauses where the owner's consultants will act as final and conclusive arbitrators of any disputes that may arise. This will mean that an individual who works for the owner will have the final say on a disagreement between the owner and the CM/GC. See Chapter 22 for more information on dispute resolution.

15. Implied warranty
    - Make sure a clause is included stating that the construction document represents the total work.
16. Coordination of drawings
    - These drawings are usually prepared by the various trades to make sure no conflicts will exist in the field (i.e., duct work, sprinkler, lights must fit within the ceiling void spaces without "hits").
    - The obligation of the design team concerning coordinated drawings should be spelled out in the contract.
17. Unforeseen conditions
    - Some form of indemnification must be stated in the contract that protects the CM/GC from any latent conditions that may be found upon exposure (i.e., earth, a wall, the roof, the floor).
    - This clause must specifically address conditions associated with foundation work borings and finding conditions other than noted in the boring logs.
    - The owner must deal with hazardous environmental conditions.
18. Owner's responsibility
    - The owner's obligation should be completely enumerated to include the items noted in Exhibit 9-4.
19. Inflation clauses
    - The contract should stipulate that if any materials escalate by some (negotiated) percentage, then adjustments would be made for the material cost increases.
20. Shop drawings
    - These drawings could have a major impact on the schedule if the design team does not approve them in a timely manner.
    - The contract should stipulate the time required for the design team to review and comment on the shop drawings.

---

1. Submittal of a complete set of drawings and specifications (construction documents)
2. Provide access to the site
3. Phase I and Phase II environmental reports
4. Survey of the site
5. Soil (or rock) boring information
6. Permits (unless stipulated otherwise)
7. Payment of CM/GC's invoices in an agreed upon period of time
8. Coordination with utility companies
9. Obtaining TCO (unless transferred responsibility to the contractor)
10. Any adjacent property approvals
11. Coordination of all non-contractor responsibilities, including communications, cable installation, and furniture and carpet installation

Exhibit 9-4
Owner's obligations.

21. Failure of the owner to make timely decisions
    - The owner should not put any time constraint on the CM/GC when the owner cannot make timely decisions.
    - Delays caused by the owner's indecision should be addressed.
    - You do not want this item to become adversarial, but the CM/GC has to protect himself from possible delays caused by the owner.
    - The best way to handle this would be to notify the owner in writing that unless a decision is made by a certain date, delays would occur.

22. Owner-provided suppliers and vendors
    - The contract should identify the owner's suppliers and vendors.
    - The work of the owner's suppliers and vendors should be included in the CM/GC's schedule.
    - Any delays by those suppliers and vendors should not affect the CM/GC's schedule. If their work affects the final schedule, then the owner must compensate the CM/GC for any resulting delay.
    - The vendors and suppliers will not interfere with the CM/GC's work. If they do, then the CM/GC must advise the owner in writing of the potential consequences.
    - The vendor and supplier will provide their own laborers and containers for cleaning up the area in which the work is being performed.

23. Notification provisions
    - The contract may state that the CM/GC must notify the owner of any problem, schedule change, or cost changes and the notification must be accomplished within a limited time.
    - Based on certain case law, the CM/GC must adhere to the notification time or possibly lose the ability to collect any money for additional work or request a time extension.

24. Order of precedence
    - Make sure the contract states the order in which the design documents are based. Which takes precedence—the drawings, specifications, sketches, or other documents produced by the design team?

Obviously, the CM/GCs may not be able to negotiate all of these favorable clauses, but they must understand the risks if certain onerous clauses are included in the contract.

SUMMARY

- The owner usually dictates the construction contract.
- Several forms of construction contracts can be used with some variations of the base types.
- Stipulated sum (lump sum)
- Cost plus

- GMP
- Construction management with no risk
- Construction management with risk
- T&M or unit pricing
- The CM/GC's lawyer must review the contents of the construction contract.
- The CM/GC must review the contract for onerous clauses and then decide how much risk he is willing to accept.

10

Insurance and Bonds

(We never have a problem.)

CONSTRUCTION INSURANCE

Construction insurance can be defined as a contract whereby the insurance company seeks to provide coverage and indemnify the construction manager/general contractor (CM/GC) against a potential peril, loss, damage, or liability that arises from the performance of the construction work. Insurance for a construction project is a policy (a form of a contract) that the CM/GC purchases which covers the contractual insurance and risk management requirements of commercial general liability, builders risk, workers' compensation, automobile, completed operations, errors and omissions, environmental liability, etc. Purchasing an insurance policy with an insurance carrier, who collectively pools the risks of many CM/GC's and construction projects together to provide coverage and protection, is a way to manage risk. Insurance for construction projects in the urban environment presents many unique challenges based on size, dollar amount, congestion in the construction area and surrounding area, and the people and property that are in the direct zone of influence of the project.

Often insurance is not properly administered in the construction industry because the project executive (PE) and project manager (PM) are very busy with the planning and construction of the project, and cannot find time to properly administer the risk management and insurance processes. This can result in serious financial consequences and exposure to risk that a company may not be able to afford. Insurance is not an area to be neglected, as one mistake in this area can have far-reaching consequences. CM/GCs who have not purchased the proper type and amount of insurance for a project have faced major problems when a risk they thought was covered in the policy was not, and they found themselves with insufficient or no insurance coverage with a catastrophic loss.

Many CM/GCs have an in-house insurance advisor/manager, or they retain the services of an external insurance consultant/advisor to assist with the proper management and administration of construction insurance, which is highly specialized and complex. The challenge is to purchase the proper insurance with the appropriate levels of coverage and deductibles, covering the projects risks for a reasonable premium and from a well-established, reputable, and highly rated insurance company. The CM/GC often must purchase a series of policies to provide comprehensive insurance coverage for the firm and the project. It is important that the policies be read, along with the description of the coverage, list of exclusions, qualifications, definitions, deductibles, amount of coverage, policy limits, excess coverage, and umbrella coverage to provide appropriate insurance coverage. Insurance policies are often relied on to help resolve claims on a construction project, and to manage project risks that the CM/GC does not want or cannot afford to be exposed to.

WAIVER OF SUBROGATION AND INDEMNIFICATION

An important feature of construction insurance policies is the waiver of subrogation clause. This clause provides that the insurance carrier, after settling and paying a claim, cannot pursue the damages paid for by the insured parties' insurance carrier, against the party responsible for the actual damages. By including this clause, the CM/GCs can insulate themselves from such subrogation claims from a subcontractor working on the project, and have the subcontractor's primary insurance carrier resolve the claim through their insurance policy.

Another provision in insurance policies is the indemnification clause, whereby subcontractors and their insurance companies have to provide the primary insurance for the work the subcontractor is performing on the project. This is a contractual requirement between the parties. If there is a lawsuit that arises against multiple parties working on the project, the subcontractor that performed the work has to indemnify and defend the CM/GC, owner, consultants, and other named parties on the insurance certificate and policy. The subcontractors' insurance company and their legal staff do this.

TYPES OF INSURANCE

Commercial General Liability Insurance

This type of insurance provides coverage for claims by third parties against the CM/GC and all additional named insured parties. A pedestrian who has an accident while passing by a construction site would be covered under this type of insurance policy.

Builders Risk Insurance

This type of insurance provides coverage against the insured's loss to the property during the construction process. A break in a water service in the building during the construction process, which damages electrical, mechanical, elevator, and plumbing systems would be covered under this type of insurance policy. It is a primary coverage rather than against a claim from a third party. One needs to define the policy period for which it is in effect, which could be when the project is completed or a date thereafter.

Errors and Omissions Insurance

This is the professional liability insurance for the CM/GC and design professionals. CM/GCs cannot easily obtain architectural or engineering services Errors and Omissions (E&O) coverage. The CM/GC must be careful with this type of insurance, especially if involved with a design-build project. Chapter 19 covers design-build projects further.

Environmental Liability Insurance

This is a specialized insurance policy to cover pollution and hazardous material damages such as asbestos, mold, lead paint, medical waste, fuel oil, etc. The coverage under the commercial general liability (CGL) policy is very limited, and a separate policy with broader environmental coverage is often obtained to deal with these matters. It is recommended that the owner directly hold the contract for environmental work and provide the insurance policy for the appropriate coverage, with the CM/GC as the additional named insured.

Workers' Compensation Insurance

Workers' Compensation (WC) coverage is a state mandatory insurance to provide coverage for the CN/GC's workers if they are injured while performing their work. It provides for lost wages, medical coverage, and loss of partial or full ability to work because of an accident on the job. See Chapter 5, Exhibit 5-28, which is a workers' compensation table for trades in major cities.

Automobile Coverage

This provides coverage for all automobiles and trucks used in connection with the construction of the project, as well as transporting personnel and material to and from the project site. The insurance company will require the names and drivers license information for all drivers operating vehicles covered under the policy. If a driver's report from the Department of Motor Vehicles is not good, that person may be excluded from coverage under the policy, and thus should not drive company vehicles.

PROOF OF INSURANCE

The contract documents will define the types and levels of insurance requirements for the project. Exhibit 10-1 is a sample insurance rider for when an owner-controlled insurance program/contractor-controlled insurance program (OCIP/CCIP) may be used on a project.

It is not sufficient to identify only the coverage amounts, named insured's, and types of policies. Full compliance with the contractual insurance requirements, waivers of subrogation, indemnifications, etc., must also be ensured. All insurance must be confirmed initially with an apostle letter from the insurance company followed by a certificate of insurance describing the coverage, insurance company issuing the policy, insurance limits, named insured, and additional insured. An apostle letter is a letter of commitment by the insurance company to provide the insurance coverage contractually required for the project, prior to the actual issuance of the certificate of insurance. Proof of insurance is usually a contract requirement, and must be confirmed prior to the start of the construction project. If a loss were to occur prior to procuring insurance and obtaining evidence of same, the ability to get coverage for the event may be limited. It is important to note that an insurance certificate may not be valid if the CM/GC or subcontractor contract is not executed. Exhibit 10-2 is a sample certificate of insurance.

Exhibit 10-1
Insurance
requirements.

All subcontractors are to include in their lump sum base bid the cost for the insurance as described below.

The CM/GC and the owner have retained an insurance agent as their Insurance Broker for the project. The CM/GC and/or owner will be providing an Owner or a Contractor Controlled Insurance Program (OCIP)/(CCIP) for the project. Your firm in order to be enrolled in the CCIP program and track the actual project site payroll.

As part of your bid, please fill out these forms and include same with your bid.

Please note that no construction work is to proceed at the site until your firm has been enrolled in this OCIP and/or CCIP, and/or you have supplied your own corporate insurance.

The OCIP/CCIP insurance program provides for the actual calculations by owner, CM/GC, and insurance broker, of the insurance credit to be deducted from your base lump sum bid. This will be based on the actual payroll and work performed. The deduct alternate which was requested in the original bid documents will not be used as an actual cost, but rather as a budget estimate of the projected credit. Subcontractor to cooperate with the owner, CM/GC, and insurance agent to classify personnel, submit certified payroll, and copies of all current insurance certificates and policies to assist with the evaluation of the insurance credit.

Subcontractor shall provide insurance as follows in your base bid:

1) Workers' Compensation and Employer's Liability.
   a) Statutory Workers' Compensation (including occupational disease) in accordance with the law and including the other states endorsement.
   b) Employer's liability insurance with a limit of at least $500,000.
2) Comprehensive General Liability (IICGLII) with a combined single limit for bodily injury, personal injury, and property damage of at lease $4,000,000 per occurrence and aggregate. The limit may be provided through a combination of primary and umbrella/excess liability policies.

   Coverage shall include the Broad Form Comprehensive General Liability Endorsement (GF 04 05 of the ISO or an equivalent). Coverage shall provide and encompass at least the following:

   a) X, C, and U hazards, where applicable;
   b) Independent contractors;
   c) Blanket written contractual liability covering the indemnification set forth herein below;
   d) Product liability and completed operations, with the provision that coverage shall extend for a period of at least twelve (12) months from project completion;

Exhibit 10-1

(Continued)

   e) CGL coverage written on an occurrence form;

   f) Endorsement naming the following entities as additional insured: CM/GC, owner, all subsidiaries, and other parties.

   g) Waiver of subrogation.

3) Comprehensive automobile liability (including all owned, leased, hired, and non-owned automobiles) with a combined single limit for bodily injury and property damage of at least $2,000,000 per occurrence. The limit may be provided through a combination of primary and umbrella/excess liability policies.

4) Umbrella and/or excess liability policies used to comply with CGL and/or auto liability limits shown above shall be warranted to be excess of limits provided by primary CGL, auto and employers liability.

5) Certificate of insurance indicating the project must be submitted, approved, and available to CM/GC, prior to commencement of work, and provide for 30-day written notice prior to cancellation, non-renewal, or material modification in any policy to:

6) A separate certificate as per owners and their building management's requirements note in attachment.

7) All insurance carriers must: (i) be licensed in the state of; and (ii) be rated at least A in Best's.

8) Subcontractor shall secure, pay for and maintain property insurance necessary for protection against loss of owned, borrowed, or rented capital equipment and tools, including any tools owned by employees and any tools, equipment, stagings, towers, and forms owned, borrowed, or rented by subcontractor. The requirements to secure and maintain such insurance is solely for the benefit of subcontractor. Failure of the subcontractor to secure such insurance or to maintain adequate levels of coverage shall not obligate CM/GC, owner or their agents and employees for any losses. If subcontractor secures such insurance, the insurance policy shall include a waiver of subrogation as follows:

   a) "It is agreed that in no event shall this insurance company have any right to recovery against CM/GC and/or owner."

9) Should subcontractor engage a sub-subcontractor, the same conditions applicable to subcontractor under these insurance requirements shall apply to each sub-subcontractor.

10) Subcontractor shall purchase and maintain an owner's and contractor's protective liability insurance policy in the name of Structure Tone Inc., and Bloomberg to cover all exposures, including bodily injury and death arising out of and in the course of this contract. Policy should be written to conform to the ISO CG 0009 form (an occurrence form coverage).

    Limits of liability shall not be less than:

    Combine and single limit

    $2,000,000 each occurrence

    $10,000,000 aggregate

11) The owner and the CM/GC are providing a wrap-up insurance policy for the project. Subcontractors are to provide all required workers' compensation and liability for off-site work, trucking, and automobile insurance. Provide a projected budget deduct for the owner's and/or CM/GC's wrap-up insurance policy.

Exhibit 10-2

Certificate of liability insurance.

Certificate of Insurance				Issue Date (MM/DD/YY)	
	THIS CERTIFICATE IS ISSUED AS A MATTER OF INFORMATION ONLY AND CONFERS NO RIGHTS UPON THE CERTIFICATE HOLDER. THIS CERTIFICATE DOES NOT AMEND, EXTEND OR ALTER THE COVERAGE AFFORDED BY THE POLICIES BELOW.				
PRODUCER	INSURERS AFFORDING COVERAGE				NAIC #
	INSURER	A			
INSURED	INSURER	B			
	INSURER	C			
Subcontractor's Name	INSURER	D			
	INSURER	E			

COVERAGES
THIS IS TO CERTIFY THAT THE POLICIES OF INSURANCE LISTED BELOW HAVE BEEN ISSUED TO THE INSURED NAMED ABOVE FOR THE POLICY PERIOD INDICATED. NOTWITHSTANDING ANY REQUIREMENT, TERM OR CONDITION OF ANY CONTRACT OR OTHER DOCUMENT WITH RESPECT TO WHICH THIS CERTIFICATE MAY BE ISSUED OR MAY PERTAIN, THE INSURANCE AFFORDED BY THE POLICIES DESCRIBED HEREIN IS SUBJECT TO ALL THE TERMS, EXCLUTIONS AND CONDITIONS OF SUCH POLICIES. LIMITS SHOWN MAY HAVE BEEN REDUCED BY PAID CLAIMS.

CO LTR.	TYPE OF INSURANCE	POLICY NUMBER	POLICY EFFECTIVE DATE (MM/DD/YY)	POLICY EXPIRATION DATE (MM/DD/YY)	Limits	
	GENERAL LIABILITY	POLICY NUMBER			EACH OCCURRENCE	$ 1,000,000
X	COMMERCIAL GENERAL LIABILITY	PERPROJECT AGGREGATE ENDORSEMENT			PRODUCTS-COMP/OP AGG.	$ 1,000,000
	CLAIMS MADE [X] OCCUR.	50' RAILROAD EXCLUSION ELIMINATED			PERSONAL & ADV INJURY	$ 1,000,000
	OWNER'S & CONTRACTOR'S PILOT				GENERAL AGGREGATE	$ 2,000,000
X	ISO FORM CG0001	(11/BB PR EQUIVALENT)			FIRE DAMAGE (Any one foto)	$ 300,000
X	CONTRACTL LIAB.				MED. EXPENSE (Any one person)	$ 5,000
	GENL AGGREGATE LIMIT APPLIES PER: POL-ICY [X] PRO-JECT LOC					
	AUTOMOBILE LIABILITY	POLICY NUMBER			COMBINED SINGLE LIMIT	$ 1,000,000
X	ANY AUTO				BODILY INJURY (Per person)	$
	ALL OWNED AUTOS					
	SCHEDULED AUTOS				BODILY INJURY (Per accident)	$
	HIRED AUTOS					
	NON-OWNED AUTOS				PROPERTY DAMAGE	$
	GARAGE					
	EXCESS LIBILITY	POLICY NUMBER			EACH OCCURRENCE	2,000,000
X	UMBRELLA FORM	PER PROJECT ENDORSEMENT INCLUDED			AGGREGATE	$ 2,000,000
	OTHER THAN UMBRELLA FORM					
	WORKERS COMPENSATION AND EMPLOYER'S LIABILITY	POLICY NUMBER			[X] STATUTORY LIMITS	
		COVERAGE APPLIES IN STATE OF JOBSITE OPERATION			EACH ACCIDENT	$ *500,000
		UNDER THIS SUBCONTRACT			DISEASE-POLICY LIMIT	$ *1,000,000
	THE PROPRIETOR, PARTNERS EXECUTIVE OFFICERS ARE [X] INCL. EXCL.	USL&H COVERAGE IS INCL-UDED WHERE NEEDED			DISEASE-EACH EMPLOYEE	$ *500,000

This must be crossed out →

SHOULD ANY OF THE ABOVE DESCRIBED POLICIES BE CANCELLED ** BEFORE THE EXPIRATION DATE THEREOF , THE ISSUING INSURER WILL ENDEAVOR TO MAIL **30** DAYS WRITTEN NOTICE TO THE CERTIFICATE HOLDER NAMED TO THE LEFT BUT FAILURE TO MAIL SUCH NOTICE SHALL IMPOSE NO OBLIGATION OR LIABILITY OF ANY KIND UPON THE COMPANY, ITS AGENTS OR REPRESENTATIVES.

AUTHORIZED REPRESENTATIVE

*** NON-RENEWED OR MATERIALLY CHANGED*

Initials_____/_____

OWNER OR CONTRACTOR CONTROLLED INSURANCE PROGRAMS

For a large construction project being built in the urban environment, a wrap-up insurance program is often evaluated for the project. The wrap-up insurance can be provided either by the owner, OCIP, or by the contractor, CCIP. When a project is in excess of $50 million, a wrap-up policy may be economically advantageous for the project.

The wrap-up policy takes the place of all of the individual insurance policies that the owner, CM/GC, and subcontractors would normally provide to cover their contractual requirements, risks, and the construction work. The concept is the same, whether the owner or contractor procures the wrap-up insurance policy for the project—to provide one umbrella insurance policy for the entire project that will cover all parties. A wrap-up insurance policy requires someone on the CM/GC's staff to administer the program. This person would sign up subcontractors prior to the start of the construction work, track insurance credits in the subcontractor's base contract and change orders, gather certified payrolls to submit to the insurance carriers to keep track of labor categories and hours expended to perform the work, send accident and incident reports to the insurance carrier, send supporting documentation regarding an accident or incident to the insurance carrier, arrange for witnesses and supervisory personnel to be called in to testify if there is any litigation, and so forth. The CM/GC needs to ensure that a qualified insurance administrator is on staff to be able to perform the required functions and provide the proper interface with the wrap-up insurance carrier. This is an additional cost to the general conditions of the project (See Chapter 17).

This one-stop insurance shopping has many benefits to the project, owner, CM/GC, and subcontractors. Exhibit 10-3 summarizes the advantages of a wrap-up insurance program provided either by the owner (OCIP) or by the contractor (CCIP).

1. Provides insurance coverage for all parties working on the site through one insurance carrier and policy (watch out for certain unions that require their members to be covered under the union's workers' compensation insurance program).
2. Combines insurance coverage, safety coordination, and claims reporting into one comprehensive process for all parties working on the site.
3. Can result in savings in insurance premiums for the overall project, especially based on the economy of scale.
4. Usually has less exclusion than coverage found with individual CM/GC, owner, subcontractor, and consultant's policies.
5. It provides a uniform and enforceable level of insurance coverage and a safety program associated with it for the project.
6. It challenges all involved parties with the opportunity for prevention and mitigation of any accidents, incidents, and potential damages.
7. Reduced litigation in the event of a claim, with only one insurance carrier involved with insurance coverage and handling of claims.

Exhibit 10-3

Advantages of a wrap-up insurance program (OCIP or CCIP).

(Continued)

Exhibit 10-3
(*Continued*)

8. A wrap-up provides a unified defense for all parties in the event of a claim.
9. Contract indemnification provisions are easier to put in place with all parties insured by one insurance carrier.
10. A wrap-up minimizes the likelihood of litigation for on-site work injuries and cross-liability suits.
11. Higher limits of coverage are usually available under a wrap-up insurance program through one policy.
12. Usually results in fewer coverage disputes, along with faster and more professional claims processing.
13. The wrap-up insurance policy usually has less qualification and exclusions of coverage.
14. Common insurance coverage and terms provide a level of comfort to all parties involved.
15. One insurance company will insure, indemnify, and hold harmless all parties.
16. One insurance premium to be paid directly by the owner and/or CM/GC.
17. Better control of the issue of insurance certificates and coverage for all parties working on the project.
18. One insurance company and legal department to defend all claims in connection with claims arising out of the performance of the work.

SURETY BONDS

A surety bond is a three-party contract, which involves the principal (e.g., the subcontractor), the surety company guaranteeing the principal's performance of the work, and the obligee (either the owner or the CM/GC) for which the bond is provided. It is important that the CM/GC understand that a surety bond is not an insurance policy, although in some respects there are some similarities. A surety bond is also a contract to cover the default in the performance of the work by the principal, and the cost incurred to remedy the deficiency. Surety bonds are often required for all municipal, public, governmental, and quasi-governmental work. Sometimes bonds are required on private projects to meet financing requirements, or if the subcontractor performing the work is questionable as to their ability to complete the work as per the contractual requirements. The bond provides an additional level of protection that the work will be completed in accordance with the contract. A subcontractor, supplier, or vendor providing new technology or equipment to the construction project may be a reason to consider a bond.

Surety bonds are expensive, depending on the record of accomplishment of the firm and its principals, financial well-being, reputation in performing similar projects, and perceived risk to the bonding company. This three-party relationship is further defined below.

Principal

This is the party who is performing the work and who directly pays for the surety bond. The bond guarantees the subcontractor performing the work. The bond allows the principal to get jobs that it would not otherwise be able to without a surety bond.

Obligee

This is the entity that is protected by and most directly benefits from a surety bond, either the owner or the CM/GC. Governmental entities are required by the Miller Act (federal law) and little Miller Acts (state laws) to have surety bonds to guarantee that taxpayer monies are being used prudently. It could also be, for example, a mechanical subcontractor who is responsible for the sheet metal portion of a job and decides to require a bond from the sheet metal sub-subcontractor who will be supplying and installing the ductwork correctly and for the specified contract amount.

Surety Company

This is typically the surety division/department of an insurance/bonding company. It receives a fee from the principal (subcontractor) for putting its money at risk and for guaranteeing to the obligee (owner or CM/GC) that the principal will perform per the contract requirements, including paying those who have provided labor and material to the project.

Most surety companies sell their bonds through independent insurance agents. Some agents are bond specialists. These agents will be able to give the CM/GC and subcontractor good advice on what the surety companies require to provide a bond. They will also have good relationships with surety underwriters in order to know which surety company would be best for the particular subcontractor. A CM/GC typically has one surety agent and one surety company with whom they deal to write all of their bonds. A surety company considers the relationship between the company and the contractor to be very important.

It is important for CM/GCs and subcontractors to know that unlike companies that write insurance policies, bonding companies that write surety bonds have the legal right to go back against any CM/GC or subcontractor who causes them a loss on a surety bond and recover the total amount of the loss and the surety's associated expenses. Surety companies sometimes require that the CM/GC and the subcontractor provide collateral for the value of the surety bond. Sometimes, however, the bonding company will be comfortable with the financial status, experience, reputation, and performance of the CM/GC and write the bond based on their relationship with the CM/GC. The CM/GC and subcontractors will have to sign an indemnity agreement specifically agreeing to this before the surety company will write the bond. Often, the surety company will require the owners of the CM/GC and subcontractors, along with their spouses, to sign personally for the indemnity agreement. Therefore, if the subcontractor causes the surety company a loss, not only would the subcontractor be liable to pay the surety back, but the owners would be responsible as well. However, surety companies are not secured creditors. At any one time, they probably have many millions of dollars of bonds written for projects that are not completed. If they were to file the indemnity agreement for the contracts they are guaranteeing, the contractors would have a very difficult time getting a bank to loan them money.

CM/GC and subcontractors usually need a few license bonds that are required by various governmental agencies in order to qualify a CM/GC and subcontractor for licensing in a particular city, county, or state. This is especially true in large urban environments. These are usually of rather small amounts and surety companies write the bonds much more freely than larger performance and payment bonds. Many of these bonds only guarantee that the work that is done by the principal is done according to the building codes—not that the work actually gets done or that suppliers of labor or material are paid.

TYPES OF BONDS

Various types of bonds can be required in a contract to protect the parties in the performance of the work. The different types of bonds are bid bond, performance bond, payment bond, maintenance bonds, and completion bonds. These bonds are discussed in the following.

Bid Bonds

Bid bonds are often required by obligees to be presented with other bid documents. This type of bond guarantees to the obligee who the low bidder is, that they will sign the contract for the project, and will present a performance and payment bond, if required. The amount of a bid bond's guarantee is usually 5 to 10% of the amount of the total bid. This is the bond amount or the potential bond penalty. Usually, surety companies do not charge for bid bonds. If the contractor was low and awarded the job, but decided not to enter into the contract, the obligee can make a claim against the bond in order to make up the difference between that low bidder's price and the price of the second bidder, or to be able to put the project up for re-bid.

Performance Bonds

Performance bonds are written for a specific project or contract. They guarantee that the contractor will perform the contract according to all of the contract documents. Therefore, the information in the contract documents is what is guaranteed: the completion time as well as interim completion goals, all of the specifications, any minority participation, as-built drawings, environmental concerns, the frequency of payments that the contractor will have to put up with, etc. The surety guarantees that the contractor will perform. Therefore, if the contractor fails to do so, the surety probably has an obligation to step in to complete the contract. However, if the principal on the bond has good reason not to complete the job, such as not getting paid by the obligee or another breach committed by the obligee, the principal must communicate this to the surety. Otherwise, the surety will have no choice but to step in and complete the job. Then, the surety has the right to go against the principal for any expenses the surety incurred in completing that project. Exhibit 10-4 contains a sample subcontractor performance bond.

Payment Bonds

Payment bonds are written for a specific project or contract. They guarantee that the CM/GC and subcontractor will pay all those who furnish labor and material for the job. Payment bonds often guarantee the payment of utility bills as well. Usually, the CM/GC

Exhibit 10-4

Sample
subcontractor
performance bond.

KNOW ALL MEN BY THESE PRESENTS: That *(name of subcontractor)* as Principal, hereinafter called Principal, and *(name of Surety Company),* as Surety, hereinafter called Surety, are held and firmly bound unto *(name of Owner/CM/GC),* as Obligee, hereinafter called Obligee, in the amount of Dollars *($ amount of the bond),* for the payment whereof Principal and Surety bind themselves, their heirs, executors, administrators, successors and assigns, jointly and severally, firmly by these presents. WHEREAS, Principal has by written agreement dated *(date of signed contract between Principal and Owner/CM/GC)* entered into a subcontract with Obligee for *(description of the project and scope of work to be performed)* in accordance with drawings and specifications prepared by *(name of Architect, Engineer, and Specialty Consultants)* which subcontract is by reference made a part hereof, and is hereinafter referred to as the subcontract. NOW, THEREFORE, THE CONDITION OF THIS OBLIGATION is such that, if Principal shall promptly and faithfully perform said subcontract, then this obligation shall be null and void; otherwise it shall remain in full force and effect. Whenever Principal shall be, and be declared by Obligee to be in default under the subcontract, the Obligee having performed Obligee's obligations there under:

(1) Surety may promptly remedy the default subject to the provisions of paragraph 3 herein, or;

(2) Obligee after reasonable notice to Surety may, or Surety upon demand of Obligee may arrange for the performance of Principal's obligation under the subcontract subject to the provisions of paragraph 3 herein;

(3) The balance of the subcontract price, as defined below, shall be credited against the reasonable cost of completing performance of the subcontract. If completed by the Obligee, and the reasonable cost exceeds the balance of the subcontract price the Surety shall pay the Obligee such excess, but in no event shall the aggregate liability of the Surety exceed the amount of this bond. If the Surety arranges completion or remedies the default, that portion of the balance of the subcontract price as may be required to complete the subcontract or remedy the default and to reimburse the Surety for its outlays shall be paid to the Surety at the times and in the manner as said sums would have been payable to Principal had there been no default under the subcontract. The term balance of the subcontract price, as used in this paragraph, shall mean the total amount payable by Obligee to Principal under the subcontract and any amendments thereto, less the amounts heretofore properly paid by Obligee under the subcontract.

Any suit under this bond must be instituted before the expiration of two years from date on which final payment under the subcontract falls due.

No right of action shall accrue on this bond to or for the use of any person or corporation other than the Obligee named herein or the heirs, executors, administrators or successors of the Obligee.

(Continued)

Exhibit 10-4
(*Continued*)

Signed and sealed this day of

IN THE PRESENCE OF:

_____ (Seal)

Principal

_____ (Seal)

_____ (Seal)

By
Attorney-in-Fact

and subcontractor, and therefore the surety, are only responsible for no more than two tiers of sub-subcontractors below them. However, the contractor must read the contract documents, and perhaps consult its construction-oriented attorney, to make sure of this depending on applicable law in the location. The CM/GC and subcontractors need to make changes in the contract language if there are any legal concerns before the contract is signed and the bond submitted. An alternative to payment or performance bonds is contractor default insurance, sometimes referred to as subguard. It is important to contact the insurance and bonding consultants to properly evaluate these types of programs. Exhibit 10-5 contains a sample payment bond.

Maintenance Bonds

Maintenance bonds guarantee that for a specified period after the project is complete, the project will remain free of defects caused by poor workmanship and defective materials. Sureties do not like to write maintenance bonds beyond 2 or 3 years after the completion of the project.

Completion Bonds

Completion bonds have obligees that do not actually have a contractual obligation to the principal. A municipality may require a completion bond that guarantees that streetscape work will be done correctly. Although, the city is not paying for the work—the owner of the building nearest the streetscape work probably is—the city still wants the work done correctly and perhaps according to the latest city ordinances and master plan for the area, requiring a certain amount and type of sidewalk and lighting. These types of bonds are unique because the CM/GC and subcontractor could be required to finish the work even if they are not being paid. The CM/GC and subcontractor are supposed to be paid by one party, but the bond's obligee is another, in this case the municipality. Surety companies are usually quite cautious about writing these bonds.

These are generalizations. Read the bonds—they put your company and its principals at risk. An obligee can ask for standardized bond forms like those developed by the American Institute of Architects (AIA) or American General Contractors Association (AGC); however, some obligees have their own bond forms that can be onerous. Sometimes the standard bond forms have changes and clauses that may cause concerns or problems as well.

KNOW ALL MEN BY THESE PRESENTS: That *(name of subcontractor)* as Principal, hereinafter called Principal, *(name of Bonding Company)* and hereinafter called Surety, are held and firmly bound unto *(name of Owner/CM/GC)* as Obligee, hereinafter called Obligee, in the amount of Dollars *($ dollar amount of the contract or bond),* for the payment whereof Principal and Surety bind themselves, their heirs, executors, administrators, successors and assigns, jointly and severally, firmly these presents. WHEREAS, Principal has by written agreement dated *(date of signed contract between subcontractor and Owner/CM/GC)* entered into a subcontract with Obligee for *(description of the project and scope of work)* in accordance with drawings and specifications prepared by *(name of Architect, Engineers, and Specialty Consultants)* which subcontract is by reference made a part hereof, and is hereafter referred to as the subcontract. NOW, THEREFORE, THE CONDITION OF THIS OBLIGATION is such that if the Principal shall promptly make payments to all claimants as hereinafter defined, for all labor and material used or reasonably required for use in the performance of the subcontract, then this obligation shall be void; otherwise it shall remain in full force and effect subject, however, to the following conditions:

(1) A claimant is defined as one having a direct contract with the Principal for labor, material, or both, used or reasonably required for use in the performance of the contract, labor and material being construed to include that part of water, gas, power, light, heat, oil, gasoline, telephone service or rental of equipment directly applicable to the subcontract.

(2) The above-named Principal and Surety hereby jointly and severally agree with the Obligee that every claimant has herein defined, who has not been paid in full before the expiration of a period of ninety (90) days after the date on which the last of such claimant's work or labor was done or performed, or materials were furnished by such claimant, may sue on this bond for the use of such claimant, prosecute the suit to final judgment for such sum or sums as may be justly due claimant, and have execution thereon. The Obligee shall not be liable for the payment of any costs or expenses of any such suit.

(3) No suit or action shall be commenced hereunder by any claimant.

   (a) After the expiration of one (1) year following the date on which Principal ceased work on said subcontract it being understood, however, that if any limitation embodied in this bond is prohibited by any law controlling the construction hereof such limitation shall be deemed to be amended so as to be equal to the minimum period of limitation permitted by such law.

   (b) Other than in a state court of competent jurisdiction in and for the county or other political subdivision of the state in which the project, or any part thereof, is situated, or in the United States District Court for the district in which the project, or any part thereof, is situated, and not elsewhere.

(4) The amount of this bond shall be reduced by and to the extent of any payment or payments made in good faith hereunder.

(Continued)

Exhibit 10-5
Sample payment bond.

Exhibit 10-5
(*Continued*)

Signed and sealed this day of

IN THE PRESENCE OF:

_____ (Seal)

Principal

_____

_____ (Seal)

_____ (Seal)

Subcontract Labor and Material Payment Bond.
This bond is issued simultaneously with another   By
bond in favor of the general contractor conditioned   Attorney-in-Fact
for the full and faithful performance of the contract.

SURETY COMPANIES

A surety company that is asked to write a bond for a CM/GC or subcontractor with whom it has never done business will probably ask many questions and request a lot of documentation. The surety company will want to get to know the contractor and its business. Most surety underwriters consider surety to be a relationship-driven business. That relationship should include an independent bonding agent. The underwriter will want to see a few years of CPA-prepared fiscal year-end financial statements, schedules of jobs in progress and completed jobs, an interim financial statement, a copy of a bank letter of credit, the company's bylaws and legal documents, and a completed questionnaire that gives the surety some basic information about the contractor. If the surety is going to require the personal indemnity of the owners, that surety will want a current personal financial statement as well. Exhibit 10-6 is a sample questionnaire from a surety company.

The CM/GC and subcontractor should look for a surety company that wants to have a relationship with them. The surety could help the CM/GC and subcontractor grow their business in a steady, prudent, and conservative manner. It might be able to help with relationships with others such as CPAs, bankers, or attorneys who are construction specialists. A surety company that has a long-term, good relationship with a CM/GC and subcontractor is more likely to ride out a bad time than a surety that does not have that close relationship. The CM/GC and subcontractor should feel comfortable contacting the surety immediately if they run into a problem on a job that could lead to a claim against the surety. The surety has obligations to the obligee, but it also has an obligation to the principal to make sure it does not pay claims for which the principal has a defense, such as the obligee has not paid them or the architect has not responded to them with information they require to proceed and complete the project. The CM/GC and subcontractor will want a stable surety company with which to do business. The bonding agent should be able to help with this. It is prudent to check with the A.M. Best rating service as well as the U.S. Treasury listings to see how the surety company is rated. Some obligees have requirements that sureties must have

1. Name of Firm: _____

2. Address:_____

   _____

3. Fiscal Year End _____

4. Phone: _____ 4a. Fax: _____

5. Contracting Specialty: _____

6. Contact Person: _____

7. Title_____

8. Year Business Started: _____

9. Type of Business __ Corp __ Part. __ Prop. __ Sub S. Corp

10. State of Incorporation: _____

11. Area of Operation _____

12. List the corporate officers, partners or proprietors of your firm:

Name	Position	Owner	Percent	Date	Name of Spouse
A)					
B)					
C)					
D)					
E)					

13. Will the above individuals and spouses personally indemnify Surety?
    __ Yes   __ No

14. If no, explain: _____

    _____

15. Is there a buy/sell agreement among the owners of the business?
    __ Yes   __ No

16. Is this agreement funded by life insurance? __ Yes   __ No

17. Corp. Indemnity? __ Yes   __ No

18. Cross Corp. Indemnity? __ Yes   __ No

19. How many people does your firm employ? _____

20. How many work crews? _____

21. Has your firm or any of its principals ever petitioned for bankruptcy, failed in business or defaulted so as to cause a loss to a Surety?   __ Yes   __ No

22. If yes, please explain: _____

    _____

23. Is your firm or any of its owners or officers currently involved in any litigation?
    __ Yes   __ No

24. If yes, please explain: _____

    _____

25. What percentage of the firm's work is normally for:

26. Government Agencies: _____ %   Private Owners: _____ %

27. What percentage of the firm's work is normally subcontracted: _____ %

28. Are bonds required of subs? ___ Yes ___ No

Exhibit 10-6

Contractor bonding questionnaire.

(Continued)

Exhibit 10-6
(Continued)

29. What trades do you normally subcontract? _____
30. What is the largest amount of uncompleted work on hand at one time in the past?
31. Amount: $ _____   Year: _____
32. What is the largest job you expect to do during the next year? $ _____
31. What is the largest uncompleted work program expected during the next year?
    $ _____
32. What is your expected annual volume next year? $ _____
33. What trades do you normally undertake with your own forces? _____
    _____

34. SIC CODE _____
35. Do you lease equipment? __ Yes   __ No   Type of lease? _____
36. What are the terms of the lease? _____
    _____

37. Name of your CPA: _____
38. Address: _____
39. Phone: _____   Contact Person: _____
40. On what basis are taxes paid? ___ Cash ___ Completed Job ___ Accrual
    ___ % of Completion
41. On what basis are financial statements prepared? ___ Cash   ___ Comp. Job
    ___ Accrual   ___ % Complete
42. How often are financial statements prepared? ___ Annually ___ Semi-annually
    ___ Quarterly ___ Monthly
43. Do you have a full time accountant on staff?   ___ Yes   ___ No
    Yrs. Experience: _____
44. Are job records kept? ___ Yes   ___ No
45. How often reviewed? _____   How often updated? _____
46. Do they show job detail? ___ Yes   ___ No
47. Frequency? _____
48. Name of your Bank: _____
    a. Address: _____
       _____
    b. Phone: _____   Contact Person: _____
    c. Amount of line of credit: $ _____
49. Expiration date: _____
50. What is interest rate? ____%
51. UCC Filing? ___ Yes   ___ No
52. How is credit secured? _____
53. Is your firm union? ___ Yes   ___ No
54. What is firm's Dun & Bradstreet Number? _____
55. On what level of assurance are financial statements prepared? ___ CPA Audit
    ___ Review   ___Compilation
56. D & B Rating: _____
57. Pay Record: _____
58. Date of Rating: _____

59. Previous Bond Companies

Exhibit 10-6
(*Continued*)

     Name   Years   Reason for Leaving

    A. _____ _____ _____

    B. _____ _____ _____

    C. _____ _____ _____

60. List five of your largest contracts:

Job Name	Location	Type Project	Contract Price	Gross Profit	Date Completed	Bonded?
A)						
B)						
C)						
D)						
E)						

61. List five of your major suppliers

Name	Address	Telephone	Contact
A)			
B)			
C)			
D)			
E)			

62. List five subcontractors (or contractors if you are a subcontractor) with whom you do business:

    A)

    B)

    C)

    D)

    E)

63. List any subsidiaries and affiliates of the contracting firm:

Name of firm	Type of Business	Ownership of Business	Annual Revenue	Gross Profit	Work Bonded
A)					
B)					
C)					
D)					
E)					

64. REMARKS:

_____

_____

_____

                    Completed by:

                    Title:

                    Date:

Exhibit 10-7

CM/GC Surety concerns.

1. Who are the owners of the business?
2. Are the owners active in the business?
3. Has there been a recent change in ownership or key upper management personnel?
4. Have there been major changes in the business?
5. Has there been rapid expansion of the business?
6. Are there sufficient and competent managerial professionals to run the organization?
7. Are there sufficient and competent technical and engineering professionals to build the project?
8. Have there been major changes in the focus of the business?
9. Has the CM/GC developed a strategic, business, marketing, and annual plan?
10. Is the company adequately insured?
11. Is the firm concentrating its marketing efforts in its market niche?
12. Does the firm have a proper and adequate accounting system and practices?
13. Does the firm have a project management system to monitor and control the project's schedule, budget, quality control, etc.?
14. Does the firm have a prior history of problems with procuring, estimating, scheduling, purchasing, constructing, administering, and close-out of a project?
15. Does the firm have a good safety and quality control program?
16. Has the firm ever been denied insurance or bonding coverage?
17. Has the firm, or any of its staff, been found guilty of criminal activities?
18. Does the firm have any pending litigation?

certain minimum ratings in order to be acceptable to provide the required bonds and will not accept bonds written by sureties who have ratings below those minimum levels.

Exhibit 10-7 lists the major concerns that a surety company has of a CM/GC.

SUMMARY

- Insurance planning is essential to managing construction risks and avoiding excess losses.
- Insurance policies are often used to limit exposure to settlement of disputes.
- Construction insurance is a very specialized area, and insurance professionals should be called on to properly manage the insurance process.
- No contractor should be allowed to work on a construction project without a certificate of insurance proving that the required insurance coverage is in effect.
- A performance bond is not an insurance policy, but a contract between three parties—the obligee, the principal, and the surety company—to be responsible for

the potential default in the performance of the work or cost incurred to remedy the deficiency in the work.

- There are many types of bonds. It is important to understand the contract terms and conditions as to what type of bonds are required.

- A bond may be requested if the GC or owner is not fully comfortable with the ability of a contractor to perform the work, and to minimize potential financial exposure because of the work not being performed to the terms of the contract.

- Termination of a contractor and calling in the bond with the bonding company should be a last resort. The contractor should first be given notice to remedy the problems.

- There should be a dispute resolution clause in the bond, such as arbitration to allow for resolution of a claim.

- Try to avoid clauses in the contract and bond that allow for termination for convenience instead of for due cause. Once a bond is issued, it cannot be canceled. It protects the obligee until the contract has been performed, unless there has been non-cardinal changes. A cardinal change may be a large change in contract price of more than 20% or an addition of an item of work not normally performed by that particular CM/GC or subcontractor.

11

Security
(Why are we missing so much copper?)

SECURITY OF THE CONSTRUCTION SITE

Security at a construction site is concerned with securing the overall site, limiting access to authorized personnel only, and proper storage of materials, safeguarding the site from any risks, and preventing theft and vandalism at the site. The security of a construction site is of utmost importance to ensure the safety of the personnel and materials on the site, and to ensure that only authorized personnel are allowed on the site. During the development of a site logistics plan for the project, as detailed in Chapter 6, the locations of site access, construction shanties and field offices, storage and staging of materials, and construction materials will be identified. During the construction process, the construction manager/general contractor (CM/GC) is usually responsible for the security of the site. Sometimes when an existing building, such as a U.S. federal courthouse, will remain in the custody and control of the owner during the construction process, the owner will continue to provide security services for the building with their own security team. The owner also usually picks up the security responsibility as they start to deliver their furniture, fixtures, and equipment, and the building is getting ready for occupancy.

SITE ACCESS BY CONSTRUCTION PERSONNEL

Construction personnel will usually enter the site from an entrance separate from the owner's entrance if they are occupying the site at the same time. All construction trade personnel should be given proper identification badges to be worn at all times. Without a badge, a person should not be allowed access to the site. The CM/GC will usually produce the badges for the project, and give them out to the subcontractor foreman, who then distributes them to the authorized personnel from their company working on the site. The names of the personnel assigned to the badge, along with all other proper identification and documentation should be gathered and returned by the subcontractor's foreman to the CM/GC for control purposes. In addition, the CM/GC may wish to consider ensuring that all construction personnel have received their safety training specific to the site prior to issuing them a security badge.

A security guard should be posted at the construction personnel gate during normal working hours (in addition to 1 hour before and after) to control site access for the majority of the construction personnel. An off-hours site access gate should also be established for personnel working late or arriving early, and for emergency off-hour access to the site. Security guards are normally supplied and controlled by the CM/GC during the construction phase of the project, as part of the general conditions of the project.

SITE ACCESS BY THE OWNER

If the owner is going to occupy the building, or a portion thereof, during the construction process (which often happens later in the construction cycle as the project is nearing completion), then a separate entrance should be established for the owner's personnel, with the owner's own security guards and procedures in place. Once the building is cohabited by both the owner and CM, it is a good idea to have one security force provide security of the site. This is usually supplied by the owner.

SECURITY OF CONSTRUCTION MATERIALS

Construction materials are delivered to a temporary loading dock for the site, which is usually located on street level, adjacent to the hoist. A security guard should be posted at this location to control the delivery of all materials, as well as the removal of all debris from the site. The guard should have a copy of the delivery schedules and hoist log to ensure that the deliveries are scheduled, and occur in the sequence as scheduled, or make the necessary adjustments as the day's construction operations may require. The security guard should record all deliveries and removals, company name, license plate and operator of the truck, and any unusual activities. These logs should be turned over to the CM/GC daily for review, scheduling, and security control purposes.

Any personnel accessing the site with construction materials should also be documented and controlled. The loading dock entrance should not become an alternate back door entrance to the construction site. The loss of construction materials is costly to the project in both money and time.

SECURITY SYSTEMS

It is often wise to have a roaming guard on the site as well, who will walk the site periodically, ensuring that everyone is wearing a proper ID badge, observe the working conditions, and report any abnormal events to the CM/GC. The security guard can also act as a fire watch off-hours, provided that they have had the proper training and are licensed by the appropriate agencies. Security guards are often required to use a security lock key system to ensure that they have made their rounds and visited all of the critical spaces.

For a large project, consideration should also be given to utilizing closed circuit television cameras to provide surveillance of the construction site 24 hours a day, 7 days a week. The cameras should be laid out and installed by a professional company to ensure adequate coverage of the site at all major points of entry and egress, as well as central locations through which personnel and materials flow. These video recordings should be kept in a permanent electronic file, in the event that an issue develops.

Consideration should be given to supplying the security guard with a radio to be able to communicate with the other security guards, construction mangers, etc. In addition, the guard should be given an emergency contact list for police, fire, ambulance, hospital and first aid station, all essential CM personnel, etc. (see Chapter 5 Exhibit 5-30). The CM should also have an emergency contact list including the names and emergency contact numbers for all major subcontractors, suppliers, and vendors working on the site. A landline telephone is essential to have at the guard's booth at the site. The use of a mobile telephone may also be considered for the guard to allow for a more timely response if there is an event in the building away from the guard's booth. Timely response and notification of an incident can make a substantial difference in minimizing the implications of a situation.

Unauthorized personnel should be directed to the CM/GC's office for access to the sites. The guard or the site safety manager should also ensure that the person is wearing proper clothing, such as shoes and a hardhat, before allowing access to the site. (See Chapter 16, Exhibit 16-7A).

CONSTRUCTION SHANTIES AND FIELD OFFICES

The CM/GC's site office and subcontractor shanties will be located in reasonable proximity to the site entrance. However, they may be on an alternate floor in the building, on the sidewalk bridge, retail space, etc. Consideration should be given to identifying a controlled path for a visitor to take from the construction personnel entrance to ensure that they do not go wandering aimlessly around the construction site.

Construction materials and equipment are often stored in the subcontractor shanties and storage areas, where they may be under lock and key. However, this does not stop people from breaking in and stealing valuable equipment. In addition, materials and equipment may be stored on the floors of the construction site where the work is being performed. Often this presents an open invitation to someone who may be looking to steal the material for reuse or resale. This is often referred to in the industry as MONGO (or the salvage value of the materials, such as copper, steel, brass, nickel, and aluminum to the local salvage yard). One sure sign of this is when the recycling truck arrives at the loading dock of the site during construction. Reels of 500MCM copper cable over 500 feet long have disappeared after delivery, to be resold for salvage value. When events like this happen, it is detrimental to the construction process because the material has to be re-ordered, resulting in additional expenses and potential delays to the project. This cost is often born by the subcontractor, as most contracts read that it is the subcontractors' overall responsibility to secure their own materials in a safe manner. If the loss is large enough, the subcontractor may consider filing a claim with their insurance carrier; however, this usually results in an increase in the insurance policy premiums and difficulty renewing policies.

SECURITY OF PRE-ORDERED MATERIALS

Often construction materials have a long lead-time, and must be procured prior to the award of subcontracts and the start of the construction process. Items such as refrigeration equipment, cooling towers, generators, UPS systems, raised flooring, etc. need to be ordered by the CM/GC and then assigned to the appropriate subcontractor upon delivery. Materials that are pre-ordered may arrive prior to the start of construction, and will require off-site storage until they can be delivered to the site and set in place. In addition, large quantities of construction materials such as light fixtures, raised floor panels, etc., may need to be stored off-site until they are ready to be installed. Consideration should be given to an off-site storage facility, usually a fully bonded and insured warehouse, which will be able to store these materials for an extended period. It is important to note that the materials should have a certificate of insurance prior to being paid for by the owner to protect everyone's interests.

CONSTRUCTION SITE SECURITY PLAN

The construction site's security plan must be coordinated with the overall site logistics plan for the project. Exhibit 11-1 indicates the overall site logistics plan, which also incorporates the site security plan.

Exhibit 11-2 is a checklist for a site security plan.

Exhibit 11-3 is a checklist for project ID card and access procedures.

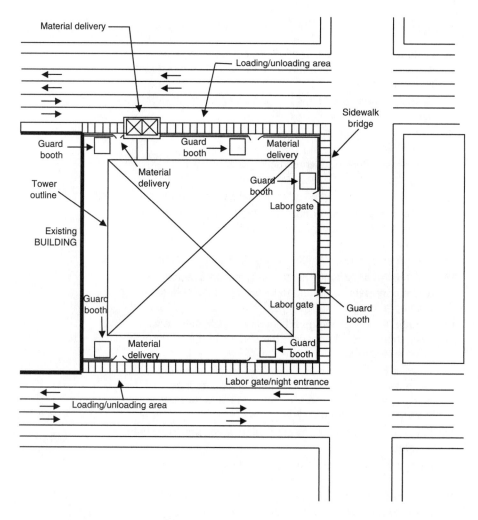

Exhibit 11-1
Site security plan.

Item	Included
1. Hire a professional security service company to provide guard service.	
2. Set up a separate entrance for construction personnel with a guard for normal working hours.	
3. Set up a separate entrance with a guard for owner's personnel.	
4. Set up a separate entrance for delivery and removal of all materials.	
5. Schedule all deliveries, and keep a log of all scheduled deliveries and applicable details.	

Exhibit 11-2
Checklist for a site security plan.

(*Continued*)

Exhibit 11-2
(*Continued*)

Item	Included
6. Issue a pass to all construction personnel on the site, with a number, the employee's name, and the subcontractor for whom they work.	
7. Keep a current list of all active personnel and passes that have been issued and are valid.	
8. All material leaving the site must have a pass signed by the CM/GC.	
9. Provide an after-hours construction entrance/exit for personnel.	
10. File an incident report of any unusual activities at the site.	
11. Obtain the contact information for the local police precinct, fire department, hospital, and first aid center in case of an emergency.	
12. Post all emergency numbers in the CM/GC's field office.	
13. Maintain an emergency contact information directory for all the principals, foremen, and PMs for all the subcontractors working at the site.	
14. Arrange for a roaming guard service to walk the site with a watchman's key to observe the entire site during normal hours and after hours.	
15. Arrange for the installation of closed-circuit television cameras at strategic locations to monitor personnel entering and leaving the site, deliveries, and removal of materials from the site.	
16. File a police report for any criminal activities and prosecute all offenders.	
17. Have the subcontractors store construction materials and tools in a safe and secure location.	
18. Be especially careful to secure tools that can easily be removed and used in the home.	
19. Security, like safety, is everyone's responsibility.	
20. Determine who will provide the security service to the site—the owner or the contractor?	
21. Use the site logistics plan to determine the construction personnel entrance, owner's entrance, construction materials, loading dock, etc., to determine the location and quantity of site access points and the number of security guards that will be required.	
22. Establish an after-hours night and weekend entrance for access and egress to the site.	
23. Provide a secure fence and gates around the site to secure it after hours.	
24. Determine what additional security measures are required, such as closed-circuit cameras, motion detectors, security lock key systems, etc., and determine if the security company that is providing guard service can provide them.	

Item	Included
25. Keep logs of all personnel and materials entering and leaving the site.	
26. All personnel should have a security badge identifying who they are, the subcontractor for whom they are working, ID number, etc.	
27. Consider a photo ID for the security badge.	
28. Ensure that all personnel issued a security badge have had the proper safety and site-specific training, and are wearing proper construction clothing and protection.	
29. Arrange for off-site storage at a bonded warehouse, which is in reasonable proximity to the construction site.	
30. Consider a Minority Woman Business Enterprise (MWBE) firm for providing the security services to assist in meeting MWBE project and corporate objectives.	
31. Do not allow construction personnel to access the site via the loading dock, other than to assist with deliveries.	
32. Coordinate the security logs with the hoist log to ensure appropriate subcontractor materials are delivered and removed.	
33. Establish a security pass system for the removal of any large quantity of materials, equipment, etc., from the site.	
34. Put an identification number on all tools and equipment.	
35. Lock up storage for tools and materials, especially those that can be used in the home.	
36. Post "No Trespassing" signs.	
37. Provide adequate site illumination for after-hours.	
38. Secure all field offices, shanties, lock ups, and trailers.	
39. Consider alarm systems for field offices and critical locations.	
40. Lock out the electrical service to the hoist, crane, and other heavy equipment after hours.	

Exhibit 11-2 (*Continued*)

Item	Included
1. All personnel working on the site shall carry an ID card at all times and be clearly visible.	
2. Consider having the employee's photograph, name, and the subcontractor with whom they are employed identified on the ID card.	
3. Incorporate the issuance of the employee's ID card with their safety training prior to starting work at the site.	
4. Each subcontractor should keep track of their authorized personnel working on the site each day, and submit it to the CM/GC and the security guard service.	

Exhibit 11-3 Checklist for project ID card and access procedures.

(*Continued*)

Exhibit 11-3

(*Continued*)

5. Any person on the site who does not comply with the above requirements will be removed from the site.
6. The CM/GC's project superintendent will carry a copy of the list of all personnel who are authorized to have access to the site.
7. Be sure to obtain ID cards from any employees who leaves the site once they have completed their work.
8. The guard security service should perform a random spot check of ID cards for all employees on a daily basis.
9. Any person visiting the site on an interim basis should be escorted to the CM/GC office for screening and issuance of a temporary ID card for the appropriate period of time.
10. Visitors should be distinguished from employees.
11. Visitors should be escorted at all times.
12. Establish access control procedures for after-hours.

SUMMARY

- During the construction process, the CM/GC is usually responsible for security of the site.
- The owner usually takes responsibility for security once furniture, fixtures, and equipment are delivered and the building is getting ready for occupancy.
- Security of a construction site is very important to ensure the safety of the personnel and security of the construction tools and materials at the site.
- The security plan for the site needs to be integrated with the site logistics plan.
- The utilization of security guards, closed-circuit cameras, logs documenting access and egress, security badges, and off-hour access control are all important to a well-run and secure site.
- The location of the contractor and subcontractor's field offices and storage shanties must be properly planned to facilitate access to the site by workers and visitors and to have the trades people coordinate and interface with each other.
- Materials and tools that can be easily removed and used in a person's home are prime targets for theft, along with salvage metals, which are referred to in the industry as MONGO.
- Theft of materials adds to the overall cost of construction in both time and money.
- It is important to note that the materials should have a certificate of insurance prior to any of the material being paid for by the owner to protect everyone's interests.
- Large quantities of construction materials such as light fixtures, raised floor panels, etc., may need to be stored off site until they are ready to be installed. A bonded warehouse is often used in this situation.

12

Renovation and Demolition
(I thought this was a structurally sound building.)

INTRODUCTION

One of the most difficult projects to work on is a major renovation of an existing building (see Exhibit 12-1, renovation flowchart). Drawings that have been developed for the project are usually accomplished without the full benefit of knowing what is behind the walls and ceilings. When dealing with older buildings (50 to 100+ years old) the existing drawings are usually not available. Thus, the design team who has to prepare the construction documents for a renovation project has to base their design on visual observations. Latent conditions would not be known until demolition was initiated. Probes of various components of the structure are taken to try to find out more about the structural elements of the building. This is usually a difficult undertaking. In some cases, the consultants may have to x-ray or boroscope some sections of the building to determine the critical components of the structure.

Whatever information is indicated on the drawings, the project manager (PM) reviewing this type of project has to be cautious in evaluating the scope. The PM may have to consider other resources to assist in the review of the project. These may include:

1. Local building department documents
2. Local archives
3. Local historical societies
4. Old newspaper articles
5. Local library

PREPARATION OF COSTS

As in other type of projects, but more so with a renovation, a site visit is mandatory. When walking through the project prior to submitting a cost, the site visit checklist in Exhibit 12-2 should be followed.

Exhibit 12-1

Renovation
flowchart.

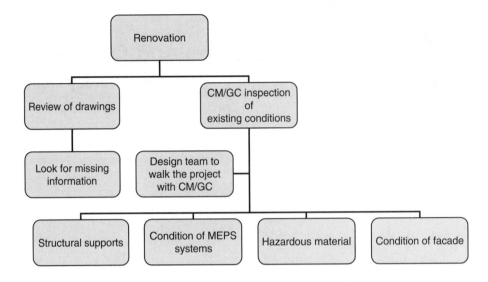

When reviewing the contract documents, read all notes contained therein. Care should be taken with any caveats the design team may have added (i.e., the construction manager/ general contractor [CM/GC] is responsible for any latent conditions found behind walls). If the building is going to be used for another purpose, then the intended load (in pounds per square foot [psf]) will have to be checked against the original load capacity (psf) for the original use. This may be found by looking at building department documents or finding out the original use and then checking the code used at the time and its allowable load. If it is impossible to obtain the information required, then a structural engineer may have to prepare some calculations to determine the load carrying capacity of the structural floor systems. As a last resort, a load test of the structure may be required. This could cause a collapse of the structure, which would not be very productive.

If the building being renovated is a landmark or is located in a landmarks district, then additional care would be required by the PM. As noted in Chapter 4, the design team must obtain all approvals from the Landmarks Commission. The PM's responsibility may involve the submittal of the actual material to the Landmarks Commission for final approval. This may delay the project, so some float time will have to be allocated for this approval process and added general conditions may be required. If any restoration work is required, then specialty subcontractors will have to be retained. They should review the scope and complexity of the project as soon as possible. The costs associated and the time allocated with restoration work could be very substantial.

If after checking the utilities, the PM finds that they cannot be used effectively, then the utility companies will have to be contacted for temporary service. In most cases, the fire department will require the installation of an active standpipe going up the building. This is required so that if a fire should occur, the fire department will have access to water in the building. The fire department will also require some form of access into the building. If a hoist is required, then the fire department could use the hoist as a means

1. Take notes on special conditions and place them on the drawings in the appropriate area.
2. Take photos of all components of the building. Note on the drawings where the pictures were taken.
3. The designer should accompany CM/GC on the site visit. If defects are noted but are not indicated on the drawings, then the PM should have the designer sign off on the defects found.
4. Any adjacent buildings should be reviewed for any structural connections and for a possible need for protection of the contiguous structure via underpinning and shoring.
5. Prior to the walkthrough, find out what the building was used for (maybe multiple uses since the time of its construction). This can sometimes be obtained through building department records.
6. Look for any potential hazardous materials such as asbestos, lead paint, old oil tanks, old transformers, or other materials that may have been used for manufacturing (if that is what it was used for). Have a hygienist review the substances found. If hazardous material is found, then it would have to be removed.
7. Inspect the roof and investigate (where possible) the condition of the base supporting material, flashing, roof membrane, penetrations, and parapets. If these items are found to be deteriorating, then a new roof and associated supplemental elements may be required. An old roof may have asbestos, so this hazardous material would have to be removed properly.
8. Inspect the exterior of the building and look for major cracks. This may indicate that major settlement has taken place. Also, see if the walls are plumb and not out of alignment.
9. Look for standing water or any water marks on the walls. This could be caused by roof leaks, pipe leaks, or water seeping into the basement from the surrounding areas.
10. Look for fire damage especially as it relates to wood structures. This could impact the structural integrity of the wood.
11. See if any floor deflections or excessive cracking have occurred. This could be caused by structural defects.
12. If it is a wooden structure, then further investigation will have to be made on the condition of the wood (use an ice pick), joists supports, and termite or carpenter ant damage. These elements would have to be replaced or "sistered" (overlapping of the wood joists).
13. Investigate (where possible) all the structural connections (wood, steel, or concrete) to make sure full structural integrity is being maintained.
14. Determine much fire proofing will be required.
15. If drywall is to be used under the wood joists, how will the ceiling system be hung (usually from the joists)?
16. All window lintels should be investigated for potential deterioration.
17. Make sure the windows can open or close even if the windows are to be replaced.
18. Inspect all the pipes to determine if deterioration has occurred. In some instances, it may be worthwhile to have the designer cut out sections of questionable pipes for review and evaluation and look through the cleanouts. Insulation on the pipes may be asbestos and thus would have to be removed.

(Continued)

Exhibit 12-2
Renovation site
visit checklist.

Exhibit 12-2

(*Continued*)

19. The electrical systems should be reviewed for total required capacity and for investigating the conditions of the conduits, wires, and circuit breaker boxes. Also check that DC power is not being used. The local utility company should be contacted to determine if any adverse conditions or violations exist.

20. All mechanical systems should be investigated, especially if some of the equipment will be retained. The duct work should be checked for possible mold intrusion.

21. The PM should check with the design team to determine if reports were done on the condition of the mechanical, electrical, plumbing, and sprinkler (MEPS) systems. If available, the reports should be obtained prior to putting pricing together.

22. If the CM/GC firm does not have mechanical, electrical, plumbing, and sprinkler expertise on staff, then a MEPS subcontractor should join the CM/GC on the site visit.

23. During winter conditions (depending upon the location), the PM will have to determine how heat can be supplied to the building so that trades people can work effectively. This has to be done in a safe and secure manner.

of access. If a hoist has not been installed or it is at one end of the building, access may be required via a staircase. Thus, the PM must make sure the stairs are in good condition to support the firefighters and their equipment. This would also be true for the trades people and the equipment that may have to be carried into the building via the stairs.

When reviewing the project, a determination will have to be made of how much cutting and patching will be required. This could include:

1. Openings for elevator shafts
2. Openings for mechanical, electrical, and plumbing chases
3. Relocation of stairs

When these new openings are required, the PM has to take into account the possible shoring of the structure until adequate supports are provided. In addition, the logistics of providing the supports have to be reviewed and evaluated. Special shoring or underpinning may be required.

After the PM has performed the proper due diligence on the scope of work involved, then the preparation of the cost can be initiated. One of the key elements of the costs for the project is the determination of what type of contingency should be added. The contingency would be based on the availability of accurate information versus estimates. For restoration projects, a 10% contingency for accurate information and a 25% contingency for a large amount of guess work.

A contingency could also be limited if a written understanding is reached between the CM/GC and the owner that the construction documents represent the scope of the project as now defined. Any latent findings not represented by the construction documents would then be handled as a change order as defined in the contract.

RENOVATION PROJECT—POST-DEMOLITION

The PM must check all dimensions on the drawings. Any discrepancies should be reviewed immediately with the design team via the request for information (RFI) procedures. Any element of the project that is now exposed must be brought to the attention of the design team. This is especially true when hidden structural systems are uncovered that are in conflict with the intent of the design documents. The PM must advise the design team and the owner immediately (in writing) of this type of discovery. Photographs must accompany the description of the systems found, as well as the drawing number (and detail, if applicable) where the system was to be built. Exhibits 12-3 and 12-4 show the poor conditions that may be encountered during a renovation. It is good practice to have the design team walk through the project after the required demolition has been accomplished. This will give the design team and the owner time to investigate all the uncovered areas and make decisions on potential revisions to the construction documents. In addition, the design team will have the opportunity to re-evaluate field conditions and possibly revise the dimensions noted on the existing construction documents.

SCHEDULE

With renovation projects, it is imperative that the PM keep track of every disruption to the schedule. If latent conditions are found that could possibly impact the schedule, then the delay allocated to this process must be documented in detail. All specific dates of when a portion of the project was stopped and when it started again have to be

Exhibit 12-3
Photo of deteriorated joists.

Exhibit 12-4

Photo of deteriorated wires and pipes.

documented. A revised schedule must then be sent to the owner and the design team, showing the impact of the delay. The revised schedule must be signed off (literally with the approved signatures) by both the owner and the design team. If this is not done, potential claims could arise. A realistic schedule for a renovation project should include contingency time for latent defects corrections.

SAFETY

As with any project, safety is of paramount importance. However, with restoration and demolition, extra caution must be taken. Besides the items enumerated in Chapter 5, the items indicated in the safety issues checklist in Exhibit 12-5 should be followed.

DEMOLITION

Demolition involves the removal of sections of a building or the complete removal of the structure. This undertaking is dangerous and must be carried out by skilled subconstructors and trades people. As with any project in an urban setting, the safety of the public is of paramount importance. Thus, the safety items discussed previously must be in place prior to starting any major demolition project.

With demolition, the logistics of removing the debris must be thoroughly thought out. The location of debris chutes must be so located that its removal is done expeditiously and without interfering with other demolition operations. Most codes will indicate the

1. All openings have to be protected by a very strong and high barricade.
2. Floors must be checked for any "soft" areas (you do not want any workers falling through the floors).
3. All hoses and wires have to be hung from a high spot so that workers will not trip on them.
4. Wet downs of floors create slippery conditions. This should be accomplished at the end of the day, when most workers have left the job.
5. Any gas that runs through the building should be shut at the main entrance valve.
6. All exposed flooring nails should be removed immediately.
7. All debris must be removed from the job site as soon as it is created.
8. Fire extinguishers and water (building riser) must be readily available.
9. Temporary lights must be hung throughout the job site. This is especially true for dark stairs and hallways.
10. Escape routes must be established in case of fire or a collapse. The routing map should be hung in all areas of the building. In addition, at the tool box meetings, the routing should be explained to all the workers (in English and Spanish).
11. Safety signs must be placed at all locations within the building.
12. A safety director is imperative for renovation and demolition projects.
13. All chemicals, gasoline, diesel fuel, and kerosene used for the project must be kept in locked fire-proof cabinets.
14. Cylinders used for torches must be secured via chains within a safe fireproof storage area.
15. On all exposed open perimeters of the building, safety cables must be strung around the exposed areas so that workers cannot fall from an unprotected area.
16. Safety netting has to be placed around the perimeter of the building for any material that may fall from the building.
17. The local building code may also require the complete enclosure of the structure so that debris and dust can be contained.
18. Weather conditions must be evaluated on a daily basis. The site may have to be secured (all material tied down) due to adverse winds, hurricanes, or tornado conditions.

Exhibit 12-5

Safety issues checklist.

type of chutes that can be utilized. In most cases, they have to be fireproof and completely enclosed (containing the debris) and emptied into either a dump truck or a large dumpster. This must be done in an area that is away from the public. Dust must be kept to a minimum in the urban setting. With contiguous buildings surrounding the building being demolished, excessive winds can be created which can cause a minor dust storm. Thus, a wet down of the areas being demolished will have to be explored and evaluated (due to possible slippery conditions for the workers).

Adjacent buildings must be thoroughly investigated to make sure no structural tie-ins have been made either initially or over the preceding years. If a structural connection is found, then the owner's structural engineer must evaluate the problem and determine its resolution.

If a building is contiguous to another building's wall, then the CM/GC will be responsible for "parging" (repairing) the wall. In addition, temporary shoring of the wall may be required if the wall is supported by the structure that is to be demolished. See Chapter 6 for more information on underpinning.

HAZARDOUS MATERIAL

When dealing with buildings that were constructed prior to 1975, the likelihood of finding some form of hazardous material used in the construction process is great. The Environmental Protection Agency (EPA) has prepared a list of hazardous materials. The list includes some of the following materials found on construction sites:

1. Asbestos
2. Lead paint
3. Mold
4. Fuel oil
5. Polychloro-biphenyls (PCBs)

Asbestos

Asbestos is one of the most difficult hazardous materials to remove. Asbestos can be removed or encapsulated. For large areas, removal seems to be the most efficient means. Occupational Safety and Health Administration (OSHA) and the EPA have set up standards for the removal and disposal of asbestos. Standards may vary from state to state, but the basic procedures are as follows:

1. A hygienist performs air quality and sample testing of the area in question.
2. If positive results for asbestos are found, then the local municipality and EPA are notified.
3. A waiting period of at least 10 days is usually required for proper notification of tenants, neighbors, and community groups.
4. After the notification period, the area is enclosed in plastic and an exhaust fan is set up to produce a negative pressure inside the space. Clean rooms and showers are set up for the workers.
5. The workers removing the asbestos must be in protective gear established by OSHA.
6. Asbestos removal for large structural areas is achieved by wetting down the asbestos and then removing it and placing the material into approved plastic bags.
7. At the end of a workday, the workers must place their clothing in approved bags (for disposal) and then take showers to remove any fibers that may have gotten on their bodies.

8. The space is constantly monitored for asbestos fibers.

9. The asbestos in the bags is then removed and placed in approved transportation vehicles. The EPA must approve the asbestos haulers.

10. The asbestos is then transported to an approved disposal site.

11. All transfer manifests and certificates from the approved dumping sites must be retained by the PM.

SUMMARY

- Renovation projects are difficult to assess due to the design teams' limited knowledge of the existing building.

- Drawings of the existing structure are usually not available so the design team has to rely on visual observations, probes, and in some cases x-rays.

- In the process of developing costs for a renovation project, a detailed site visit is required by the PM.

- Safety for renovation and demolition projects must be thought out so that the public and the trades people are properly protected.

- After partial demolition, the design team must prepare a field investigation survey to determine if any changes are required to the original construction documents.

- The CM/GC must constantly update the construction schedule to make sure any delay caused by the field survey is reflected in the schedule. The owner and design team must sign off on the revised schedules.

- A logistic plan of how the debris is to be removed is a necessity.

- Protection of adjacent buildings must be evaluated.

- If any hazardous material is discovered, it must be disposed of in a manner approved of by OSHA and the EPA.

- When dealing with asbestos, special precautions and procedures must be followed in order to remove this hazardous material.

- Windborne dust and debris created by the proximity of urban buildings will have to be dealt with on an individual basis.

13 Meetings and Communications
(Someone besides me has to understand.)

COMMUNICATIONS IN AN EVER-CHANGING ENVIRONMENT

Communication is the art of conveying to another party your thoughts, feelings, information, and intentions through both verbal and non-verbal signals. Proper and timely communications are essential to the success of a construction manager/general contractor (CM/GC) and to a construction project. Communications are even more challenging while constructing a project in the urban environment, in that there are more project stakeholders given the congestion and diversification encountered. Communications often utilize many different media such as telephone calls, e-mails, letters, meetings, memos, requests for information (RFI), field information memos (FIM), reports, logs, and transmittals, each with their own unique way of conveying information.

The environment in which a corporation operates changes rapidly depending upon the dynamics of the politics, regulations, technical issues, personnel, material and resource availabilities, and competitive situations. Unless a company and its personnel have their ears to the ground and are tuned in to the environment, the company will quickly find itself behind the times and behind the eight ball. Change is a constant in life, and the construction industry is no different in that regard. One needs to be able to understand and manage the environment in which he or she is operating in order to manage the change process. Change is a given, so either you must manage change or change will manage you. In the latter scenario, you may find yourself in an unfavorable situation and environment.

VERBAL AND NON-VERBAL COMMUNICATIONS

The spoken word, that is, verbal communication, articulates directly what it is we are trying to express. Often people feel that what is said (the content) is more important than how it is said (the delivery or the form). However, the non-verbal aspect of communication is the most important because the mannerisms, gestures, tone of voice, spirit, and inflections of the person speaking often communicate more than the spoken word. One needs to pay particular attention to his or her gestures and expressions, as they will convey more about the intent of the conversation than the spoken word.

PROJECT STAKEHOLDERS AND THEIR SPECIAL INTERESTS

One must also consider the diverse group of stakeholders that are involved and their unique focus on the project. In the urban environment there are more stakeholders given the size and diversification of larger cities, special interest groups that form in this environment, and the amount of large construction projects that occur. Chapter 1, Exhibit 1-7 contains a list of the project stakeholders for reference.

Each stakeholder has his or her own needs, wants, and desires as they relate to the project, and often stakeholders do not speak the same language or have the same focus or interest. It is therefore very important to understand each stakeholder's position, values, opinions, etc., as each will be different. To have good communications, you must understand the audience you are addressing and deliver the message accordingly.

COMMUNICATIONS PLAN

Having in place a communications plan and network is essential for the efficient and effective operation of a company and the successful completion of a project. Exhibit 13-1 presents a checklist of items to consider in a communications plan. This should be used as a guideline. A communications plan should be customized for the specific requirements of a corporation and the project.

MEETINGS

Meetings, provided that they are properly organized and managed, can be very useful for exchanging information and updating all personnel involved on the major issues, their status, and resolution. Often meetings are called and held without a formal purpose, agenda, or focus. Such meetings can be very time-consuming, counter-productive, and frustrating. Meetings are an important forum for communication among the entire project team and all of its stakeholders in which they share and convey the open items, status of the project, open change orders, shop drawing status, RFIs, bulletin status, field conflicts, requisitions, payments, and new business issues. Exhibit 13-2 contains guidelines for conducting a successful meeting.

Item	Included
1. Telephone system and directory	
2. Facsimile system and directory	
3. Internet connectivity and directory	
4. Voice mail system	
5. After hours answering service	
6. Emergency answering service and response	
7. Directory of corporate employees	
8. Directory of firms involved with a project	
9. Emergency contact list	
10. E-mail system and directory	
11. Project management website and website manager	
12. Project management software and software manager	
13. Records management policy, procedures and records manager	
14. Photographs with dates, times, and locations	
15. Progress reports	
16. Financial reports	
17. Photographs	
18. Video clips	
19. Meetings	

Exhibit 13-1

Checklist for a communications plan.

Item	Included
1. Always prepare an agenda for the meeting.	
2. Publish the agenda to all parties.	
3. Invite only the parties who are required to attend the meeting.	
4. Appoint a chairperson of the meeting to direct and manage the meeting.	
5. Stick to the topics on the agenda.	
6. Have someone other than the meeting chairperson take the minutes.	
7. Publish a set of meeting minutes within 48 hours of the meeting.	
8. Distribute the meeting minutes via E-mail or hard copy.	
9. Ensure that each item discussed at the meeting is properly recorded, the person who will be working on the resolution of the item is identified, the expected due date is determined, and the status of the item is clarified.	
10. Delegate personnel responsible for the resolution of the open items; have them report on the status and resolution of the matter.	

Exhibit 13-2

Guidelines for conducting a successful meeting.

(Continued)

Exhibit 13-2

(*Continued*)

Item	Included

11. Try to avoid solving the problem at the overall project meeting, when a large diverse group of personnel are in attendance.

12. Schedule working meetings with the personnel or committees who will be working on an open item, and have them report back on the resolution of the matter.

13. Challenge the meeting participants to do their jobs and resolve open items in a timely and professional matter.

14. Respect the opinion of others, and treat people in the same manner as you expect to be treated.

15. Conduct yourself in a professional manner.

16. Conflict or differences of opinion are healthy and beneficial, if they are properly managed.

17. Build on the synergy of the group and their diverse professional backgrounds and experience.

18. Schedule meetings with the projected completion of milestones not necessarily every week, unless required. Do not schedule a meeting just for the sake of having a meeting.

19. Organize each meeting into structured sections so that personnel have the option of attending only the portion of the meeting that is pertinent to them.

20. Prevent someone else from taking over the meeting for their own purpose or agenda.

21. Prevent people from grandstanding and getting on a soapbox.

22. Monitor the meeting's pace; keep the meeting moving along.

23. Start the meeting at the designated start time.

24. Try to limit the meeting to no more than 1 hour to make efficient and effective use of time.

25. Resolve conflicts between team members. Remember that everyone is entitled to his or her opinion, which may be a different focus on the same item within the project.

26. Keep the meeting focused.

27. Follow up after the meeting with the responsible parties on important open items to ensure proper and timely resolution.

28. Prior to the meeting solicit the opinions and support of key personnel on complex and politically sensitive situations.

29. For follow-up meeting or those of a similar nature, schedule the meetings on the same time and day of the week.

30. Any disagreement as to the minutes should be sent via E-mail prior to the next meeting and it should be reviewed and noted at the beginning of the next meeting.

TYPES OF MEETINGS

Some common types of CM/GC home office meetings are listed in Exhibit 13-3.

A sample list of construction project meetings is shown Exhibit 13-4.

It is important to note that the administration of a project with all of the different meetings, special administrative requirements, and reports can sometimes be more complex and challenging than the actual construction of the project itself. The project manager (PM) and staff must therefore have both excellent construction technology and administrative management skills in order to be successful.

DEFINITION AND PURPOSE OF THE MOST COMMON MEETINGS

Brief summaries of the purpose and focus of some of the most frequently held meetings are as follows:

Project Meeting

This meeting is focused on the overall status of the project and seeks to share with all concerned parties the major issues, opportunities, and challenges facing the project and the project team. The architect or owner representative will usually conduct this meeting

Type of Meeting	Frequency	Next Sch.Mtg.
1. Strategic planning meeting.		
2. Marketing meeting.		
3. General management meeting.		
4. Departmental meetings.		
5. Townhouse employee meetings.		
6. Emergency meetings for critical problems and situations.		
7. Organizing the response to a Request for Proposal (RFP) for a large complex project.		
8. Professional managerial and technical training meetings.		
9. Major changes and announcements within the company.		
10. Upper management meeting.		
11. Corporate goals and objectives meeting.		

Exhibit 13-3

Types of CM/GC home office meetings held.

Exhibit 13-4

Sample list of
construction
project meetings.

Type of Meeting	Frequency	Next Scheduled Meeting
1. Master project meeting		
2. Budget meeting		
3. Schedule meeting		
4. Bid and award meeting		
5. Value engineering meeting		
6. Safety meeting		
7. Coordination meeting		
8. Pre-construction meeting		
9. Design meeting		
10. Pre-award meeting		
11. Project close-out meeting		
12. Requisition review meeting		
13. Site walk-through meeting		
14. Tool box meeting		
15. Project team meeting		
16. Subcontractor meeting		
17. Kick-off meeting		
18. Insurance meeting		
19. Close-out meeting		
20. Project post-completion and lessons learned meeting		
21. QC meeting		
22. Owner's furniture, fixtures, and equipment meeting		
23. Owner's move-in coordination meeting		

during the design phase, and the chair of the meeting will be turned over to the CM/GC once construction has commenced. This meeting is usually held weekly. Exhibit 13-5 is a sample project meeting agenda.

Budget Meeting

This meeting, which includes the entire project team, is focused on developing the scope of work and a program for the project in order to establish a meaningful budget on which the owner can rely. Budgets usually are developed in the conceptual phase of the project initially, then at the design development stage (25% documents), the construction document development phase (50% documents), the bid and award phase (80% documents), and at the formulation of a guaranteed maximum price (GMP) (usually 80% buyout of trades, as the documents are seldom if ever 100% complete). The budget should be developed with 100% of the information and program requirements of the project. See Chapter 17, Exhibit 17-2 for the cost-estimating and budgeting process. Exhibit 13-6 contains a sample agenda for a budget meeting.

Agenda Item	Party Responsible	Due Date	Comments
1. Names of participants			
2. Review status of the construction documents			
3. Review procurement of long lead items			
4. Review bid and award status			
5. Review construction status			
6. Review project schedule			
7. Review project budget			
8. Review status of open change orders			
9. Review status of outstanding requisitions			
10. Review project status reports			
11. Review status of progress payments			
12. Review filings and permits			
13. Review insurance program			
14. Review shop drawing status			
15. Review open Requests for Information (RFI)			
16. Review open Field Information Memos (FIM)			
17. Review status of controlled inspections			
18. Review status of QC program and punch list			
19. Review project incident and safety reports			
20. Update open items list			
21. Review any new business			

Exhibit 13-5

Sample agenda for a project meeting.

Schedule Meeting

Schedule meetings should parallel the timing and process of the budget meetings. All items that are budgeted must be scheduled and vice versa. The schedule should identify all major project milestones, preconstruction period, bid and award period, construction period, project phasing requirements, project close-out, and owner's items such as furniture deliveries and move-in. Exhibit 13-7 is a sample agenda of a schedule meeting.

Bid and Award Meeting

These meetings are held with each of the major subcontractors or contractors who are bidding on the project. Once the bids are opened and leveled, the three responsive bidders (not necessarily the lowest bidders) are called in to review their bids, the scope of the work,

Exhibit 13-6

Sample agenda for a budget meeting.

Item	Responsible Party	Due Date	Comments
1. Names of participants.			
2. Review owner's program.			
3. Review status of the design documents.			
4. Determine best approach to budget the project as the design and construction documents develop.			
5. Review latest trade budget developed by CM/GC			
a. By trade			
b. By breakdown of work within each trade			
c. Review allowances within each trade			
6. Identify items requiring clarification and further breakdowns.			
7. Identify items that have a budget allowance and review program and design.			
8. Define the general conditions and review the budget for same.			
9. Define the CM/GC fee and review budget for same.			
10. Define insurance requirements and review budget for same.			
11. Define bonding requirements and review budget for same.			
12. Identify what overtime may be required for the project and budget same, either in each individual trade line item, or as a separate overtime line item for each trade.			
13. Identify costs for project logistics, such as cranes, derricks, hoists, scaffolding, and sidewalk bridges. Establish appropriate line items within the budget or within each trade for same, depending on who will be providing them.			
14. Establish a design contingency for the owner and the consultants to use for design changes.			

Item	Responsible Party	Due Date	Comments	**Exhibit 13-6** (*Continued*)
15. Establish the construction contingency, separate from the design contingency, to cover items encountered during the construction process.				
16. Establish the formulation for a guaranteed maximum price (GMP), if one is contractually required.				
17. Update budget open items list.				
18. Anticipated cost review.				
19. Review budget qualifications and exclusions.				
20. Evaluate value engineering options if the project is over budget.				
21. Evaluate special administrative programs and the cost to administer and comply with same.				
22. Review any new business.				

project requirements, schedule, insurance, phasing, and out of sequence work to ensure that the bid is complete and to make adjustments as required to properly level the bids and complete the analysis of the same. Subcontractors are often asked to sign the leveling sheet to ensure that they have the full scope of work and to avoid future misunderstandings thereafter. Chapter 16 explains the bid leveling process and contains examples of bid leveling sheets. A sample agenda for a bid and award meeting appears in Exhibit 13-8.

Coordination Meeting

Coordination meetings are held by the CM/GC with all of the pertinent subcontractors to coordinate the timing, location, and sequence of work to be installed for the project. The sheet metal/mechanical subcontractor usually starts the coordination process, given that the ductwork and piping is large and is the least flexible component to be installed. Once the sheet metal and piping right of ways have been established, the other trades such as the plumber, electrician, and sprinkler subcontractors will place their work on the coordination drawing and identify any conflicts that may exist. It is important that no construction materials be installed prior to the processing and approval of the coordination by all of the trades. If a subcontractor installs construction materials prior to the approval of the coordination drawing, then they are liable for the relocation if a conflict exists. It is a lot easier to move it on paper, with a pencil and eraser, than in the field once installed. Three-dimensional AutoCAD drawings greatly facilitate the coordination

Exhibit 13-7

Sample agenda for a schedule meeting.

Agenda Item	Party Responsible	Due Date	Comments
1. Names of participants.			

2. Review status of the project schedules.
    Pre-construction schedule.
    Bid and award schedule.
    Construction schedule.
    Project close-out schedule.			

3. Identify any items of work not on the schedule.			
4. Identify the critical path of the schedule.			
5. Identify and track long lead items, along with their procurement, manufacturing, and delivery.			
6. Identify corrective action plans if the schedule is not being adhered to.			
7. Identify all activities for the project, including work performed by the consultants (delivery of project documentation), by the owner (delivery of the owner's directly purchased materials, i.e., furniture systems, carpet, accessories, security, A/V systems, and telecommunications systems), and by other parties working at the site (other base building CM/GC, other tenant CM/GC, and specialty contractors) and incorporate them into the overall project schedule.			
8. Identify milestones in the schedule that need to be met in order for the project to be on schedule (approvals, permanent power, enclosure, roof, elevators, controlled environment for finishes, hoist removal, derrick and crane removals).			
9. Identify the target date for obtaining a temporary/final certificate of occupancy.			
10. Review resource loading of schedule.			
11. Review cost loading of schedule.			
12. Review any open scheduling items.			
13. Review any new business.			

Agenda Item	Responsible Party	Due Date	Comments
1. Names of participants.			
2. Determine the pre-qualification process for subcontractors bidding the project.			
3. Review the subcontractors' bidders list and finalize it.			
4. Develop a sample bid package for all parties to review.			
5. Develop a bid response form.			
6. Determine who will receive the bids and how, where, and by whom will they be received and opened.			
7. Determine the role of the owners or outside consulting auditing firms in the bid and award process, and plan accordingly.			
8. Develop a list of alternative prices to be included in the bid.			
9. Develop a list of unit prices to be included in the bid.			
10. Develop a bid leveling sheet to document the receipt of the bids.			
11. Identify the long lead items, and bid them to equipment manufacturers directly to ensure that they are available at the project site in accordance with the project schedule.			
12. Prioritize the sequence in which the construction subcontractor trades will be bid, as they cannot all be bid on at once by the CM/GC project team. Develop a specific bid and award schedule for each trade.			
13. Review the bid leveling process to determine the responsive subcontractor.			
14. Analyze any exclusions and qualifications that the subcontractor's bid may contain to eliminate them, and ensure an award for a complete scope of work.			

Exhibit 13-8

Sample agenda for a bid and award meeting.

(*Continued*)

Exhibit 13-8
(Continued)

Agenda Item	Responsible Party	Due Date	Comments
15. Identify items that may fall in multiple subcontractors' trade jurisdictions, and ensure that they are purchased in a coordinated manner to purchase and install the items only once.			
16. Review recommendations for award of subcontracts.			
17. Advise the owner and their consultants of the expected turnaround approval time for each subcontractor trade award.			
18. Establish the approval authority required for the award of subcontracts.			
19. Advise the owner of subcontractor trade awards and project commitments against the budget.			
20. Review status of the bid and award process against the schedule and budget.			
21. Review alternate bid prices taken and their status.			
22. Update open items list.			
23. Review any new business.			

process. Chapter 16 contains additional information on the coordination process. Exhibit 13-9 contains a sample agenda for a coordination meeting.

Safety Meeting

Safety meetings are held with all of the subcontractors working on the site and their personnel. The meeting that each subcontractor has with its workers is referred to as a tool box meeting. Chapter 5 discusses safety and tool box meetings in more detail. Exhibit 13-10 contains a sample agenda for a safety meeting.

Pre-Construction Meeting

Pre-construction meetings are held at the beginning of the project as described in more detail in Chapter 1. Exhibit 13-11 has a sample agenda for a pre-construction meeting.

Contract Review Meeting

It is important that the owner and CM/GC define their contractual relationship early on in the project life cycle, as described in Chapter 9. Exhibit 13-12 contains a sample agenda for a contract review meeting.

Agenda Item	Responsible Party	Due Date	Comments
1. Names of participants.			
2. Review the coordination process and which subcontractor trade will lead the coordination effort.			
3. Schedule coordination meetings and submission of coordination documentation to conform to the requirements of the overall project construction schedule.			
4. Review the routing and status of the coordination documents among the trades.			
5. Establish where possible and feasible a right of way for each of the trades installing their work in the ceilings, under the floor, in shaftways, and on the roof.			
6. Identify conflicts or "hits" among the trades where conflicts exist in the installation of their work. Determine which subcontractor work will take precedence over the others in terms of the right of way.			
7. Have all subcontractors involved sign off on the coordination drawings.			
8. Complete the coordination process prior to the start of the actual construction work.			
9. Keep track of all coordination layouts.			
10. Review any new business.			

Exhibit 13-9

Sample agenda for a coordination meeting.

Agenda Item	Responsible Party	Due Date	Comments
1. Names of participants.			
2. Review project-specific safety plan prepared by CM/GC.			
3. Review project-specific safety plan prepared by each subcontractor following the CM/GC safety plan.			
4. Review indoctrination process for all new employees regarding safety, prior to starting work on the project.			

Exhibit 13-10

Sample agenda for a safety meeting.

(*Continued*)

Exhibit 13-10

(*Continued*)

Agenda Item	Responsible Party	Due Date	Comments
5. Review schedule and topics for tool box meetings.			
6. Review safety signs to be posted around the construction site.			
7. Review minutes of all safety meetings held by the subcontractors with their employees.			
8. Review project incident and safety reports.			
9. Review safety walk-through schedules with all involved personnel.			
10. Schedule site visits with the safety person from the insurance company.			
11. Schedule site visits with the corporate safety manager of the CM/GC's home office.			
12. Develop appropriate training programs for all workers, based on new safety regulations and any accidents or incidents encountered on the project.			
13. Discuss filing of all accident and incident reports, along with all supporting documentation.			
14. Review filing and posting of all required safety reports with all governing authorities.			
15. Review any new business.			

Exhibit 13-11

Sample agenda for a pre-construction meeting.

Agenda Item	Responsible Party	Due Date	Comments
1. Names of attendees.			
2. Review of project program requirements.			
3. Review of overall project schedule and milestones.			
4. Review of overall project budget.			
5. Review of responsibilities of the architect.			
6. Review of the responsibilities of the engineers.			

Agenda Item	Responsible Party	Due Date	Comments
7. Review of the responsibilities of the specialty consultants.			
8. Review of the owner's governing body.			
9. Review of the responsibilities of the CM/GC.			
10. Review of any other agencies involved with the project.			
11. Review of usual scope of work.			
12. Review of Green LEED project status.			
13. Review of special administrative programs.			
14. Review of project filings and permits.			
15. Review of occupancy requirements and project phasing.			
16. Review of long lead items and their procurement.			
17. Review of value engineering items.			
18. Review of insurance program.			
19. Review of bonding program.			
20. Review of bid and award strategy.			
21. Review pre-qualification procedures.			
22. Develop bidders list.			
23. Schedule.			
24. Review the project's logistical plans.			
25. Review QC program.			
26. Review safety program.			
27. Identify project risks and the risk management plan for the project.			
28. Review project administrative programs.			
29. Review project reports and documentation.			
30. Review submission and approval procedures.			

Exhibit 13-11
(Continued)

Agenda Item	Responsible Party	Due Date	Comments
1. Names of participants.			
2. Review the form of contract to be utilized.			
3. Review the major terms and conditions of the contract.			
4. Review payment terms and conditions.			

Exhibit 13-12

Sample agenda for a contract review meeting.

(Continued)

Exhibit 13-12

(*Continued*)

Agenda Item	Responsible Party	Due Date	Comments
5. Review insurance and bonding requirements.			
6. Review waivers of subrogation and hold harmless clauses.			
7. Review who will be the person responsible for determining the intent of the contract and construction documents.			
8. Review the dispute resolution process.			
9. Review the retainer amount and reduction of same.			
10. Review the terms for the fee, general conditions, and project contingencies.			
11. Review the terms for damages.			
12. Review the project description, scope, and schedule.			
13. Review any guarantees for the project budget, schedule, and interim milestones.			
14. If a guaranteed maximum price (GMP) is required, review the formulation and timing for arriving at the GMP.			
15. Review any shared savings clauses of the construction contingency within the GMP.			
16. Review any fees and general conditions for additional services, such as coordinating the owner's work.			
17. Review any new business.			

SUMMARY

- Communications among the project team are essential for the timely and proper exchange of information and data about the project to all involved.
- Communications are both verbal and non-verbal.
- Non-verbal communication often says more than verbal communication about the message that the person is trying to deliver.

- A diversified group of stakeholders and team members are associated with each project, and each person involved brings his or her area of expertise to the project, with a slightly different focus on the various items encountered in the project.

- Meetings are an important part of the communication process, and many different types of meetings will be held, each with different agendas and goals.

- Do not schedule a meeting just for the sake of holding a meeting.

- A communication plan is necessary for the efficient and effective operation of the company and the successful completion of a project.

- Do not schedule a meeting without a defined purpose and identified goals and objectives.

14 Project Documentation, Logs, and Reports

(The consultants have to know what we want.)

PROJECT DOCUMENTATION AND RECORD KEEPING

Project documentation provides a record of the history and progress of the project, agreements reached, directions given, and decisions made. Proper project documentation provides a record of the project and is often used to settle disputes, contract claims, court mediation, arbitration, insurance claims, accident investigations, and is used for future reference. In the urban environment, projects tend to be very large in size, duration, and complexity, thus requiring the appropriate project documentation to capture all of the various activities and events that occur during the project, which often lasts for several years.

Project documentation provides written and electronic material regarding many different types of activities and events. Exhibit 14-1 contains a sample list of the activities and events that occur during the project life cycle that need to be documented.

Exhibit 14-1A is a project information form, which organizes all of the pertinent initial information for the project.

Not only is it important for the construction manager/general contractors (CM/GC) to construct the project on budget, on schedule, and quality controlled, but they must document the process as well. At times, this can be more challenging than the actual construction process. Many clients have large staffs of accountants, auditors, and lawyers, who often also sit on the Board of Directors. These people have a lot of influence on the selection process of the CM/GC, and the scorecard of their performance. It is important to understand what information and documentation they will require throughout the project, in addition to the contractually required documentation. Not producing the proper and timely documentation, information, and reports can lead to delayed payments, adverse dealings with the owner's corporate staff, and being found in default of

Exhibit 14-1

Sample list of activities to be documented during a project.

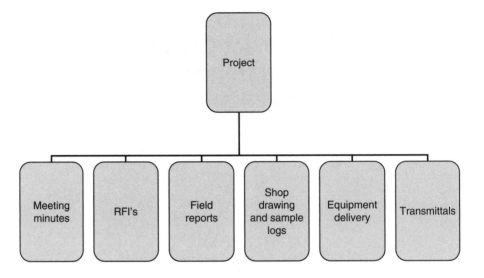

the contract for administrative non-compliance. It is important that all documentation, information, and reporting requirements be properly identified at the beginning of the project, and systems developed to properly gather the data and develop it into timely and accurate information, upon which management can be kept informed and take timely action.

The CM/GC must have a professional staff that can construct the project as well as administer, manage, and document the process. It is not often that a person is qualified and experienced to do both simultaneously. In addition, there are usually so many demands placed on the staff that they do not have the time to focus on both the construction and administrative requirements of the project. Therefore, many CM/GCs split this responsibility among the overall project staff, and have personnel who are assigned to the administration and documentation of the project in the office, and others who are assigned to construct the project working with the subcontractors, suppliers, and vendors in the field.

There are many meetings held during the course of the project in its different phases, which are discussed in detailed in Chapter 13. In these meetings important decisions will be made about the project schedule, budget, phasing, special programs, and resolution of problems. All of this is critical to the success of the project, and must be documented properly. Unfortunately, people tend to have amnesia when it comes to critical issues that have presented a challenge or problem to the project, especially when it comes to the finances and schedule of the project.

The projects records needs to be properly preserved and archived after the construction process is over. Accident claims and lawsuits for damages have at times arisen several years after the construction of the project, and must be defended. If you are summoned into a court of law, and are asked to produce the projects records and documentation and cannot, the legal system often views this as a willful lack of proper record keeping.

Exhibit 14-1A
Project
information.

Company Name

Address

City, State, ZIP

Phone Number

Bid date: _____

Bid time: _____

Project location: _____

Legal description: _____

Project name: _____

Projected budget: _____

Project funding: _____

Owner: _____   Contact person: _____

Billing
address: _____   Phone number: _____

_____

_____

Architect: _____   Contact person: _____

Phone number: _____

Drawings/Specifications:

Complete ☐   Incomplete ☐   Latest revision date

Schedule

Start Date _____   Duration _____

Finish Date _____

Site Information:

Good access	Yes ☐	No ☐	Water available	Yes ☐	No ☐
On-site obstructions	Yes ☐	No ☐	Power available	Yes ☐	No ☐
Poor soils	Yes ☐	No ☐	Telephone available	Yes ☐	No ☐
Drainage problems	Yes ☐	No ☐	Natural gas available	Yes ☐	No ☐
High security risk	Yes ☐	No ☐	Office available	Yes ☐	No ☐
Demolition needed	Yes ☐	No ☐	Debris service available	Yes ☐	No ☐

TYPES OF PROJECT RECORDS AND REPORTS

The following is a list of project records and reports, along with a brief description of each and their importance and relevance to the project:

1. Project meeting minutes—Meeting minutes document the progress of the project and all of the important decisions pertaining to it, from its inception to its completion. Meetings are usually held on a weekly basis. (See Chapter 13, Exhibit 13-5, which contains a sample agenda for project meeting.)

2. List of field information memos (FIMs)—A FIM is issued by the design team to deal with field conditions and modifications that may be required, that are not large

Exhibit 14-2

Field information
memo (FIM).

Project: _____

Date Originated: _____

FIM No.: _____

Originating Party: _____

Description of field condition

[ ]

Responding Party: _____
Date responded to: _____

Description of resolution:

[ ]

in nature, and that require immediate attention in order to allow the project to progress. Exhibit 14-2 is a sample FIM.

3. List of contract drawings and specifications—This is a definitive list of drawings and specifications that define the scope of work for the contractual requirements of the project.

4. List of addendums—Addendums are often issued during the bid and award period to clarify details and respond to information requested by the bidders.

5. List of bulletins—Bulletins are issued after the award of the project. Bulletins define a change in the work, which must be priced and submitted as a change order for review and approval.

6. List of sketches—Sketches are sometimes utilized as an interim means of initiating a change in the work by the owner or design team, which is not extensive to warrant the issuance of a bulletin. Sketches are usually then incorporated into the next bulletin when issued.

7. Shop-drawing log—Both the construction manager and design team to track the submission of all shop drawings by each subcontractor specialty will maintain shop-drawing logs. Shop drawings are the basis upon which the subcontractor will

#	Section Number	Work Item	Required Submission	Received		Sent to		Returned		Additional Comments
				Date	Contractor	Date	Reviewer	Date	Comments	

Exhibit 14-3
Shop drawing-log.

perform the actual layout and detailing of the construction work of the project, and be approved by the design team. Exhibit 14-3 contains a sample shop-drawing log.

8. Project schedule—The project schedule will detail the sequence of activities and events that are required to construct the project. The progress of the project against the original schedule and contractual requirements is important to the success of the project (see Chapter 15).

9. Project budget and committed cost—The project budget establishes the baseline of cost for each unique trade to construct the project. The committed cost report tracks the costs as committed against the baseline budget to measure variances in the budget versus actual costs (see Chapter 17).

10. Anticipated project cost—The anticipated cost report tracks the projection of the overall cost of the project through its completion, taking into account committed costs, pending change orders, claims, and other costs to complete the project (see Chapter 17).

11. Approved change orders—The approved change order log details the change orders that have been reviewed and approved by the owner for work to be incorporated into the project (see Chapter 17).

12. Pending change orders—The pending change order log tracks the change orders that have been submitted and are under review or change orders that may be submitted for review and approval (see Chapter 17).

13. Safety meetings—Safety meetings address the safety concerns and issues for the project. Safety meetings are usually held weekly to keep everyone aware of safety concerns, accidents, and accident prevention (see Chapter 5).

14. Coordination meetings—coordination meetings are held between the construction manager and the subcontractors to coordinate the installation of work into the overall project, within the space limitations and clearances allowed (see Chapter 13).

15. Tool box meetings—Tool box meetings are an extension of the project safety meeting between the foreman and the workers for each trade performing their specialized work (see Chapter 5 and Chapter 13).

16. Subcontractor meetings—The CM usually schedules a meeting with all subcontractors working on the project to review the status of their firm's unique work, review problem areas, discuss the work for the following week, and address overall coordination and management of the trades on the project (see Chapter 13).

17. Accident reports—The accident report documents an accident that has happened on the project. The report should capture the pertinent information as to the nature of the accident and the personnel involved (see Chapter 5).

18. Incident reports—The incident report documents an incident that happened on the project, such as a demonstration at the work site, a work stoppage, an administrative government visit, or events in the area that have affected work at the job site (see Chapter 5).

19. Project photographs—Photographs are usually taken on a weekly basis to document the progress of the project. Photographs are usually taken of the overall

Exhibit 14-4
Project photo-
video log.

Construction Company   *Date:* _____

Address   *Owner:* _____

City, State, ZIP   *Contractor:* _____

Phone Number   *Project number:* _____

       Project name: _____

Item No.	Video / Photo No	Type		Description	Date	Taken By
		Progress	Defect			

project site, along with photographs of each floor. As is often said, "A picture is worth a thousand words." Sometimes a video of the project is also taken to show real-time footage of the construction process. Exhibit 14-4 is a sample photo-video log.

20. Progress reports—Progress reports are usually issued on a weekly, semi-weekly, or monthly basis to keep the project team informed of the progress of the work. The progress report usually contains status of budget, schedule, open items, change orders, bids and awards, pending change orders, etc. A sample progress report is included in Exhibit 14-5.

21. Daily log—A daily log is maintained by the superintendent in the field to track the progress of the work that day, where work is being performed, which subcontractors

Exhibit 14-5

General progress report.

PROJECT: _____

OWNER: _____

CONTRACTOR: _____

PROJECT MANAGER: _____

PROJECT NO. _____

CONTRACT NO. _____

DATE: _____

Amount of Original Contract	$	
Approved Change Orders to Date		
Anticipated Over-run or (Under-run) in Uncompleted Work		
Actual Over-run or (Under-run) in Completed Work		
Rental Revenue from Contract Owner not included above		
Estimated Total Amount of Principal Contract	$	
Other Contract Work not Included in Principal Contract		
Total Estimated Contract Volume	$	
Contract Revenue to Date .... $		
Less: Contract Advances for Materials on Hand		
Contract Advances for Plant and Move-in..........		
Revenue Reduction for Uncompleted Work		
Other		
Total Amount of Work Completed to Date: ...	$	
Uncompleted Contract Volume	$	
Percent Complete Based on Original Contract .... _____ %		
Percent Complete Based on Total Estimated Contract .... _____ %		
Time Allotted by Original Contract .... _____ Days		
Extension of Contract Time .... _____ Days		
Total Contract Time .... _____ Days		
Contract Time Elapsed .... _____ Days		
Percent of Original Contract Time Elapsed .... _____ %		
Percent of Total Time, Including Extensions, Elapsed .... _____ %		
Date Contract was Physically Completed –If Completed .... _____		
Expected Date of Physical Completion –If Not Complete .... _____		

Exhibit 14-6

Daily log.

Contractor	Day		Deta	
	Job Name		Job No	
Field Notes			*Weather*	
			Temp	
			Craftsmen	No
			☐Superintendent	
			☐Clerk	
			☐Bricklayers	
			☐Carpenters	
			☐Cement Masons	
			☐Electricians	
			☐Iron Workers	
			☐Laborers	
			☐Operating Engrs	
			☐Plumbers	
			☐Pipe Fitters	
			☐Sheet Metal	
			Total	
			Equipment	Hrs

Contract Extras	Authorized By	Amount	Material Purchases
Equipment Rented Today	From	**Rate**	

Approved _____

are working, number of personnel that are working, weather conditions, and any usual events that may have occurred that day. Exhibit 14-6 contains a sample daily log.

22. Daily construction report—This report documents in detail the type of work that is being performed by each trade and the overall progress of the project, in more detail than the daily log. Exhibit 14-7 contains a daily construction report.

23. Manpower reports—These reports are utilized to document the number of workers in all capacities (foreman, journeyman, apprentices, minority and disadvantaged personnel, etc.) that are performing work on the project each day.

24. Daily equipment use log—This log keeps track of all of the major equipment, mostly for logistical purposes, such as the tower cranes, hoist, pumps, bulldozers, lifts, lulls, heaters, and crawling cranes. Exhibit 14-8 contains a daily equipment use log.

Exhibit 14-7

Daily construction report.

	DATE	S	M	T	W	TH	F	S
	DAY							

Project

Job No.

Client

Contractor

Project Manager

WEATHER
TEMP
WIND
HUMIDITY

AVERAGE FIELD FORCE

Name of Contractor	Non-manual	Manual	Remarks

VISITORS

Name of Contractor	Non-manual	Manual	Remarks

EQUIPMENT AT THE SITE

CONSTRUCTION ACTIVITIES

DISTRIBUTION
1. Project Manager
2. Field Office
3. File
4. Owner

Signature:

PAGE 1 OF _____ PAGES

Title:

25. Equipment delivery and installation log—This log tracks the major deliveries of all construction materials and equipment to the site, and when and were they are installed. Exhibit 14-9 shows an equipment delivery and installation log.

26. Off-site storage log—This log is utilized to keep track of all materials and equipment stored off-site awaiting delivery to the project site. Materials are often stored in bonded warehouses in close proximity to the construction site, so that when they are needed they can be readily delivered. Exhibit 14-10 is an off-site storage log.

27. Request for information (RFI) log—The RFI log is used to track the status of all RFIs and their answers. RFIs can be used to verify verbal information and direction,

Exhibit 14-8

Daily equipment use log.

Construction Company
Address
City, State, ZIP
Phone Number

Date: _____

Prepared by: _____

Contractor: _____

Project name: _____

Weather: _____

Machine Number	Machine	Rate per Hr	W = Working I = Idle	Cost Code				Total Hours	Total Cost
			W						
			I						
			W						
			I						
			W						
			I						
			W						
			I						
			W						
			I						
			W						
			I						
			W						
			I						
			W						
			I						
			W						
			I						
			W						
			I						
			W						
			I						
			W						
			I						
			W						
			I						
			W						
			I						
			W						
			I						
			W						
			I						
			W						
			I						
			W						
			I						
			W						
			I						
			W						
			I						
			W						
			I						

confirm decisions, request dimensions, and clarify field conditions. Exhibit 14-11 contains a sample RFI and Exhibit 14-12 shows a sample RFI log.

28. Inspection and approval reports—These reports document the status of all controlled inspections and approvals required for the project either by the project specifications or by governing jurisdictional agencies. It is important to track this information to ensure that the appropriate party who will perform the controlled inspections and approvals is available in a timely manner, as well as to make the work available for inspection so as not to impede the progress of the work (see Chapter 3).

29. Filings and permits—The building department and regulatory filings for the project all have associated permits, which need to be tracked to ensure that they are not outdated, are renewed as required, and are signed off at each phase of the work (see Chapter 4).

Exhibit 14-9

Equipment delivery
installation log.

Company Name   *Date:* _____

Address   *Building:* _____

City, State, ZIP   *Location:* _____

Phone Number   *Project number:* _____

                                         Project name: _____

Date	Invoice No	Description	Date Rcvd	Date Used	Location

30. Insurance documents—All contractors and subcontractors working on the project must be enrolled with the proper insurance requirements for the project. Certificates of insurance are issued to each entity, verifying the coverage required for the project, expiration date, and all of the additional and named insured. It is extremely important that no firm be allowed to work on the project until it has the proper insurance and documentation. Certificates of insurance are usually valid for 1 year, but may not be a calendar year, and can vary from firm to firm. It is important to track expiration and renewal dates on the insurance policies (see Chapter 10).

31. Bonding documents—If the project requires a bond to be provided by the construction firm, it must be obtained prior to the start of work. Different types of bonds may be required based on the bid and contractual documents (see Chapter 10).

Exhibit 14-10
Off-site storage log.

Company Name

Address

City, State, ZIP

Phone Number

Date: _____

Building: _____

Location: _____

Project No.: _____

Project
Name: _____

Date	Invoice No.	Description	Qty Stored	Qty Rcvd	Qty Used	Location Used	Qty Remaining

32. Requisitions and payments—Requisitions are usually processed on a monthly basis, with payment made within 30 days of receipt of the requisition. Money is the lifeblood of many construction firms, which are in the construction business and not the financing business in which they sometimes find themselves when they are not paid in a timely manner. Requisition and payment logs help keep track of requisitions, total contract amount, payments to date, retainage, etc. (see Chapter 20).

33. Certified payroll—Projects often require certified payrolls to document the utilization of minorities, women, and disadvantaged workers, as well as equal opportunity employment (EOE) programs for personnel and firms working on the project. Many of these certified payrolls are required to satisfy the terms of the contract and administrative programs associated with the project.

34. OSHA reporting—OSHA requires that a report be filed annually summarizing the firm's accidents by category and year, and that the report must be posted at every job site on which the firm is working. In addition, OSHA requires that an accident report be filed for certain types of accidents, especially an accident involving a fatality. Non-compliance can lead to violations and fines (see Chapter 5).

35. Test reports—Test reports contain all controlled inspection reports required by governing authorities and the contract (see Chapter 3).

36. Weather reports—Weather reports document the conditions on a given day that the work is being performed, and may be utilized if there is a weather-related accident, a delay claim for adverse weather, extraordinary weather conditions that affect the project, etc.

Exhibit 14-11
RFI.

Project Name:_____

Project #_____RFI #_____

Date Requested:_____Requested By:_____

Send To :_____ Attn:_____

Description:_____

Specification Reference:_____Drawing Reference:_____

_____

Question:

```



```

Possible Solution(s):

```



```

Check ONE: __ Critical __ Important __ Routine

Date Required:_____ Signed:_____

Answer:

```




```

Scope Change? __ Yes __ No

Date Replied:_____ Signed:_____

37. Punch list—The punch list for the project documents all deficiencies in the work performed that need to be corrected (see Chapter 21).

38. CM/GC submittals—When the CM/GC submits documents, samples, drawings, equipment cuts, and reports, a letter of transmittal to document the items being sent accompanies the items. A sample letter of transmittal is shown in Exhibit 14-13.

Project:
Address:
Job No.:
Status Date:

RFI No.	Date Originated	Originator	Date Responded	Person Responding	Status

Exhibit 14-12
RFI log.

Exhibit 14-13
Letter of transmittal.

From:

To:

| Transmittal Number: |
| Date: | Job Number: |
| Project Name: |
| Attention: |
| Re: |

We are sending you...   Attached   Under separate cover via _____ the following items:

___ Shop drawings   ___ Prints   ___ Samples   ___ Specifications
___ Copy of letter   ___ Change order   _____

Prepared by: _____

Copies	Date	Number	Description	Action Code	

These are transmitted as checked below:

___ For information
___ For approval   A. ____ Approved as submitted   1. ___ Resubmit   __ copies for approval
___ For your use   B. ____ Approved as noted   2. ___ Submit   __ copies for distribution
___ As requested   C. ____ Returned for corrections   3. ___ Return   __ corrected prints
___ For review and comment   D.

For bids due _____ ____ Prints returned after loan to us
             Month / Day   *Year*   *Time*

Remarks

Copy to: _____   Signed: _____
        _____   Printed Name: _____
        _____

If enclosures are not as noted, please notify us at once.

The CM/GC will often use a submittal form, a sample of which is contained in Exhibit 14-14. The submittals are kept track of in a submittal log, a sample of which is in Exhibit 14-15.

39. Waste management report—This report keeps track of all debris removed from the site. In today's environment of Green LEEDS projects, recycling of construction waste is very important. In addition, when hazardous materials are removed from the site, such as asbestos, lead, petrochemical products, PCBs, etc., a waste management report is prepared to document the nature of the material and how it is being disposed of. Exhibit 14-16 contains a sample waste management report.

40. Field investigative report—When an event happens at the construction site that warrants an investigation, a field investigation report is prepared to document the event and the findings of the investigation. Exhibit 14-17 contains a field investigative report. At times the architect may issue a field change report to modify the work as it is happening in the field. Exhibit 14-18 is a sample field change report.

41. Compliance with special administrative programs—Many projects constructed in the urban environment participate in special incentive programs, which provide sales tax reductions, property tax reductions, federal income tax credits, access to inexpensive power sources, etc. In addition, many large projects have an owner controlled insurance program (OCIP) or a contractor controlled insurance program (CCIP), which require proper administration for compliance (see Chapter 10). Exhibit 14-19 contains a list of special administrative programs often encountered when constructing a project in the urban environment.

42. Letter of completion—At the completion of the project, the architect will issue a letter of completion attesting that the project has been built in accordance with the intent of the construction documents and that the punch list is completed. This signifies substantial completion of the project. Exhibit 14-20 has a sample letter of completion.

43. Procurement log—The CM/GC and its subcontractors will be purchasing various items of construction materials for use in the construction process. The procurement log keeps track of all of these materials, when they are needed, and where they are stored, to ensure that they are available when they are to be incorporated into the construction project.

ELECTRONIC RECORD KEEPING AND PROJECT WEBSITES

In today's electronic age, computer- and Internet-based project management systems and programs are used to facilitate the communication process. Specifically they assist in the sharing of information, drawings, specifications, bid packages, budgets, schedules, logistical plans, project history, project performance, responses to RFIs, status of shop drawings, construction documentation logs, change order logs, payment logs, approval logs, meeting minutes, and punch list. A private corporation

Project:

Job No.

Project Mgr:

Contractor:

Date Received	Trans-mittal No.	Description	Subcontractor / Ref. Spec. Section	Con-tractor Trans. No.	No. Copies Rec'd.	Action: No Exceptions Taken	Action: Make Correc-tions Noted	Action: Revise & Resubmit	Action: Rejected	Date Received	No. Copies Returned	Remarks

Exhibit 14-14
Sample of submittal log.

Project Title:
Project No.

Project Manager:
Contractor:

SECTION NO.	ARTICLE NO.	SPECIFICATIONS SECTION TITLE		DATA REQUIRED								DATE OF SUBMITTAL	DATE REJECTED	DATE RESUBMITTED	DATE ACCEPTED	NOTES
		DIV.	(Indicate Division No. if applicable)	SAMPLES	SHOP DWGS.	MATL. OR PARTS LIST	DESCRIPTIVE DATA	MFRGS LITERATURE	CERTIFICATES	OPERATION INSTR.	TESTS					

Exhibit 14-15

Submittal control log.

building its new corporate headquarters utilized a project management system to assist with the management of the overall project. Some CM/GC firms have also developed their own in-house programs that are customized to their unique policies and procedures and ways of doing business. It is important that whatever computer-based system you elect to utilize that is able to adapt to the unique requirements of the project.

Exhibit 14-16

Waste
management
report.

Project Name:

Address:

Report Date:

Waste Description	Date Removed	Volume (cu yd)	Weight (lb)	Destination		Is waste hazardous?
				Town	Landfill / Recycling / Re-use	

One of the advantages of a computer-based project management system is the ability to integrate all of the project requirements and information into one system. This allows the project team to simultaneously input, evaluate, track, and monitor all of the aspects of a multi-disciplinary, technically complex, large project. The schedule for the project can be integrated with the budget and the resources (both manpower and materials), to allow you to cost and resource load the schedule with additional information. This gives you a better overall assessment of the status of the project, if you are ahead or behind schedule, what resources are required to meet the schedule, how you can make up for delays before it may become too late to react, and project the final schedule and cost with more information and intelligence. The S Curve can

DATE	S	M	T	W	TH	F	S
DAY							

Project   No.

Contractor

Subject

To

WEATHER
TEMP
WIND
HUMIDITY

Exhibit 14-17
Field investigation report.

DISTRIBUTION
    1. Project Manager
    2. Field Office
    3. File
    4. Owner   Signature: Title:

be developed to monitor and control this process (Chapter 20 contains further information on the S Curve).

The actual performance of the project, time, money, and resources can be tracked against the planned activities of the original project schedule, budget, and resources to pinpoint variances and deviations from the original plan. The PM should evaluate the project's overall performance by reviewing in-depth the budget, schedule, open items, and resources at least monthly to assess the overall project status. This will show the PM potential schedule, budget, or resource problems. Early detection and timely action are the keys to mitigating a problem before it becomes too late.

Exhibit 14-18

Field change report.

Construction Company
Address
City, State, ZIP
Phone Number

Date: _____
Project: _____
Project number: _____
Location: _____

_____
Revision No: _____

Reference Data					
Specification section No: _____		Page No: _____		Para No: _____	
Drawing No: _____		Detail: _____		Other _____	
Sketch No: _____		Dated: _____			

1) *Description of Problem:*

_____
_____
_____
_____

2) *Proposed Solution:*

_____
_____
_____

3) *Affected Contractor(s):*

_____

4) *Cost and Schedule Impact:*

_____
_____

Exhibit 14-19

List of sample administrative programs.

OCIP	Owner-Controlled Insurance Policy (See Chapter 10)
CCIP	Contractor-Controlled Insurance Policy (See Chapter 10)
MWBE	Minority- & Women-Owned Business Enterprises
SDB	Small Disadvantaged Business
LMB	Local Minority Business
EO50	Executive Order No. 50 (federal law governing definition of economically disadvantaged persons)
	Local Property Tax Abatement and Deferral Programs
	Local Tax Exempt Status for the Construction Project

Date

Addressee's contact information

Exhibit 14-20
Letter of
completion.

Re: CONTRACTNUMBER:
       PROJECT NUMBER:
       PROJECT NAME:

Dear Sirs:

To the best of our knowledge and in accordance to the standard of practice, the above-referenced project has been completed in all respects in accordance with the Contract Documents.

Item:	Date:
Substantial Completion	
Start of Warranties & Guarantees	
Phase I Completion	
Phase II Completion	

_____   _____

Architect's Signature   Date

_____   _____

Construction Manager/General Contractor   Date

SUMMARY

- The successful completion of a project is not only constructing it on schedule, budget, and with quality, but also the proper administration, record keeping, documentation, and distribution of timely information about the status of the project. At times this can be more challenging and complex than the actual construction process, especially when you are dealing with a sophisticated owner and a large, complex project.

- Corporate clients today have large accounting and audit departments, which often sit on the Board of Directors. Their unique reporting requirements must be identified early on in the project, and systems developed to track and report the required information.

- Without the proper documentation of the history of the project, you may not be able to properly defend yourself with the approval of change orders, events affecting the project beyond your control, litigation, claims, etc.

- When an accident or incident occurs, it is important to document the event and to gather all pertinent information that may be required to defend the situation in a lawsuit or claim, which often occurs years later, long after the project team is gone from the project.

- Records management, both during and after the completion of the project, is important. All records should be categorized and retained to comply with the contract requirements, and settlement of all claims, disputes, warranty, and guarantee issues.

- All documentation, reports, and information required for the project must be identified at the very beginning of the project, in order for the proper systems to be put in place to gather the data, present it in meaningful and timely reports of information, and keep management informed and allow them to make timely and proper decisions.

15

Schedules
(I thought back-of-envelope dates were sufficient.)

WHAT IS A SCHEDULE?

A schedule is the placement of a sequence of activities in a time-related logical order. A schedule is also a dynamic tool that is used to measure, monitor, and control the progress of a project. Thus, if we were changing a rear flat tire on a car, the following activities would occur (with related time):

1. Stop car—3 seconds
2. Inspect tires—15 seconds
3. Open trunk—5 seconds
4. Place warning signal—15 seconds
5. Remove spare tire and jack—2 minutes
6. Block front wheels—1 minute
7. Jack-up car—5 minutes
8. Remove lug bolts—2minutes
9. Remove tire—1 minute
10. Put tire on side of road—1 minute
11. Put spare on car—1 minute
12. Tighten lugs 85%—1 minute
13. Lower car to sit on wheel—30 seconds
14. Tighten lugs to 100%—2 minutes
15. Lower car completely—1 minute
16. Put jack and tire in trunk—2 minutes
17. Close trunk—10 seconds

Exhibit 15-1

Scheduling flow chart.

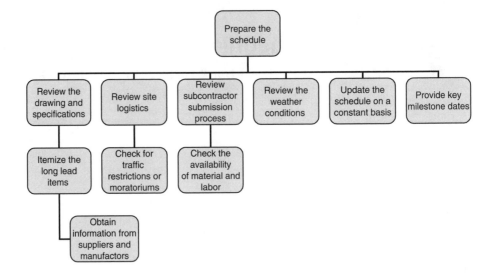

Here you see how the changing of a tire has to follow a logical order of activities so that the primary objective (putting on a new tire) can be achieved. Each activity is a task that must be accomplished, with an associated time, in order to complete the project. For the construction process, the same sequence of activities is followed with the objective of completing a project within a certain period. The American General Contractors Association states that activities have five elements:

1. Consumes time—How long will the activity take?
2. Consumes resources—Takes a certain number of trades people and material to perform an activity.
3. Has a start and end time—When can the activity start and when will it end?
4. Can be assigned to a trade—Placing steel would be done by the ironworkers' trade.
5. Can be measured—The actual activity can be measured against the time originally anticipated.

See Exhibit 15-1 for a schedule flow chart.

THE IMPORTANCE OF SCHEDULES

Schedules are important to project managers (PMs), general contractors (GCs), construction managers (CMs), and owners for the following reasons:

1. Schedules enable PMs to plan the efficient use of equipment, material, and labor.
2. Schedules help CM/GCs determine the overall time-related cost of the general conditions portion of the contract. (General conditions are the CM/GC's actual out-of-pocket expenses.)

3. Schedules are used by the owner to determine the project completion date, cash flow projections, and to secure construction loan financing.

4. Schedules enable the PMs, subcontractors, and suppliers to determine if there is sufficient labor and materials available for the project.

5. Schedule variances (delays) from the approved baseline schedule must be shown to the owner for possible adjustments.

6. Schedules must be updated on a daily basis. This will help to avoid any claims by showing the sequence of the activities that may have caused a schedule to be extended.

7. Schedules are not created in a vacuum. The person preparing the schedules must understand the construction process.

8. Schedules that are dictated by the owner must be thoroughly reviewed by the CM/GC. If the dates are unrealistic, the owner must be notified immediately. Notify the owner in writing with the details of why the date is unrealistic.

9. If overtime work and paying premium cost for materials is required to meet the schedule, the owner should be notified accordingly.

10. Schedules can track the resources (labor and material costs) of a project.

11. Schedules can tell you the percentage of various components of a project that are complete.

TYPES OF SCHEDULES

Two basic types of schedules are used in the construction industry: (1) a bar chart and (2) the critical path method (CPM).

A bar chart as depicted in Exhibit 15-2 is a simple representation of the various tasks for a project. The left side of the chart shows the task and the top of the chart indicates the time (and associated calendar) required to complete the task. The bar indicates the length of time the task will take. To make the bar chart schedule more realistic, a detailed breakdown of an activity can be shown (Exhibit 15-3).

In its simple form, the bar chart schedule makes it easier for the subcontractors to understand the relationship of various tasks. From this bar chart, the subcontractors can schedule their work and order the material, product, or equipment required to complete the work. The bar chart has several limitations, including:

1. Details of the various tasks are limited.

2. It does not show the connection and interaction between tasks.

3. It is difficult to display any delay impact.

4. It does not indicate the critical flow that must take place in order to complete the project on time.

A CPM (Exhibit 15-4) schedule shows all the interactions between the various tasks. In addition, it indicates which tasks are critical and thus the PM can follow the "yellow

Activity ID	Activity Description	Dura	Resource ID	Budgeted Cost
1000	Project Start	0		0.00
1010	Submit shop drawings	1	SUBCONT	0.00
1020	Approval of shop drawings	1	CONSULT	0.00
1030	Fabricate the reinforcing steel	4	SUBCONT	0.00
1040	Ship the reinforcing steel to site	1	SUBCONT	0.00
1050	Bend reinforcing bars	1	LATHER	3584.00
1060	Submit concrete design mix	1	SUBCONT	0.00
1070	Approval of design mix	2	CONSULT	0.00
1080	Number of cubic yards required	1	SUBCONT	0.00
1090	Order wood shoring & formwork	1	SUBCONT	0.00
1100	Delivery of wood to job site	2	SUBCONT	0.00
1110	Cut wood to required sizes	4	CARPNTR	10944.00
1120	Erect shoring and formwork	2	CARPNTR	7296.00
1130	Spray the formwork	1	LABORER	156.00
1140	Place any electrical conduit	1	ELECTRCN	904.00
1150	Set reinforcing chairs	1	LATHER	3584.00
1160	Place reinforcing steel	1	LATHER	3584.00
1170	Set elevation of top of concrete	1	SURVEYOR	0.00
1180	Order concrete and pump truck	1	SUBCONT	0.00
1190	Set pump truck	1	SUBCONT	0.00
1191	Concrete trucks arrive	1	SUBCONT	0.00
1192	Pour concrete into pump truck	1	LABORER	312.00
1200	Start pouring concrete	1	SUBCONT	0.00
1210	Perform slump test and take cylinders	1	TEST LAB	0.00
1220	Shovel and vibrate the concrete	1	MASON	1248.00
1230	Finish off the concrete	2	MASON	1760.00
1240	Cure the concrete	2	LABORER	312.00
1250	Strip off the formwork	2	LABORER	2496.00
1260	Leave the shoring	7		0.00

© Primavera Systems, Inc.

Start Date	02-04-08
Finish Date	02-28-08
Data Date	02-04-08
Run Date	04-24-08 10:01

CONC

Concrete Pour
Progress Run : 02-04-08

1. Detailed Schedule

Early Bar
Progress Bar
Critical Activity

Exhibit 15-2
Base bar chart.

Activity ID	Activity Description	Dur	Planned Start	Planned Finish			
					APR 2008	MAY	
0000	Place Reinforcing Steel	8*	04-29-08 07:00	04-29-08 14:59	Place Reinforcing Steel		
0010	Lift Reinforcing Steel Bundles Via Crane	1	04-29-08 07:00	04-29-08 07:59	Lift Reinforcing Steel Bundles Via Crane		
0020	Review shop drawings to indicate location	1	04-29-08 08:00	04-29-08 08:59	Review shop drawings to indicate location		
0030	Place reinforcing steel on top of chairs	3	04-29-08 09:00	04-29-08 11:59	Place reinforcing steel on top of chairs		
0040	Tie bars to chairs and croos reinforcing steel	3	04-29-08 12:00	04-29-08 14:59	Tie bars to chairs and croos reinforcing steel		

Start Date   04-29-08 07:00
Finish Date   04-29-08 14:59
Data Date   04-29-08 07:00
Run Date   04-29-08 11:51

Early Bar
Progress Bar
Critical Activity

EX00

1. Detailed Schedule

Activity Expansion

Date	Revision	Checked	Approved

Exhibit 15-3
Activity breakdown.

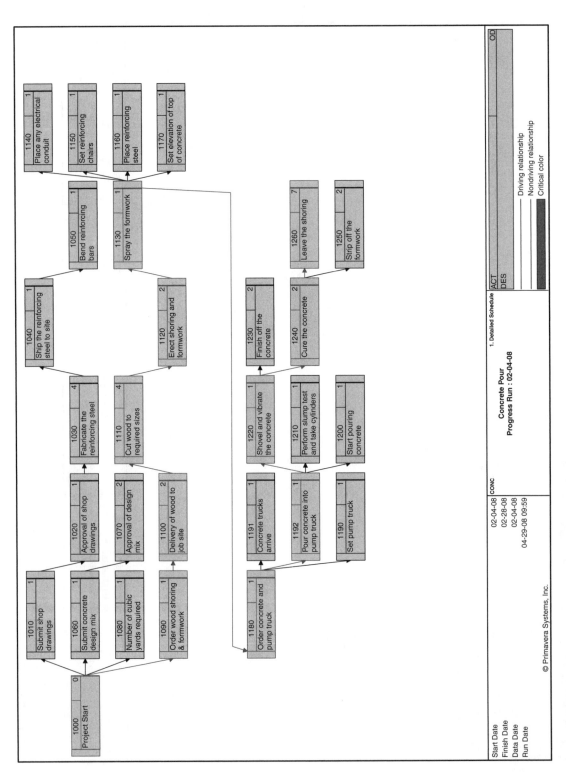

Exhibit 15-4

CPM chart.

© Primavera Systems, Inc.

brick road" from initiation to completion. Graphic displays of the major and minor operations taking place during the life of the project are noted in the schedule.

The CPM shows the tasks that cannot start until other tasks are completed and shows how some tasks can continue simultaneously with other tasks. It represents concise information about the sequence of construction and shows how to predict the required time to complete the project. One of the major advantages of the CPM is that it indicates real time dates of when a task will start and when it will completed. The PM can then:

1. Identify critical activities (i.e., why CPM).
2. Schedule subcontractors, material, and equipment delivery dates.
3. Evaluate alternative construction methods.
4. Follow the progress of the project and record all changes.

As with all great methods, the CPM has certain disadvantages:

- It is a complex method of preparing a schedule.
- It usually requires a scheduling expert for developing a schedule.
- Every activity must be updated daily.
- It is usually difficult for the subcontractors to understand. The CPM shows a multitude of tasks with which a subcontractor may not be involved. This tends to confuse subcontractors as to their responsibilities and when they need to be carried out.

DEVELOPMENT OF THE CPM SCHEDULE

In order for the CPM schedule to be effective, the PM and the scheduler must assess the project and then determine the best approach for completing all the work as established by the construction documents. This will entail the following:

1. Review all construction documents.
2. Inspect the site and the surrounding areas (i.e., existing buildings, traffic patterns).
3. Establish tasks and duration.
4. Construct dependency of the various activities.
5. Determine critical path and float times.
6. Review the initial draft with the subcontractors and suppliers.
7. Establish the first schedule approved by all parties as the "baseline" schedule.
8. Create procedures for making changes to the baseline schedule. All changes will have to be documented and revision dates noted on the schedules.

Critical CPM terminology includes the following:

Activity flow: Sequence of work from one task to another.

Order of activity: An indication of which work event follows another.

Duration: Time required to complete a task.

Nodes: Graphic representation of specific tasks.

Early start: The earliest date an activity can start with the completion of the required preceding activities.

Late start: The latest date an activity can start and not affect the start of succeeding activities.

Early completion: Completion of a task prior to its initial scheduled date.

Late completion: Completion of a task later then originally scheduled.

Float: The time duration between early start and late start for an activity.

See Exhibit 15-5 for the definitions related to the CPM chart.

The float time is a critical element of the project due to its extra time capability. How should the CM/GC deal with this aspect of the CPM schedule? In all projects, some type of contingency must be included within the schedule. In some cases, the float time may be inherent within the task itself. This is usually true of bar chart schedules. However, the CPM is so detailed that the contingency is included for unforeseen circumstances or conditions. The question then comes up of who owns the float—the owner or the CM/GC? This must be established in the contract or else early in the project. If it is not stated or agreed to, then a potential dispute could arise.

SCHEDULING PROGRAMS

Several scheduling programs are available for the construction process. These include:

1. Microsoft Project (office.microsoft.com/en-us/project/default.aspx)
2. Primavera Project Planner (primavera.com)
3. Suretrak Project Manager by Primavera Systems ("Primavera Light") (primavera.com/products/suretrak.asp)

Scheduling programs cannot tell you how to sequence the activities of a project. Thus, the sequence established for, for example, concrete placement (see section immediately following) would have to be developed by a knowledgeable PM. The scheduling program will take the information provided by the PM and then graphically show the activities of their interaction with other relevant activities.

The scheduling program has other capabilities besides showing a graphic representation of the activities and the time associated with each one:

1. Resources (material/manpower) for the activity (Exhibit 15-6)
2. Cost related to the activity (Exhibit 15-7)
3. Baseline (initial) schedule (Exhibit 15-2)
4. Progress of the schedule and percentage complete of each activity (Exhibit 15-8)

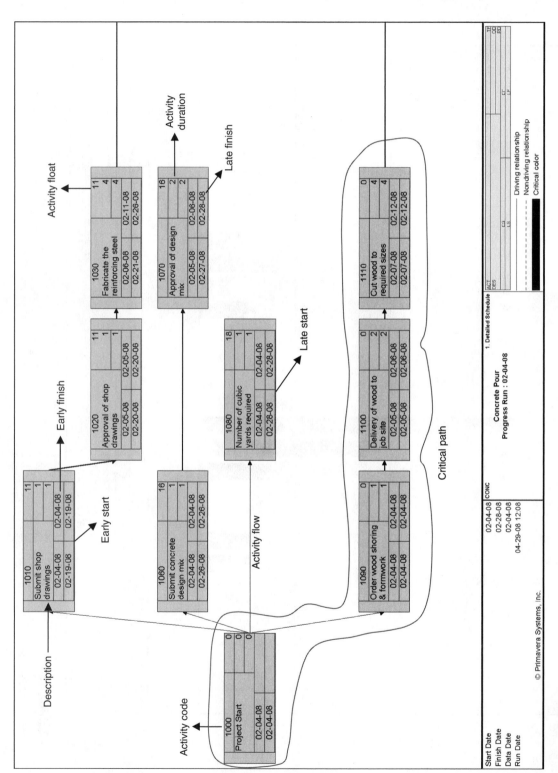

Exhibit 15-5
Definition for CPM chart.

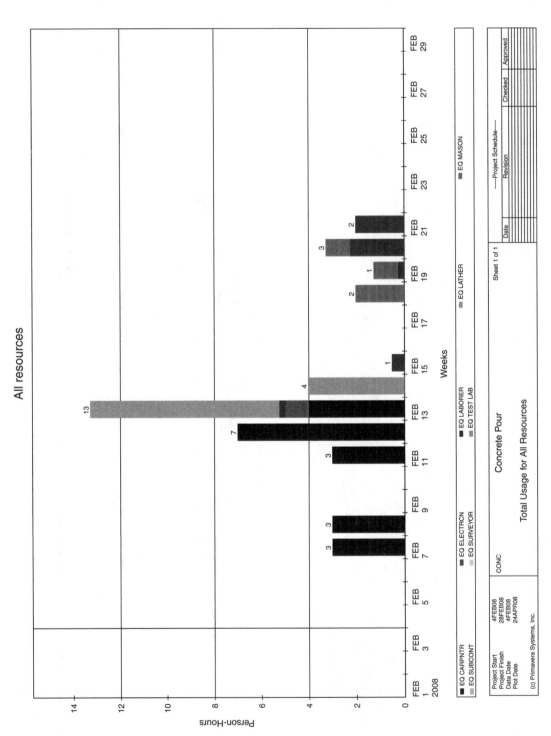

Exhibit 15-6
Resources chart.

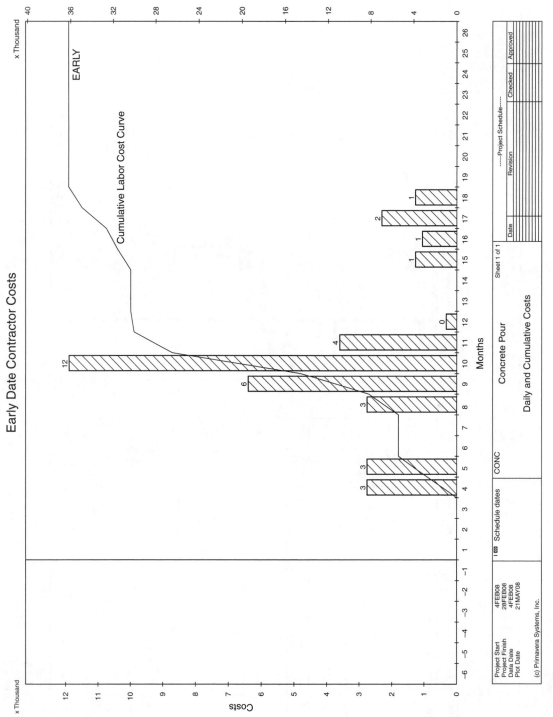

Exhibit 15-7
Cost.

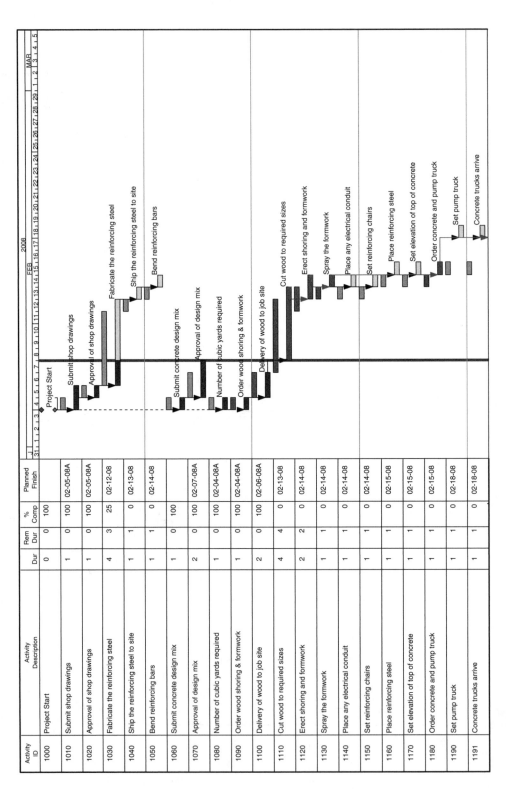

Exhibit 15-8

Progress schedule.

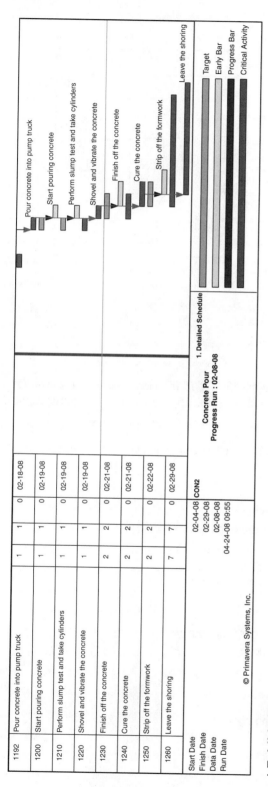

ID	Activity				
1192	Pour concrete into pump truck	1	1	0	02-18-08
1200	Start pouring concrete	1	1	0	02-19-08
1210	Perform slump test and take cylinders	1	1	0	02-19-08
1220	Shovel and vibrate the concrete	1	1	0	02-19-08
1230	Finish off the concrete	2	2	0	02-21-08
1240	Cure the concrete	2	2	0	02-21-08
1250	Strip off the formwork	2	2	0	02-22-08
1260	Leave the shoring	7	7	0	02-29-08

Start Date 02-04-08 CON2
Finish Date 02-29-08
Data Date 02-08-08
Run Date 04-24-08 09:55

Concrete Pour
Progress Run : 02-08-08

1. Detailed Schedule

© Primavera Systems, Inc.

Target
Early Bar
Progress Bar
Critical Activity

Exhibit 15-8
(Continued)

Exhibit 15-9

Concrete pour
drawing.

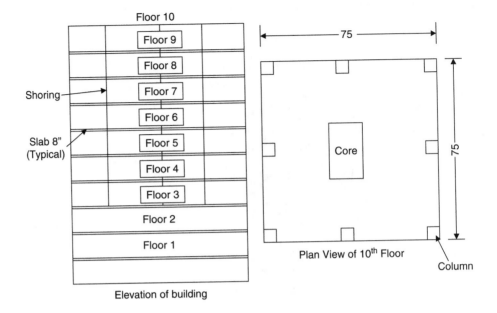

A Concrete Pour

If we look at Exhibit 15-9, we see that the 75 ft × 75 ft concrete deck of a tenth-floor slab has to be poured. The activities noted in Exhibit 15-10 are required to construct the 75 ft × 75 ft floor area.

Exhibits 15-2 and 15-4 are the bar chart and CPM charts that depict the concrete pour.

Exhibit 15-10

Concrete pour
for a 75 × 75 floor
area.

1. Submit shop drawings for the 10th floor.
2. Approve the reinforcing steel shop drawings.
3. Fabricate the reinforcing steel.
4. Ship the reinforcing steel to the site.
5. Bend the reinforcing steel per the shop drawings.
6. Submit the concrete design mix.
7. Approve the design mix.
8. Determine the total cubic yards of concrete required.
9. Order wood shoring and form work.
10. Deliver the wood to the job site.
11. Cut wood to required sizes.
12. Erect shoring and form work.
13. Spray the form work with releasing agent.
14. Place any electrical conduit.
15. Set reinforcing chairs.
16. Place reinforcing steel.

17. Set elevation for top of concrete.
18. Order concrete and pump truck.
19. Set pump truck.
20. Concrete trucks arrive.
21. Pour concrete into the pump truck.
22. Start pouring the concrete.
23. Perform slump test and take cylinders.
24. Shovel and vibrate the concrete.
25. Finish the concrete.
26. Cure the concrete.
27. Strip off the concrete.
28. Leave the shoring in place.

Exhibit 15-10
(*Continued*)

PREPARATION OF A SCHEDULE

When preparing a schedule, the checklist in Exhibit 15-11 should be reviewed.

Factors Affecting Schedules

The following factors affect schedules:

- Availability of labor
- Availability of material
- Long lead items
- Weather conditions
- Proper winter protection
- Holidays
- Hunting season (especially in urban areas where the trades people have to travel long distances to hunting reserves)
- Restriction of activity mandated by a municipality (i.e., parades, VIPs coming to the city)
- Labor disputes
- Changes requested by the owner or architect
- Safety issues
- Insurance company evaluation of the site
- Municipal safety agency site inspections (including OSHA)
- Fires with damage
- Strikes
- Unforeseen site conditions
- Archeology finds

Exhibit 15-11

Preparation of a schedule checklist.

1. Be very familiar with the project construction documents.
2. Have the site inspected.
3. Know the logistic plan for the site.
4. Obtain input from the subcontractors who will be performing the activities for the project. This may entail reviewing the operations with several subcontractors. This may affect the sequence of activities and related time for completion.
5. Review key elements of the project to determine the availability of material and manpower.
6. Establish key milestone dates (i.e., when will curtain wall arrive at the site).
7. Review all of the subcontractors' activities to make sure conflicts are avoided.
8. Update the schedule on a constant basis and notify the owner immediately of any potential delays.
9. Track all long lead items weekly.
10. Look at start and end days for activities that may take place during poor weather conditions.
11. If a problem occurs, then isolate the particular problem and determine how the event can impact the total schedule.

- Poor subcontractors who are behind schedule
- Delayed payment schedules
- Delay in receiving certain permits
- Not receiving paper work in a timely manner in order to obtain a temporary certificate of occupancy (TCO)
- Problems with neighbors
- Disputes
- Change orders
- Cost escalation
- Wrong scheduling logic
- High volume of construction taking place within the locale in which you are working
- Bonding company requirements
- Security breaches
- Acceleration of the project by the owner

FAST TRACKING

Fast tracking is a method for starting construction prior to having all the construction documents (plans and specifications) completed. See Exhibit 15-12 for the fast tracking flow chart. Its main purpose is to start construction as soon as possible so that time is not lost having to wait for the completion of all documents. Most major

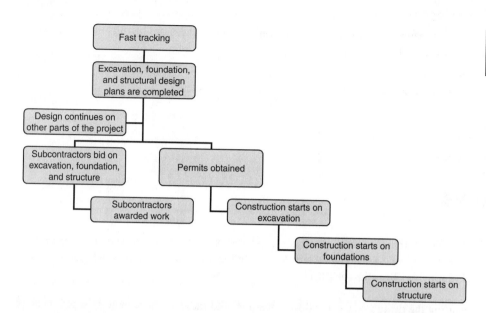

Exhibit 15-12
Fast tracking flow diagram.

private construction projects in an urban environment start construction using the fast track method. The owner putting up a project wants his positive cash flow to start as soon as construction is completed and the building is ready for occupancy. The initial work on the project would start as soon as the foundation and structural system drawings are completed. During this early construction phase, the design team will try to complete the remaining construction documents. Depending upon the complexity of the project, the design team may concentrate on completing the mechanical, electrical, plumbing, sprinkler, façade, and elevator design drawings. Long lead items (discussed in the next section) will also be ordered, so that those items will not affect the schedule. Many advantages are accrued by fast tracking and they include:

1. Capability of ordering long lead items during the early phases of a project
2. Expedites the construction process
3. Positive cash flow would start sooner for an owner
4. Can avoid winter or other adverse weather conditions
5. Can save on inflationary cost of materials, products, and equipment
6. Can obtain critical subcontractors early in the process

Along with the advantages, certain disadvantages must also be evaluated:

1. Total cost of the project is unknown. However, with a guaranteed maximum price (GMP; see Chapter 9) a price can be developed using allowances for unknown materials, products, and equipment, and providing a contingency for additional unknown conditions.
2. Coordination of all trades can be compromised
3. Potential for more change orders

4. May have to obtain various municipal approvals and more permits then what would normally be required

5. Contract with the owner has to be clearly defined:

   • Percentage of construction documents completed

   • Allowances

   • Contingency amount and who owns it

LONG LEAD ITEMS

Long lead items are those materials, products, or equipment that when initially ordered may fall outside a designated milestone date for when they should be at the job site. See Exhibit 15-13 for the long lead items flow chart.

During the design phase, the design team should indicate those materials, products, or equipment where potential scheduling problems could exist. The PM and the purchasing department should review the design documents and prepare a list of the potential long lead items. All critical items should be reviewed with the owner with the recommendation that these items are purchased immediately. These items, once purchased, would then be assigned to the approved subcontractor responsible for the installation. Once the subcontractors are selected, confirmation should be obtained from them as to the availability of additional potential critical items not purchased initially. The subcontractors should also give notice of any other items that either are no longer fabricated or will create a potential scheduling problem. To track the process of the long lead item from shop log submittal to design team approval the long lead item tracking log depicted in Exhibit 15-14 should be utilized. After the long lead items have been ordered, they

Exhibit 15-13

Long lead item flow diagram.

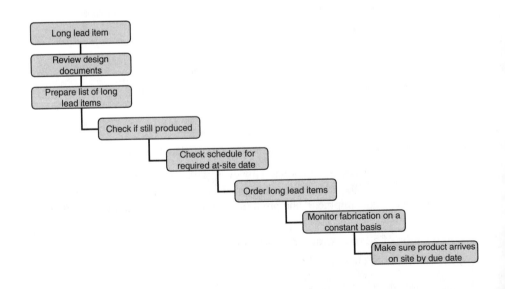

Long Lead Item	Date Required At Job Site	Shop Drawing Submittal Date	Shop Drawing Approval Date	Shop Drawing Returned to Sub	Start Fabrication	Complete Fabrication	Ship Date
1							
2							
3							
4							
5							
6							
7							
8							
9							
10							
11							
12							
13							
14							
15							
16							
17							
18							
19							
20							

Exhibit 15-14
Long lead items tracking schedule.

Material	Job Required Date	Confirmation By Name	Date	Confirmation By Name	Date	Confirmation By Name	Date
1							
2							
3							
4							
5							
6							
7							
8							
9							
10							
11							
12							
13							
14							
15							
16							
17							
18							
19							
20							

Exhibit 15-15
Tracking long lead items log.

1. Steel for a structure
2. Concrete
3. Pre-cast components
4. Reinforcing steel
5. Copper piping and wire
6. Curtain walls
7. Elevators
8. Escalators
9. Some landscaping material
10. Paving (plants close down for the winter)
11. Hardware
12. Special light fixtures
13. Millwork
14. Wood doors
15. Special ceilings (i.e., metal)
16. Vinyl wall coverings
17. Carpet
18. Mechanical units (especially chillers and cooling towers)
19. Transformers and switchgear
20. Glass doors
21. Stone for walls, floors, and paving
22. Special fabrications (railings, architectural finishes, etc.)
23. Roofing material
24. Products coming from a foreign country
25. Materials that are going to foreign countries

Exhibit 15-16

Potential long lead items.

should be tracked on a continuous basis. A follow-up log is indicated in Exhibit 15-15. The list of potential long lead items is noted in Exhibit 15-16.

TWO-WEEK LOOK AHEAD

In order for the CM/GC and the subcontractors to keep track of the schedule on a continuous basis, it is a good idea to break down the schedule into two-week increments. In this way, all the participants can focus on the activities that have to be accomplished within the next two weeks. In addition, any milestone dates would be noted on the two-week look ahead schedule. The milestone dates would be those activities that must be accomplished in order to meet the schedule end date. This is not to say that you forget about the main schedule because you never want to lose sight of the big picture—the completion of the project and the related milestone dates. Exhibit 15-17 shows the elements for preparing a two-week look ahead schedule.

Exhibit 15-17

Two-week look ahead schedule.

		DATE: FROM_____TO_____	
	Trade	Activity	Work to be accomplished
1			
2			
3			
4			
5			
6			
7			
8			
9			
10			
11			
12			
13			
14			
15			
16			
17			
18			
19			
20			

SUMMARY

- A schedule indicates activities and associated time to perform the activity.
- There are two major types of schedules: bar charts and CPM.
- Bar charts are a simple representation of activities.
- CPM indicates the interaction of various activities.
- A logical sequence of activities has to be developed in order to prepare a meaningful schedule.
- Various external factors have an important impact on the schedule.
- Fast tracking is a construction method for starting a project without having 100% completed construction documents.
- Long lead items for the project must be tracked on a continuous basis so that they can arrive at the site when required.
- A two-week look-ahead schedule should be prepared so that all the construction participants can focus on these future activities and any critical milestone dates.

16

Subcontractors and Bidding
(The people who actually build the project and want to be paid.)

WHAT IS A SUBCONTRACTOR?

A subcontractor is an organization that has a particular specialty to perform a critical element of a construction project. Usually, all the hired workers at this type of organization have the ability to perform the technical aspect of their specialty. See Exhibit 16-1 for the subcontractor's flow chart.

At one time, contractors performed most of the work that was required for a project. This included electrical work, mechanical work, and plumbing work as well as the basic elements of the project. However, by proceeding on this basis, the contractor not only had to maintain a staff of these specialty persons but these hired workers had to be extremely skilled. This was a costly proposition, especially if only limited work was available. As a result, specialty subcontractors evolved as a substitute for the contractors' own skilled work force. Here the contractor could choose from a number of subcontractors and obtain a competitive price with skilled workers, as required for a project. It should be noted the subcontractors' work constitutes approximately 80 to 90% of the total cost of the project. A typical subcontractor organization is noted in Exhibit 16-2.

As you can see in Exhibit 16-2, the subcontractor has to deal with secondary and tertiary fabricators and suppliers. Therefore, strong oversight by the project manager (PM) is required because of the number of entities involved with the subcontractor's work.

The number of subcontractors has expanded over the years as projects have become more sophisticated. Some of the more common types of subcontractors are indicated in Exhibit 16-3.

Exhibit 16-1

Subcontractor and bidding-flow chart.

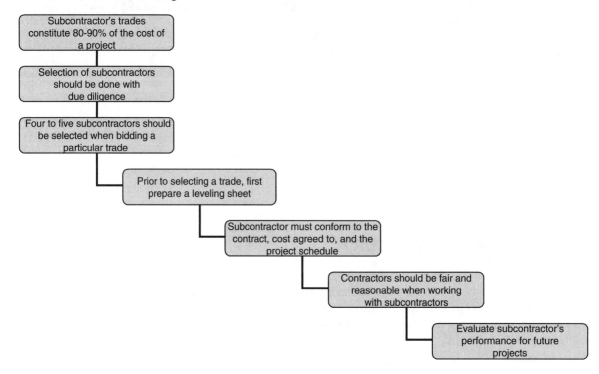

Exhibit 16-2

Subcontractor's organization.

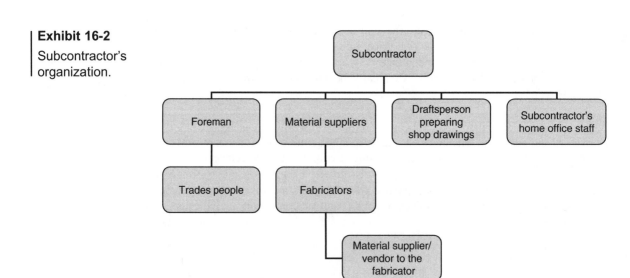

Exhibit 16-3

Subcontractor's list.

1. Concrete
2. Steel
3. Excavation
4. Demolition
5. Mechanical
6. Electrical
7. Plumbing
8. Sprinkler
9. Elevators (and other conveying systems)
10. Façade
11. Roofing
12. Drywall
13. Plastering
14. Flooring (i.e., carpet, hardwood, Vinyl Tile [VT])
15. Ceiling
16. Painting
17. Landscaping
18. Millworkers
19. Paving
20. Brickwork
21. Stone work
22. Communications and data
23. Glaziers
24. Controls
25. Pile work
26. Sheeting
27. Blasting
28. Security
29. Hoists
30. Ornamental metal
31. Hazardous material removers
32. Underpinning
33. Cranes
34. Site work
35. Site utilities

THE SELECTION PROCESS

Most established contractors have a list of qualified subcontractors that they have used over the years. However, it is always good to bring in new and qualified subcontractors so that better prices may be obtained and to make sure the existing subcontractors do not become complacent. This may be dictated by market conditions or the requirement to bid to more than three or four subcontractors as requested by the contract. In addition, the construction manager/general contractor (CM/GC) may want minority- or women-owned businesses to be involved in the project. When looking for qualified subcontractors, the following factors should be considered:

1. Reputation (i.e., Do they meet schedules? What is the quality of their work like? Do they use specified material?)
2. References from owners, architects, engineers, and other contractors
3. Quality of work
4. Financials (Dunn and Bradstreet reports)
5. Current work load
6. Number of skilled workers
7. Foremen-to-worker ratio
8. Safety records (for the past 3 years), EMR safety rating index
9. Bonding capability and rating
10. Insurance capability and rating
11. Union vs. non-union
12. If union, have they paid all of their dues?
13. Experience (i.e., have they built similar types of projects?)
14. Are they known to issue many unsubstantiated change orders?
15. Claim history
16. Ability to work with the other subcontractors
17. Can they perform when out-of-sequence work is required?
18. Do they have any outstanding lawsuits (including mechanic liens)?
19. Have they pleaded guilty or been found guilty of criminal offenses?
20. Are they under any current criminal investigation?
21. How do they handle punch list work?
22. Do they submit all the required completion documents in a timely manner?
23. Owner's name
24. Are they on any magazine index (i.e., ENR)?

A typical subcontractor's qualification sheet is noted in Exhibit 16-4.

When a subcontractor is selected to be on the CM/GC's bid list, the PM has to make sure that they will perform on schedule and adhere to the quality of work that is specified. A bad subcontractor can destroy a project by doing any of the following:

1. Going bankrupt or have insufficient funds to complete the work.
2. Hiring/employing insufficient skilled workers.
3. Providing poor quality work.
4. Creating delays in obtaining materials and product, including long lead items.

Date: _____

Major Trade: _____

1) Company name: _____

   Address: _____

   City: _____ State: _____ Zip code: _____

   Telephone: _____ Fax: _____

   Contact person: _____

2) Years in business: _____

3) Total number of employees: _____ Office: _____ Field: _____ Shop: _____

4) List geographic areas covered: _____

   _____

   _____

5) List trades and state(s) in which company holds licenses:

   _____ License Number: _____

   _____ License Number: _____

   _____ License Number: _____

   _____ License Number: _____

6) List major services and/or products provided in order of expertise:

   1. ____ % Subcontracted: _____ 5. ____ % Subcontracted: _____

   2. ____ % Subcontracted: _____ 6. ____ % Subcontracted: _____

   3. ____ % Subcontracted: _____ 7. ____ % Subcontracted: _____

   4. ____ % Subcontracted: _____ 8. ____ % Subcontracted: _____

7) Bank reference: _____

   Contact: _____ Phone: _____

   Bank Reference: _____

   Contact: _____ Phone: _____

8) List three trade references (contact & phone):

   1. _____ Phone: _____

   2. _____ Phone: _____

   3. _____ Phone: _____

9) List three major projects presently under construction:

   Project: _____ Owner: _____

   Contact: _____ Phone: _____

   Start date: _____ Finish date: _____ Contract amount: _____

   Project: _____ Owner: _____

   Contact: _____ Phone: _____

   Start date: _____ Finish date: _____ Contract amount: _____

   Project: _____ Owner: _____

   Contact: _____ Phone: _____

   Start date: _____ Finish date: _____ Contract amount: _____

Exhibit 16-4

Subcontractor's qualification sheet.

(Continued)

Exhibit 16-4

(*Continued*)

10) List revenue for the past three years:

	2 _____	2 _____	2 _____
Private work	$ _____	$ _____	$ _____
Public work	$ _____	$ _____	$ _____

11) Current backlog of uncompleted work: $_____

12) Attach financial statement. Include most recent copy on auditor's letterhead.

13) Have you ever failed to complete a project? Yes_____ No_____
    If yes, explain: _____
    _____
    _____

14) Are you signatory to any labor agreement?   Yes_____ No_____
    Which trades? _____

15) Do you have a written safety program?   Yes_____ No_____
    (If yes, attach copy of program)

16) Do you require your field employees to be OSHA 10-hour certified?
    Yes___   No___
    (If No, please describe safety training you provide)

17) Have you been cited by OSHA within the last four years?   Yes___ No___
    If yes, explain: _____
    _____
    _____

18) Insurance—What are your standard limits of insurance coverage?
    A) General Liability
       Limit: $_____ Insurance Co. _____
       Broker: _____ Phone: _____
    B) Umbrella
       Limit: $_____ Insurance Co. _____
       Broker: _____ Phone: _____
    C) Design/build
       Limit: $_____ Insurance Co. _____
       Broker: _____ Phone: _____
    D) Workers' compensation
       Limit: $_____ Insurance Co. _____
       Broker: _____ Phone: _____
    E) Are you bondable? Yes_____ No _____
       Surety Co. _____
       Single job limit $_____ Aggregate $_____

19) Workers' compensation modification rating (EMR for last three years)
    2 _____   2 _____   2 _____
    _____   _____   _____

20) Do you offer health insurance for your employees? Yes _____ No _____
    If yes: Full coverage ____ partial coverage ____ No reimbursement ____

21) Do you have a forman employee training program? Yes _____ No ____

22) What means do you use for employee training (indicate all that apply):

In-house training programs	Yes ____	No ____
Gould institute (ABC)	Yes ____	No ____
Trade school/apprenticeship program	Yes ____	No ____
Continuing education programs	Yes ____	No ____
Seminars and workshops	Yes ____	No ____

23) Do you qualify as a SONWBA—approved minority business enterprise (MBE)?

Yes_____   No_____

Do you qualify as a SONWBA—approved women business enterprise (WBE)?

Yes_____   No_____

Do you qualify as a small business enterprise?

Yes_____   No_____

24) Does your company have any particular specialized areas of expertise?

_____

_____

_____

25) Is your firm a member of any trade/business associations?

_____

_____

_____

26) Minimum size of job your firm would like to perform $_____

27) Maximum size of job your firm would like to perform $_____

Submitted by: _____

Title: _____

Exhibit 16-4
(*Continued*)

5. Performing out-of-sequence work due to a reduction of productivity.

6. Having a poor safety record.

7. Causing a possible collapse if structural integrity is in question.

8. Affecting other subcontractors' performance by not being cooperative.

9. Creating claims with possible litigation.

THE BIDDING PROCESS

It is a good practice to bid to at least four or five subcontractors in each trade because some of the initial subcontractors may not want to bid and you want to obtain a good representation for pricing purposes. In addition, you want to make sure that sufficient time is given to the subcontractors for the preparation of their bids (which are based on the contract documents, addendums, and bulletins). The documents that should be submitted as part of the CM/GC's bid are indicated in Exhibit 16-5.

Exhibit 16-5

Subcontractor's bid documents.

1. Drawings for the project
2. Specifications for the project
3. List of alternatives
4. Preliminary schedule
5. Logistics plan
6. Safety requirements
7. Owner/contractor contract and general conditions
8. Subcontractor's contract with general conditions
9. Bonding requirements
10. Insurance requirements
11. Bid due date, time, and location
12. Bid validity date
13. Bid sheet to fill out (and person to whom it will be submitted with phone number)
14. List of any questions
15. Potential long-lead items
16. All inclusive labor rates
17. Unit prices, if requested
18. Request organization chart with names noted
19. Contact name of the subcontractor estimator
20. Length of time bid will be valid
21. Special requirements

It is important to scope out the work for each specialty subcontractor. A checklist of all the elements of the project, indicating which subcontractor will perform the work, should be prepared. The CM/GC wants to buy a trade once and never twice.

LEVELING PROCESS

When the bids come in, the PM and the contractor's estimator will evaluate the bids to make sure that they are all "apples to apples" comparison of scope so that the lowest responsible bidder (which conforms to all the requirements of the construction documents) can be determined. This would include looking at missing scope items, substitutions, exclusions, clarifications, overtime work, and any qualifications. In addition, review the secondary subcontractor's work that would be performed for the primary subcontractor (i.e., a controls subcontractor doing work for the primary HVAC subcontractor). This process is called leveling. Leveling sheets are prepared which enumerate the major components of a particular trade. See Exhibits 16-6 and 16-7, which are leveling sheets for the painting and HVAC trades. Each subcontractor's scope of work is checked against this leveling sheet. If an item is missing, the PM or the estimator will contact the subcontractor's estimator and obtain the required information. When time is of the essence, the PM will fill in the missing information based on historical data or

	Sub 1	Sub 2	Sub 3	Sub4
Base Bid Price	$ 100,000	$ 125,000	$ 108,000	$ 115,000
The painting work is based upon the mechanical drawings: M-1, M-2, M-3, M-4, M-5, M-6 and M-7 dated 7/20/08. All electrical drawings: E-1, E-2, E-3, E-4, E-5, E-6, E-7 dated 7/20/08.				
All plumbing drawings: P-1, P-2, P-3, P-4, P-5 dated 7/20/08. All structural drawings: S-1, S-2, S-3, S-4, S-5, S-6, S-7 dated 7/20/08. All architectural drawings: A-1, A-2, A-3, A-4, A-5, A-6 A-7, A-8, A-9, A-10, A-11, A-12, A-13, A-14, A-15 dated 7/15/08 as prepared by the architect and including but not limited to the following:				
1. Paint subcontractor confirms the receipt of all drawings and specification as listed above.	included	included	included	included
2. Paint subcontractor is responsible to include all overtime, fabrication, and painting to meet the completion date of June 1, 2010.	included	included	included	included
3. Paint subcontractor is responsible for furnishing and installing all required scaffolds, lifts, and associated equipment to complete the work.	$ 10,000	included	included	included
4. Paint subcontractor has read the subcontractor contract and takes no exceptions.	included	included	included	included
5. Paint all gypsum wall board throughout the project.	included	included	included	included
6. Paint all furring at all core walls throughout the project.	included	included	included	included
7. Paint all gypsum wall board ceilings.	included	included	included	included
8. Paint all gypsum column enclosures throughout the project.	included	included	included	included
9. Paint all metal hollow doors and frames per the door schedule and paint specification.	included	included	included	included
10. Paint all wood closet doors.	included	included	included	included
11. Paint all gypsum wall board at all interior stair soffits.	included	included	included	included
12. Paint all window pockets as noted on the architectural plans.	included	included	included	included
13. Paint all light coves.	included	included	included	included
14. Paint high ceiling piping in mechanical rooms and open ceiling areas.	included	included	included	included
15. Paint floor sealer dustproofing at all mechanical and electrical closets.	included	included	included	included
16. Paint subcontractor is to provide 10 man days for comebacks for touch-up painting.	included	included	included	included
17. Provide 2 cans of paint and sealer used at the end of the job.	included	included	$ 5,000	included
18. Paint subcontractor shall advise project manager of any problems prior to the painting of an area.	included	included	included	included
19. Paint subcontractor shall complete all related punch list items within one (1) week after receipt.	included	included	included	included
20. Paint subcontractor at the end of the job shall indicate on the architectural drawings the color and manufacturer of the paint used in all painted areas.	$ 2,000	included	included	included
21. Sales tax include in the bid.	$ 8,000	included	included	included
TOTAL BID PRICE	$ 120,000	$ 125,000	$ 123,000	$ 115,000
Recommend Subcontractor-Subcontractor No. 4 @ $115,000				

Exhibit 16-6

Merit shop costs–September, 2007.

	Sub 1	Sub 2	Sub 3	Sub4
Base Bid Price	$ 319,500	$ 231,000	$ 290,300	$ 316,000
The HVAC Work is based upon the mechanical drawings: M-1, M-2, M-3, M-4, M-5, M-6 and M-7 dated 7/20/08. All electrical drawings: E-1, E-2, E-3, E-4, E-5, E-6, E-7 dated 7/20/08.				
All plumbing drawings: P-1, P-2, P-3,P-4, P-5 dated 7/20/08. All structural drawings: S-1, S-2, S-3, S-4, S-5, S-6, S-7 dated 7/20/08. All architectural drawings: A-1, A-2, A-3, A-4, A-5, A-6				
A-7, A-8, A-9, A-10, A-11, A-12, A-13, A-14, A-15 dated 7/15/08 as prepared by the architect and including but not limited to the following:				
1. Furnish and install sheet metal	included	included	included	included
2. Furnish and install acoustical lining	included	included	included	included
3. Furnish and install insulation	included	included	included	included
4. Furnish and install ceiling diffusers and supply registers	included	included	included	included
5. Furnish and install return air grills	included	included	included	included
6. Furnish and install linear diffusers	included	included	included	included
7. Furnish and install AC-1, AC-2, AC-3, AC-4	included	included	included	included
8. Furnish and install smoke detectors	included	included	included	included
9. Furnish and install return air fans	included	included	included	included
10. Furnish and install condensate pumps	included	included	included	included
11. Furnish and install drip pans	included	included	included	included
12. Furnish and install water detectors	included	included	included	included
13. Furnish and install condenser water supply and return piping	included	included	included	included
14. Furnish and install refrigerant piping	included	included	included	included
15. Furnish and install wire mesh screens	included	included	included	included
16. Furnish and install 10 gauge fire damper sleeves	included	included	included	included
17. Overtime-noisy work	included	$ 8,900	included	included
18. Overtime-deliveries	included	$ 6,800	included	included
19. Overtime-welding, brazing, and burning	included	$ 11,000	included	included
20. Furnish water and air balancing and test reports	included	included	included	included
21. Furnish hydrostatic testing	included	included	included	included
22. Furnish chemical cleaning	included	included	included	included
23. Furnish and install controls and control wiring	included	included	included	included
24. Furnish start-up and commissioning	included	included	included	included
25. No sales tax on labor per certificate of capital improvement	included	included	included	included
TOTAL BID PRICE	$ 319,500	$ 257,700	$ 290,301	$ 316,000
Recommended Subcontractor-Subcontractor No. 2 @ $257,700				

Exhibit 16-7

Subcontractor HVAC leveling.

use the dollar amount used by another subcontractor. After this process is completed, the lowest actual bidder can be determined. In some cases, the apparent highest bidder actually turns out to be the lowest bidder after the leveling process had been completed. If great differences occur between the qualified lowest bidder and the next qualified bidder, then the PM may require further evaluation of the bids.

If the bids come in higher than anticipated, then the PM must look at the subcontractor's breakdown and try to determine the reason for the difference. At this stage, it is also advisable to work with the subcontractor to determine ways in which costs can be reduced. If substitutions were suggested, then any change would have to be evaluated with the design team. Arbitrary changes would be a violation of the contract documents. The design team may have to be involved with the cost reduction process through value engineering. Value engineering is a process to evaluate the use of less expensive materials or alternative procedures for performing the work without materially affecting the scope, program, or functionality of the project.

The subcontractor should sign the leveling sheet so that in the future there will be no misunderstanding between the parties involved. This would also include the agreed-upon exclusions and any exceptions noted by the subcontractor. The CM/GC wants to receive the best and final offer (BAFO) from the subcontractors. Avoid a bid war or auction between subcontractors; this can only hurt the CM/GC as the work proceeds. If the subcontractor accepts the job at a lower cost then he actually needs for the project, then the subcontractor may look for ways to save on materials or work force or develop a claim for the difference.

NEGOTIATIONS

When two or three subcontractors are very close with their bids, a sit-down session is held with each subcontractor to determine who would be the best-qualified subcontractor. Negotiations take place concerning scope and price of the work. Factors such as workload commitment, past experience, and schedule are evaluated prior to making a selection.

SUBCONTRACTOR CONTRACT

After a subcontractor is selected, he or she must sign the contract between the CM/GC and the subcontractor. The provisions of the owner/contractor contract have to be known by the subcontractor. The subcontractor should agree to these terms prior to any award. The risks that the CM/GC must take will be passed along to the subcontractors. In addition, the following provisions should be considered:

1. Are there requirements for performing out-of-sequence work?
2. The subcontractor must have a safety-certified foreman.
3. The subcontractor must have a complete complement of safety equipment.

4. The subcontractor must attend all tool box meetings.

5. The subcontractor must protect all workers and the public by placing barricades and warning signs, protecting openings, having flag persons and any other safety enclosures.

6. The subcontractor's workers who do not adhere to the contractor's safety program will be told to leave the project (after one warning).

7. The subcontractors must comply with the site safety and security plan.

8. The subcontractors must follow the logistics plan for the project.

9. Subcontractors must attend all subcontractor meetings.

10. Subcontractors must provide additional trades people if the contractor's superintendent determines that the subcontractor is not supplying a sufficient amount of workers.

11. The subcontractor will be back-charged for work that delayed any other trade or the contractor.

12. The subcontractor must adhere to the schedule even if overtime work is required (at the subcontractor's expense).

13. If the baseline schedule changes, then the subcontractor must abide by those changes.

14. The subcontractor must track all long lead items on a constant basis.

15. The subcontractor must notify the contractor when overtime work is required.

16. The subcontractor must not provide substitutions of material or construction procedures unless approved by the contractor/architect/engineer/owner.

17. The subcontractor must not issue a mechanic's lien on the project.

18. The subcontractor must provide the required insurance limits with additional named insured.

19. The subcontractor must provide a bonding certificate if required by the contractor.

20. The subcontractor must obtain all special permits and certifications required by the municipality, drawings, specifications, and state.

21. The subcontractor must set up a fireproof shanty.

22. The subcontractor must adhere to cleanliness requirements of the area in which the subcontractor is working.

23. The subcontractor must follow all guidelines for working with other subcontractors in the coordination of all work.

24. The subcontractor must provide the requested warranty and guarantees at the end of the project.

25. At the end of the project, the subcontractor must provide the following:
    - Attic stock (additional material)
    - As-built drawings
    - Maintenance manuals
    - Training (if required)

- Test reports
- Certifications

26. The subcontractor must submit any change order work with detailed back up (usually on a time and material basis).

27. Labor rates must be submitted and approved prior to award.

28. Administrative paperwork that may be required for local incentive programs must be submitted in a timely manner.

29. If requested by the CM/GC, subcontractor change order work must be done even if not all the costs have been agreed upon between the parties (usually a requirement in the owner/CM/GC's contract).

30. Any disputes between the subcontractor and the contractor will be resolved by a committee established by the contractor, owner, and architect. The other method is via arbitration or, if management so desires, by litigation.

31. Any liquidated damages imposed by the owner will be prorated to the subcontractor based on a percentage of the total project cost.

32. The subcontractor must have adequate foremen to supervise the workers and maintain quality control on the product installed.

Subcontractors and the Urban Environment

When working in an urban environment the subcontractor has additional obligations that must be considered:

1. Due to the limited amount of space for storage, the subcontractor may have to store material at an off-site location until it is to be installed. The subcontractor may be required to have sufficient additional insurance to protect the product in case of damage.

2. The shanties that are initially installed for tool storage, changing areas, and office may have to be relocated several times as the job progresses. This is necessitated by limited space and the need to complete areas in which the shanties were initially located.

3. The subcontractor is not only responsible for following federal and state codes but it must also abide by the local municipal rules and regulations of the building department and other local departments that may have jurisdiction over the construction process.

4. Insurance limit requirements will have to be increased for potential injuries to the public.

5. Safety requirements will have to be enhanced above and beyond state and OSHA mandated requirements. Some local municipal requirements may be more stringent then the state or OSHA regulations. Besides the safety of the workers, the safety of the public becomes paramount.

6. Coordination of the work with other subcontractors is required especially when working with a multitude of trades in limited areas. The sequence of the subcontractor's work may dictate a change to how equipment or material will be installed. The PM must determine this.

7. Schedules have to be established with the subcontractors for the lifting of their material, either via hoists or by cranes.

8. With limited space in which to work and the potential vastness of the project, subcontractors should be responsible for cleaning up the areas in which they are working.

STARTING THE WORK

As soon as the subcontractor is awarded the job, he must immediately start the mobilization process. This means lining up the trades people and the foreman who will be working on the project. In addition, the necessary material must be ordered. If shop drawings are required, then the preparation of those documents must start immediately.

Once the site becomes available, the subcontractor has to set up his shanty in the area assigned by the contractor. Phones and electrical connections have to be coordinated with the local phone company and the site electrician. A subcontractor contact sheet (see Exhibit 16-7A) should be prepared for notification and billing purposes.

Exhibit 16-7A
Subcontractor contact sheet.

	Trade	Name of Company	Contact Persons	Phone	FAX	Cell Phone	Email	Address
1								
2								
3								
4								
5								
6								
7								
8								
9								
10								
11								
12								
13								
14								
15								
16								
17								
18								
19								
20								
21								
22								
23								
24								
25								

SUBCONTRACTOR SITE MEETINGS

The PM should have site subcontractor meetings as soon as possible. The information to review is indicated in Exhibit 16-8.

PRODUCTIVITY

It is the responsibility of the CM/GC to ensure that maximum productivity is obtained from the subcontractors. The following is a list of things that the CM/GC can do to increase productivity on the job site:

1. Keep the project clean.
2. Keep the project safe (post signs, protect openings, and erect nets, fences, and wire railings).
3. Review the site safety plan with all workers at every weekly tool box meeting.
4. Review procedures for exiting the site in case of an emergency at every weekly tool box meeting.
5. Have all new workers become familiar with the site safety requirements.
6. Make sure that all workers have proper clothing and safety equipment.
7. Have a good first aid station.
8. Provide adequate potable water locations (especially during the summer months).

1. Logistics plan
2. Site safety plan
3. First aid station
4. Emergency procedures
5. Site safety rules
6. Security procedures
7. Location for "shanties"
8. Schedule
9. Deliveries
10. Shop drawing submittals
11. RFI procedures
12. Payment procedures
13. Coordination with other trades
14. Establishment of meetings
15. Administrative requirements
16. Long lead items
17. Quality control issues and proper testing procedures.
18. Change order review

Exhibit 16-8
Subcontractors site meetings.

9. Provide clean and adequate men and women's toilets (close proximity to where the subcontractors are working).

10. Provide a sufficient number of hoists to handle the workers in the morning and at lunchtime.

11. Provide scheduling for the handling of material coming to the job site.

12. Review the project schedules with the foreman at least once a week.

13. Request the foreman to prepare a 2-week look-ahead schedule.

14. Have a subcontractor meeting at least once a week.

15. Make sure that the site is secured so that critical material is not stolen.

16. Review RFIs and shop drawings with the foreman.

17. Shop drawings and other submittals must be delivered to the contractor according to an established schedule.

18. Unapproved documents must be resubmitted within a 2-day period.

19. Make sure that the foreman and the trades people are using approved (stamped) shop drawings.

20. Have the superintendent walk with the foreman every day and review his or her area of responsibility.

21. Start the quality control (QC) and punch list process as early as possible.

22. Make sure there is a good topping-out party (when the last structural member is installed).

23. Pay the subcontractors in an expeditious manner.

24. Make sure that all of the trades people understand English, especially as it relates to safety instructions.

25. Beware of potential slacking off during hunting season, on Fridays during the summertime, on days prior to and after holidays, and on paydays.

26. Make sure that the area in which a subcontractor is to start working is available for trades people and materials.

27. If it is a union project, then read the contracts to understand some of the practices that may affect productivity. Also, know the expiration dates of the various trades contracts.

28. Have a safety incentive program (see Chapter 5).

SUBCONTRACTOR PROCESS FOR CHANGE ORDER WORK

When a subcontractor submits a change order, the first thing the PM should do is to check the construction documents to make sure the request is valid. It is good practice to have the subcontractor submit the costs in as much detail as possible. Part of the responsibility of the PM is to make sure that the costs submitted by the subcontractor are reasonable. The form of the submittal should be indicated in the subcontractor's contract. The time and material (T&O) breakdown is a good means of presenting the change order request. It gives all the participants a good view of the cost of the change order. With knowledge of the scope of the change, the PM can make a determination on the validity of the change. Besides costs, other factors must be evaluated:

1. Time to complete the work

2. Schedule impact

3. Impact of other trades

4. Starting the work without a formal approval from the owner

5. Use of hoists and cranes

6. Overtime

Mark-ups for subcontractor change order work is usually in the range of 15 to 20%.

SUBCONTRACTOR LABOR COSTS

Subcontractors' labor costs vary throughout the United States. *Engineering News-Record* (ENR) publishes data on the hourly wage rates for various trades by major city. An example of hourly trade rates is noted in Exhibit 16-9 (non-union) and Exhibit 16-10 (union). Future years' wage rates can be obtained from enr.com. The ENR information only includes base rate and fringes. The actual amount of a union trade worker's rate is much more extensive and is noted in Exhibit 16-11. The breakdown is for a carpenter working in New York City. In addition to the basic wage and fringe benefits, costs for workers' compensation, liability insurance, mandated social security, Medicare, and overhead and profit must be added.

STATE	BRICKLAYERS		CARPENTERS		CEMENT MASONS		ELECTRICIANS	
	RATE ($)	FRINGE (%)	RATE ($)	FRINGE (%)	RATE ($)	FRINGE (%)	RATE ($)	FRINGE (%)
NEW ENGLAND	18.87	26.0	21.29	20.1	21.91	17.75	22.91	20.6
NEW YORK/NEW JERSEY	22.90	24.8	21.28	20.9	21.16	19.9	22.44	22.3
MIDDLE ATLANTIC (2)	20.95	22.0	19.46	21.2	19.61	20.6	21.60	21.6
SOUTHEAST (3)	19.22	19.0	18.49	20.3	17.86	20.0	21.51	20.6
GREAT LAKES (4)	21.15	20.9	19.65	20.7	19.83	20.4	21.86	21.3
SOUTH CENTRAL (5)	20.06	18.7	18.55	20.1	18.51	19.5	22.03	20.3
CENTRAL (6)	21.98	19.0	18.78	20.0	18.91	18.9	22.12	20.5
CENTRAL MOUNTAIN (7)	23.70	18.5	19.28	20.8	19.7	19.4	22.39	20.1
WESTERN (8)	23.15	19.0	20.04	21.3	20.24	20.7	23.33	20.4
WESTERN (9)	24.21	18.3	20.36	20.4	21.67	19.1	23.72	21.3

	HVY. EQUIP. OPERATORS		LABORERS		PLUMBERS		IRONWORKERS	
	RATE ($)	FRINGE (%)	RATE ($)	FRINGE (%)	RATE ($)	FRINGE (%)	RATE ($)	FRINGE (%)
NEW ENGLAND (1)	23.11	9.9	14.59	17.7	21.81	24.3	20.49	19.2
NEW YORK/NEW JERSEY	23.06	21.2	14.88	19.8	20.08	20.4	19.43	20.2
MIDDLE ATLANTIC (2)	21.36	21.6	13.42	21.2	21.72	22.1	20.9	20.0
SOUTHEAST (3)	20.80	20.2	12.37	19.2	18.49	22.8	21.09	19.4
GREAT LAKES (4)	21.76	21.4	13.99	21.3	21.27	23.4	21.04	20.02
SOUTH CENTRAL (5)	21.35	20.1	12.44	18.7	20.19	18.0	21.15	19.3
CENTRAL (6)	20.59	19.7	12.50	19.6	19.69	21.9	20.92	20.0
CENTRAL MOUNTAIN (7)	20.98	20.6	13.20	20.8	23.3	18.6	21.50	19.7
WESTERN (8)	22.30	20.7	14.02	20.7	24.57	21.3	21.41	19.6
WESTERN (9)	22.5	20.1	14.45	19.6	23.77	19.2	20.51	18.6

SOURCE: PERSONNEL ADMINISTRATION SERVICES, INC. SALINE, MI (PAS, Inc. http://www.pas1.com). WAGE RATES SHOWN ARE AVERAGE HOURLY BASE RATES EXCLUDING FRINGE BASE.
(1) = CT, MA, ME, NH, RI, VT   (2) = DE, MD, PA, WV, DC   (3) = AL, FL, GA, KY, MISS, NC, SC, TN
(4) = IL, IN, MICH, MINN, OH, WI   (5) = ARK, LA, NM, OKLA, TX   (6) = IA, KA, MO, NEB
(7) = COLO, MONT, ND, SD, UTAH, WYO   (8) = AZ, CA, HAW, NEV (9) = ALAS, IDA, ORE, WASH

Exhibit 16-9
Merit shop costs.
(September, 2007)

Exhibit 16-10

Hourly union pay scales–September 2007.

	Atlanta	Boston	Chicago	Los Angeles	New York	San Francisco	Seattle
Bricklayers	$ 26.80	$ 61.63	$ 53.81	$ 44.07	$ 61.62	$ 51.45	$ 42.47
Carpenters	$ 26.45	$ 53.92	$ 54.27	$ 42.47	$ 70.52	$ 49.31	$ 41.27
Cement Masons	$ 25.09	$ 58.01	$ 54.11	$ 42.45	$ 64.33	$ 44.45	$ 42.53
Electricians	$ 33.76	$ 63.00	$ 60.80	$ 50.26	$ 79.16	$ 66.89	$ 48.13
Elevators Contract.	$ 42.09	$ 59.60	$ 62.36	$ 59.46	$ 66.43	$ 66.20	$ 55.02
Glazlers	$ 25.30	$ 48.16	$ 55.37	$ 45.31	$ 61.76	$ 51.45	$ 42.41
Insulation Workers	NA	$ 52.85	$ 55.33	$ 46.40	$ 68.76	$ 55.23	$ 44.68
Iron Workers-Reinf.	$ 31.96	$ 54.28	$ 58.06	$ 51.63	$ 74.25	$ 51.63	$ 46.25
Iron Workers-Struct.	$ 36.92	$ 54.28	$ 62.94	$ 51.63	$ 79.53	$ 51.63	$ 46.25
Laborers-Building	$ 19.62	$ 42.25	$ 47.27	$ 37.88	$ 54.95	$ 37.28	$ 34.90
Laborers-Highway	$ 17.27	$ 42.25	$ 47.27	$ 42.97	$ 51.98	$ 37.28	$ 34.90
Millwrights	$ 32.77	$ 50.35	$ 51.05	$ 51.45	$ 74.08	$ 51.20	$ 42.68
Operating Engineers							
Crane	$ 32.31	$ 53.74	$ 56.78	$ 51.45	$ 82.15	$ 51.32	$ 44.57
Heavy Equipment	NA	$ 53.74	$ 55.48	$ 51.45	$ 80.15	$ 51.32	$ -
Small Equipment	NA	$ 47.69	$ 51.18	$ 51.12	$ 79.15	$ 46.45	$ -
Painters	$ 25.70	$ 53.95	$ 50.04	$ 36.91	$ 53.51	$ 49.07	$ 33.02
Pipefitters	$ 35.36	$ 60.90	$ 57.61	$ 49.47	$ 77.32	$ 51.70	$ 55.39
Plasterers	$ 25.70	$ 58.01	$ 51.65	$ 39.70	$ 58.56	$ 50.76	$ 41.30
Plumbers	$ 35.36	$ 60.70	$ 57.60	$ 36.84	$ 73.45	$ 71.09	$ 55.39
Roofers	$ 22.45	$ 49.43	$ 45.37	$ 36.42	$ 56.12	$ 40.87	$ 35.78
Sheet Metal Workers	$ 36.61	$ 60.92	$ 57.79	$ 49.36	$ 73.18	$ 62.45	$ 50.87
Teamster Truck Drivers	NA	$ 38.65	$ 49.73	$ 41.56	$ 59.21	$ 44.02	$ 47.25

Rate per hour includes Base Rate + Fringes Benefits NA-Not Available
Source:ENR Construction Economics Department

Exhibit 16-11

Carpenter's union rate.

CARPENTER RATE (until 6/30/08)			
Base			$ 41.71
Union Fringe			
	Welfare Fund	10.00	
	Pension	9.81	
	Annuity	6.60	
	AJREIF	0.70	
	Vacation Fund	6.41	
	Supplemental Fund	0.04	
	Labor-Management Fund	0.20	
	IBC	0.06	
	Supplemental Pension	1.00	
	AIP	0.20	
	Sum Total		$ 35.02
FICA		3.68	
S/FUI		4.76	
Workers Compensation		8.95	
Liability Insurance		4.81	
	Sub Total		$ 22.20
TOTAL			$ 98.93

Note: Overhead and Profit not included

The cost for workers' compensation is critical to the subcontractor's labor rate. This cost is mandated by every U.S. state and changes on a constant basis. Individual subcontractor workers' compensation rates are dependent on the safety record of the subcontractor for a 3-year period. A bad safety record would cause the base rate to increase. A good safety record would cause the rate to decrease. See Chapter 5 and Chapter 10 to obtain additional information on workers' compensation rates.

DEALING WITH PROBLEM SUBCONTRACTORS

Under certain circumstances, some subcontractors may create problems for the CM/GC. This may include:

1. Lack of sufficient trades people to complete their portion of the work
2. Adversely affecting other trades with back charges that may be required (see Exhibit 16-12)

Company Name

Address

City, State, ZIP

Phone Number

To: _____

_____

_____

Date: _____

Project: _____

Project No: _____

Location: _____

Notice No: _____

Exhibit 16-12
Potential back charge notice.

The deficiencies listed below remain uncorrected. If you fail to correct them by _____, they will be corrected by others and you will incur a back charge cost.

Deficiencies fall into the following categories, details are given below.

Defective Work _____   Debris Removal _____   Warranty _____

Incomplete Work _____   Other _____

Damage _____   Other _____

Description

3. Being behind schedule

4. A poor safety record

5. Not keeping their work area clean

6. Not attending subcontractor meetings

7. QC not up to the design documents standards

8. Placing a lien on the project (with no major reason)

9. Not conforming to the agreement with the CM/GC

When any of these factors occur, the CM/GC must act immediately so that the total project will not be adversely affected. A meeting must be set up with the principals of the subcontractor's firm. The CM/GC should explain to them that changes must be made immediately or else a 3-day notice letter will be submitted (see Exhibit 16-13). If the subcontractor has legitimate concerns, then the CM/GC should listen and make the proper corrections. The project would be better served if the subcontractors all work together and complete their work according to the agreed upon schedule and the quality that is required. No one wants to get involved with a dispute that could affect the total project.

Exhibit 16-13
Subcontractor's
3-day notice.

April 28, 2008

To: Subcontractor HVAC
      1 Main Street
      Urban, USA

Re: Urban City Center—Three-Day Notice

As per your subcontractor contract agreement, Section 10.1 dated January 3, 2008, you are to provide sufficient tradespeople to maintain the agreed to Construction Schedule (dated January 3, 2008). As of April 28, 2008, all the ductwork was to be installed and insulated. The attached calculations indicate that only 80% of the ductwork has been completed.

You have until May 1, 2008 to complete the ductwork installation and insulation work. If not completed, Urban Contractors will bring in another subcontractor to complete the work.

Project Manager
Urban Contractors

SUBCONTRACTOR CONCERNS

The Subcontractor Trade Association (STA) has listed several industry practices that adversely affect the subcontractor's ability to be a viable part of the project team. These practices include:

1. Delayed payments—This may require the subcontractors to add financing costs to their bid price.

2. Release of retainer—If the job is bonded, then the need to hold back any payment is questionable. Any retainer should be interest bearing. The retainer should be capped at 5%, at least specifically after 50% of the work is completed (not the 10% as now standard in the industry).

3. Change orders—Payments are being delayed for any change order work. Subcontractor must proceed with the work even if a price has not been agreed upon. This could create a greater risk to the subcontractor.

4. Mobilization—This process can be up to 10% of a subcontractor's cost for start up and fabrication of components. This could put the subcontractor 90 to 120 days behind the payment schedule.

5. Suppliers, manufacturers, and warehousing of material—These parties expect payment within 30 days or else they will not ship the required material. Subcontractors will not be paid for this material until it is installed or otherwise inspected in the warehouse. The payment for these items can only be included in the billing period at the end of the month. All this relates to delay of payment to the subcontractor.

6. Fast tracking—This requires overtime work with expenditure of higher labor costs. Payments are delayed and sufficient time is not given for the adequate review and approval of additional costs.

The bottom line of the aforementioned concerns and proposed changes according to the STA (from the special advertisement supplement to *New York Construction*) is "it would increase the subs working capital, bonding capacity and to meet market needs benefiting subcontractors, contractors, and developers alike."

Contractors should work with subcontractors to find ways to enhance payment schedules and to try to equalize risk for all parties. As noted previously, subcontractors account for approximately 80 to 90% of a project's costs and thus everything should be done to create a harmonious working environment between the CM/GC and the subcontractors.

Sarbanes-Oxley Act

The Sarbanes-Oxley Act is a federal statue that deals with ethics within public corporations. This act was established after the Enron debacle. Most subcontractors are not public corporations, but the owners know that ethics in the construction industry are questionable. Therefore, in order to reinforce ethics within the subcontractor community, the American Subcontractors Association (ASA) approved the Model Code of Ethics for a Construction Subcontractor on September 17, 2006. Some of the items listed in the Model Code of Ethics include:

1. Practice of bid shopping
2. Practice of bid peddling
3. Job site safety
4. Job working conditions
5. Prompt payment to suppliers
6. Image of the industry

The total Model Code of Ethics for a Construction Subcontractor can be downloaded from the ASA Website at www.asaonline.com.

SUMMARY

- Subcontractors are the key to a successful project. Thus, it behooves the CM/GC to use the best-qualified subcontractors available.
- Subcontractors work entails 80 to 90% of the cost of a project.
- Bids should be submitted to four or five subcontractor for each trade. This is to obtain a good representation of bids.
- All bids must be "leveled." This is to make sure that all prices submitted include the full scope of the work as indicated in the construction documents.
- The subcontractors' contract should not only include provisions from the owner/CM/GC contract but also provisions for maintaining quality work, a safe site, proper coordination, and following the agreed upon schedule.
- Change order work must include detail backup information.
- The PM must have weekly subcontractor meetings to maintain a proper working environment for a successful project.
- The Subcontractor Trade Association has many concerns with the construction industry. Most of them relate to the slow pace of payment and to the philosophy determining the requirement for a retainer.
- The American Subcontractors Association has prepared a Model Code of Ethics for a Construction Subcontractor as an answer to the Sarbanes-Oxley Act.
- CM/GC must work with the subcontractors in order to achieve a cohesive working construction team.

17

Costs
(It is always about money.)

COST ESTIMATING

When we start discussing costs, the first item that must be looked at is the cost estimate. What is a cost estimate? It is a best guess of the cost of the project based on available information. The information required includes the percentage of construction documents completed; total project scope; types of mechanical, electrical, plumbing, and sprinkler systems; level of finishes, foundation types; curtain wall design; and cost data from a variety of reliable sources. The more information you have, the closer the estimate will be to the actual bid cost of the project. See Exhibit 17-1 for a cost estimate flowchart.

WHAT ARE COST ESTIMATES USED FOR?

Cost estimates are used for a variety of reasons, some of which are as follows:

1. As an indication of what a project would cost at a particular stage of the project. The owner needs to know what the project will cost so that a budget can be developed.

2. In order to track costs on a continuous basis, the project manager (PM) should review the cost of the project from the initiation of the design process. Thus, a cost estimate should be prepared at conceptual design, schematic design, design development, and 75% of construction documents stage.

3. As a preparation of a pro forma for a development project. A pro forma is used by developers and financial institutions to determine the financial feasibility of a project.

4. Value engineering for a project to determine how costs can be reduced. This may occur after each design phase (schematic design, design development, and construction document phases).

Exhibit 17-1

Cost estimate flowchart.

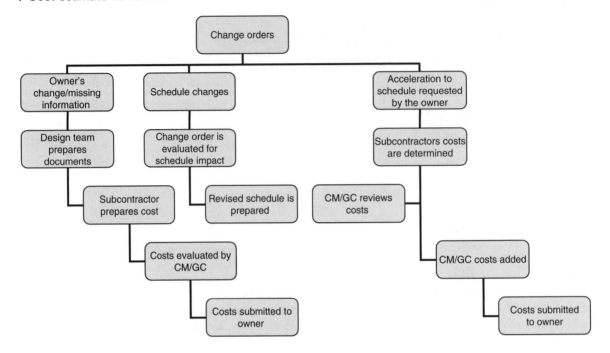

5. To evaluate alternatives for a project. This may include different types of curtain wall systems or other suppliers of mechanical equipment (in addition to the ones specified).

6. To assist in the evaluation of the bids submitted by the subcontractors.

7. As a review of change orders.

8. To evaluate the affect of a delay in the schedule.

9. To determine the cost for performing out of sequence work.

10. To indicate the cash flow of a project so that the owner will know how much money has to be available for payment to the construction manager/general contractor (CM/GC).

TYPES OF COST ESTIMATES

There are several methods used for preparing cost estimates. However, in most cases a combination of the estimate techniques discussed in the following will be used.

Cost Per Square Foot

In this method, information from previous jobs is used to determine the cost of a new project. The construction cost of a completed project is taken and then divided by the square footage of the project in question. This figure is then used to determine the cost of the new project. This method is usually used for conceptual and schematic design phases. When this method is used, the parameters for the existing project versus the new project must be consistent:

- How will the square footage be determined—using outside dimensions, using inside dimensions or some other measuring method, like the BOMA (Building Owners Managers Association) standard?

- Are the two projects similar in nature—office buildings versus residential buildings, industrial versus laboratories, retail versus medical?

- Is the location the same? If not, then the costs may need to be adjusted. This can be accomplished by using *Engineering News-Record* cost location modifiers (see the ENR Website: www.enr.com).

- Will the base structural systems be similar—concrete, steel, wood, block, or a combination?

- Is the quality of products used similar? If the existing building used glass for the curtain wall, will the new building also use a glass curtain wall?

- Will the systems be similar? If piles were used in the existing building, will this type of foundation be used in the new building?

- What costs have been included in the base costs? Is furniture included? Is overhead and profit included? Is sales tax included?

- Has an inflation factor been added to the cost of the existing building? This is needed because the existing building was completed by a definitive date and the new building may not be completed until 2 or 3 years beyond that point.

Take Off

In this method, drawings are available so that the various components of the drawings can be "counted" and a cost assigned to each element.

- The cost assigned for each element is determined by historical data kept by the CM/GC's company.

- It is also prudent to check with subcontractors to confirm the costs of the elements being priced.

- Accuracy of the estimate is dependent on the details shown in the drawings and associated descriptions.

- The unit prices must be consistent with the existing environment in which you are working.

Prices from Subcontractors

In this method, drawings are sent to subcontractors for pricing. When using this technique, a direct source of information is obtained. This is usually one of the best methods for developing cost estimates because the subcontractors know their costs for materials and for installing the product. However, the subcontractors may not have the time to give the CM/GC complete and adequate cost information. Based on their historical data, the subcontractors are able to include costs for items that may have been missed by either the cost per square foot method or the take off method.

Labor and Material Take Off

In this method, you estimate the amount of labor and material for the project. This type of cost estimating method is usually relegated to change order work.

- The person using this method must have a good understanding of the productivity rates for the various trades.
- The labor rates for the locale being worked in must be known and constantly updated dependent upon expiration dates of labor contracts and increases given for non-union labor.
- Material prices must be known and the impact of outside forces (such as a large amount of construction taking place in China, which is using vast amounts of world-wide material) must be understood.

Component

This method is similar to the cost per square foot technique, except the major components (foundations, structural systems, roofing, curtain wall, HVAC, electrical, plumbing, sprinkler, and architectural) of an existing project are used to build the cost for the new project. The cost would then be turned into a cost per square foot method.

- This technique would be used when a good understanding of the new project has been established.
- The major components of the project include the following:
    1. Foundation (i.e., spread footings, piles, mat foundation, and caisson)
    2. Structure (i.e., steel, concrete, wood)
    3. Envelope (i.e., glass curtain wall, stone façade, single-ply roof system, built-up roof)
    4. HVAC (i.e., tons for chillers, tons for cooling tower, VAV systems)
    5. Electrical (i.e., kilowatts of electricity anticipated)
    6. Plumbing
    7. Sprinkler
    8. Interior work
- The cautions listed with the cost per square foot method must also be used for this technique.
- This method can be used to check the validity of costs developed with other cost estimating techniques.

HARD AND SOFT COSTS

When costs are mentioned, most people think of the basic cost of the project only. These are the costs associated with the "bricks and mortar" of the project. However, an owner is concerned with two costs for the project—hard costs and soft costs. The hard costs are costs associated with the CM/GCs costs (general conditions, all subcontractors' costs, overhead and profit, insurance and bond costs, and applicable permit costs) to construct the project. Exhibit 17-2 is a list of items associated with the hard

CSI	ITEM	COST	$/S.F.
	Site Preparation		
	Site Utilities		
	Site Improvements		
	Roads		
	Traffic Signals		
	Retention Pond		
	Security Gate		
	Landscaping		
	Paving		
	Site Lighting		
	Foundation		
	Structure		
	Façade		
	Roof		
	Security System		
	Electrical		
	Lighting		
	Date Cabling		
	Cable T.V.		
	Telephone Switch and Racks		
	HVAC		
	BMS		
	Plumbing		
	Toilet Accessories		
	Sprinkler		
	Partitions		
	Acoustical Ceiling		
	Doors		
	Hardware		
	Painting		
	Carpet		
	Wall Covering		
	Stone		
	Tile		
	Millwork		
	Window Treatment		
	Elevator		
	Life Safety		
	Signing		
	Fence		
	Dock Accessories		
	SUBTOTAL		
	Contingency (10%)		
	SUBTOTAL		
	Contractors General Conditions		
	SUBTOTAL		
	Contractors Insurance and Bond		
	Contractor's Fee		
	TOTAL		

Exhibit 17-2

Hard costs.

Exhibit 17-3

Soft cost.

Item	Estimated Cost	Actual Cost
Land Cost		
Title Insurance		
Legal Fees		
Accountant		
Broker's Fee		
Financing Fees		
Insurance		
Real Estate Taxes		
Surveyor		
Phase I		
Phase II		
Zoning Fees		
Architect		
Civil Engineer		
Mechanical Engineer		
Landscaping Architect		
Traffic Engineer		
Environmental Consultant		
Soils Engineer		
Site Borings		
Site Fence		
Project Manager		
Permits		
Water Hook-up Charges		
Electrical Utility		
Sanitary Hook-up Charges		
Gas District		
Temporary Utility Charges		
Models and Renderings		
Office Expenses		
Travel and Entertainment		
Signs		
Testing and Inspections		
Promotional Events		
Security		
Cable Installation		
Photographs		
Management Fees		
Miscellaneous		
SUBTOTAL		
Contingency		
TOTAL		

costs for the project and Exhibit 17-3 is a list of items associated with the soft costs for a project.

When discussing costs with an owner, the definition of the costs required must be enumerated. As noted by the breakdown, the hard costs can be provided by the CM/GC but the soft costs may have to be provided by the owner. Soft costs are defined as the costs usually incurred by the owner on a construction project such as the professional services of architects, engineers, specialty consultants, interest charges, utility charges, insurances, and owners representative.

OBTAINING COST DATA

Numerous sources are available for obtaining costs for the preparation of a cost estimate. The following are some of the sources that should be considered:

1. Database from previously completed projects. This information must be constantly updated and put into a spreadsheet that indicates the basic parameters of the project.
2. Subcontractors
3. Suppliers
4. On-line services (i.e., www.get-a-quote.net)
5. Cost consultants
6. ENR estimating services. In each issue, information is provided by city for the costs for various products.
7. Other technical magazines
8. Means cost guides
9. Marshall and Swift estimator
10. Newspaper articles (be careful when using this information)

TIMING

The length of time for a project to be completed has a profound affect on the cost of the project. Obviously the greater the length of the project the higher the cost will be. The costs associated with timing can be defined as follows:

1. General conditions. These are the direct costs to the CM/GC, which consist of payroll, debris removal, office expenses, blue prints, and permits. Exhibit 17-4 is a breakdown of typical general condition items for a project.
2. Inflation considerations. For a 2-year project, for example, inflationary factors would have to be considered, such as availability of material and associated price increases and labor contract negotiations. Commodity prices are continuing to increase due to world demand for various materials. The price of oil affects every component of the construction industry. Thus, the price of oil has to be continuously watched when preparing costs for a project.
3. Time of year the project is going to be constructed. During the winter, heat has to be provided for labor and some material installations (such as concrete), which will add costs to the project. The rainy season and hurricane season could delay the schedule, which would also affect the costs.
4. Holidays such as Christmas and New Year's can affect the productivity of the workers with time being taken off from the project.
5. In certain cities, moratoriums are imposed during peak street travel times, such as holiday seasons, parades and street fairs, and when VIPs come into the city.
6. Municipal approved special events (i.e. parades, street fairs), that may occur near or around the building being constructed may impose restrictions on the site.

Exhibit 17-4

General conditions.

Description	Monthly Rate	% of Time	Total $	Comments
Account Executive				
Project Manager				
Assistant Project Manager				
General Field Superintendent				
Field Superintendent				
Assistant Superintendent				
Administrative Assistant				
Project Estimator				
Scheduler				
Hoist Engineer				
Crane Engineer				
Operations				
Purchasing				
MEPS				
Field Office Clerk				
Safety Director				
Safety Manager				
Accountant				
Administration (insurance)				
Surveyor				
Legal				
Temporary Probes or Borings				
Labor Regular Time				
Labor Overtime				
Labor Foreman				
Materials Handling Driver				
Containers				
Final Clean Up				
Permits				
Equipment/Misc. Tools				
Travel				
Blueprints				
Field Office Construction				
Field Office Furnature				
Temporary Field Office Utilities				
Computers/Network/Programs				
FAX Machine				
Copier				
Telephone				
Petty Cash				
Office Supplies				
Postage				
Cleaning Supplies				
Messenger				
Fed Ex				
First Aid Kits and Fire Extin.				
Two-Way Radios				
Cell Phones				
Temporary Toilets				
Job Photographs				
Security Service				
Temporary Signage				
Insurance				
Builders All Risk				
Bonds				
Special Union Requirements				
Hoist Rentals				
Crane Rentals				
Other Rentals				

MINIMIZING UNKNOWN FACTORS

When preparing a cost estimate, many unknown factors must be anticipated. This is more so when working in an urban environment where logistics alone takes a huge bite of the costs. In order for the cost estimate to be valid, the factors enumerated in Exhibit 17-5 must be reviewed.

1. Incomplete drawings. Missing information has to be reviewed and costs for allowances will have to be added to the estimate. 2. Site conditions. If soil borings are not available, then it may be possible to go to the local geotechnical engineer and obtain some local information. The U.S. Coast and Geodetic Survey Organization is another source for obtaining site condition information. 3. Logistics. A preliminary logistics plan will have to be prepared showing how all the material and labor can access the project. Location of hoists and cranes will have to be noted on the plan. See Chapter 6. 4. Constructability of the components. How some of the components will be put together will have to be analyzed by experienced construction PMs. 5. Lead time of products. Suppliers will have to be contacted to determine when critical components can be shipped to the site. This will have to be evaluated in terms of schedule and possible out-of-sequence work. 6. Availability of trades people. Depending on the amount of work going on in the area, the availability of sufficient labor will have to be determined. If not, then costs may have to be added for overtime work. 7. Shoring and possible dewatering. The site information will hopefully give some information on the stability of the site's excavation and the location of the groundwater table. This cost could be substantial and must be reviewed with the foundation costs. See Chapter 6. 8. Safety. Safety is a big factor in the construction industry. Costs for site safety, nets, bridges, and protection of shaft openings have to be considered. 9. Utilities. The location of the utilities around the site must be analyzed. In addition, contacting the various utility companies must be done in an expeditious manner. The cost imposed by the utility companies must be included. 10. Underpinning. The adjacent buildings have to be reviewed to determine the possible impact on the proposed building. See Chapter 6. 11. Insurance requirements. What will be the cost of the various insurance requirements established by the owner? This would be especially true for the builders all risk policy requirement. 12. Bonds. What type of bonds will be required and what will the cost be? 13. Permits. How many permits must the CM/GC obtain and what will the cost be? 14. Sheds, scaffolds, and bridges. The need for these temporary structures must be reviewed in the context of the logistics plan. If required, the cost associated with these temporary structures must be added to the cost estimate. See Chapter 6.	**Exhibit 17-5** Unknown factors to consider when preparing a cost estimate.

(Continued)

Exhibit 17-5

(Continued)

15. Sales tax. Has the proper sales tax been added to the project as required by the local municipality or state financial department?
16. Accelerated schedule. If an accelerated schedule is being considered, then the costs for overtime and fast tracking material must be added.
17. TCO (Temporary Certificate of Occupancy) or CO (Certificate of Occupancy). In certain instances, costs associated with obtaining the TCO or the CO must be considered. The local municipality may require filling fees and certain stipulations imposed on the subcontractors.
18. Testing. The cost for testing has to be added to the estimate, especially if the requirement is mandated that the CM/GC perform all the tests by using an outside agency.

GETTING THE PROPER INFORMATION

As stated previously, the more information obtained, the better the quality of the cost estimate. If the information does not give all the answers, then methods that are more aggressive must be considered. The following are some additional ways to obtain "better" information:

1. Request that the design team prepare additional sketches and outline specifications.
2. Have design scope meetings with the design team.
3. The design team should prepare 3-D models.
4. Visit the site numerous times to get a good feel of what is required.
5. Have meetings with the owner's representative and other project participants. Review some of the assumptions that are being made for the project.
6. Have experienced PMs review the available information and solicit their opinions.
7. Look for construction articles on similar types of projects.
8. Consider performing probes of the soil to obtain a better understanding of the soil conditions (if site borings have not been accomplished).
9. Obtain geotechnical maps of the area (or talk to geotechnical engineers who have performed work in the area being considered).
10. Discuss the project with manufacturers and suppliers to determine the best and least expensive way to achieve the design team's objectives.

CONTINGENCY

Every cost estimate needs a contingency. A contingency is to take care of all the unknown factors that tend to creep into a project as the design is being fully developed. Usually two types of contingencies are added to a project. The first one is the design contingency. This should take care of some of the unknowns as the costs are being developed for the various design phases. The design contingency starts out high for the conceptual design phase and should be close to zero when the costs for the 75% construction document stage are being developed. The design contingency should start out at no less then 15%.

The other contingency is for construction change orders. Because of the nature of construction and the number of parties involved in the design process, change orders are inevitable. The change orders may include scope changes (i.e., for conflicts of existing documents) or out-of-scope changes (i.e., adding more space to the project); thus, depending on the complexity of the project a change order contingency of between 3 and 10% should be considered. In Chapter 9, contingency is reviewed in the context of a GMP (guaranteed maximum price) contract.

Contingency should also be considered for the soft cost portion of the project. All contingencies should be tracked against what is actually being spent on the project.

QUALIFICATIONS

Qualifications are a series of statements that define what is included and what is excluded from the cost estimate. The qualifications will also state the general scope of the project and some of the basic information used to develop the cost estimate.

1. Drawings and sketches
2. Outline specifications
3. Start and end dates for construction
4. Long lead items
5. Special components (i.e., marble floors)
6. Exclusions (i.e., no overtime)
7. Any assumptions used

PREPARATION OF THE COST ESTIMATE

There are several ways to prepare cost estimates. Prior to the latest technology transformation, a cost estimator prepared all the estimates by hand and then used an adding machine to come up with the total estimate for the project. Now there are more expeditious ways to prepare the costs.

Spreadsheet using Microsoft Excel

- Makes placing all the information into one format very simple.
- Can make changes to one part of the estimate and will automatically recalculate new totals.

- Can easily evaluate value engineering changes.
- Can review different scenarios for options.

Special Programs for Estimating

Some of the software has a digital capability where a "wand" can read the length of material shown on a drawing. The information obtained by the wand can be converted to a cost (based on information given to the program in regard to productivity).

CAD (Computer Aided Design) Programs

- Can be used to accurately measure the various components of a project.
- With new 3-D capability programs, the estimator can "see" how intricate details come together. This will assist the estimator in defining what cost should be allocated to the detail.

CSI (Construction Specification Institute) Format

- Gives the estimator a numbering system to follow for all elements of a construction project.
- Should be used when developing the cost estimate.
- The full CSI format is shown in Exhibit 17-6.
- Division 01-General Requirements is detailed in Exhibit 17-7.
- Division 21-Fire Suppression is detailed in Exhibit 17-8.
- Division 23-HVAC is detailed in Exhibit 17-9.
- Acts as a checklist to make sure all aspects of the project have been allocated for cost.
- A typical cost breakdown based on the CSI format is shown in Exhibit 17-10.
- A typical take off sheet for HVAC work is noted in Exhibit 17-11.
- For more information on CSI, visit the Website www.csinet.org.

General

- After the estimate is completed, a check of the numbers should be made. One method would be to determine the percentage of cost for the various trades and compare it against similar projects.
- Another person familiar with the cost estimating process should review the costs and the assumptions made.
- Make sure that correct units have been used (square feet of roofing, square yards of carpet, cubic yards of concrete, etc.).
- Spot-check the math even with computerized spreadsheets.

Division Numbers and Titles

Exhibit 17-6

Construction
specification
institute–full
format.

PROCUREMENT AND CONTRACTING REQUIREMENTS GROUP
Division 00   Procurement and Contracting Requirements

SPECIFICATIONS GROUP

GENERAL REQUIREMENTS SUBGROUP
Division 01   General Requirements

FACILITY CONSTRUCTION SUBGROUP
Division 02   Existing Conditions
Division 03   Concrete
Division 04   Masonry
Division 05   Metals
Division 06   Wood, Plastics, and Composites
Division 07   Thermal and Moisture Protection
Division 08   Openings
Division 09   Finishes
Division 10   Specialties
Division 11   Equipment
Division 12   Furnishings
Division 13   Special Construction
Division 14   Conveying Equipment
Division 15   Reserved
Division 16   Reserved
Division 17   Reserved
Division 18   Reserved
Division 19   Reserved

FACILITY SERVICES SUBGROUP
Division 20   Reserved
Division 21   Fire Suppression
Division 22   Plumbing
Division 23   Heating, Ventilating, and Air Conditioning
Division 24   Reserved
Division 25   Integrated Automation
Division 26   Electrical
Division 27   Communications
Division 28   Electronic Safety and Security
Division 29   Reserved

SITE AND INFRASTRUCTURE SUBGROUP
Division 30   Reserved
Division 31   Earthwork

(Continued)

Exhibit 17-6
(*Continued*)

Division 32	**Exterior improvements**
Division 33	**Utilities**
Division 34	**Transportation**
Division 35	**Waterway and Marine Construction**
Division 36	**Reserved**
Division 37	**Reserved**
Division 38	**Reserved**
Division 39	**Reserved**

PROCESS EQUIPMENT SUBGROUP

Division 40	**Process Integration**
Division 41	**Material Processing and Handling Equipment**
Division 42	**Process Heating, Cooling, and Drying Equipment**
Division 43	**Process Gas and Liquid Handling, Purification, and Storage Equipment**
Division 44	**Pollution Control Equipment**
Division 45	**Industry-Specific Manufacturing Equipment**
Division 46	**Reserved**
Division 47	**Reserved**
Division 48	**Electrical Power Generation**
Division 49	**Reserved**

Exhibit 17-7
Division 01 general requirements.

01 00 00	**GENERAL REQUIREMENTS**
01 10 00	**SUMMARY**
01 11 00	**Summary of Work**
01 12 00	**Multiple Contract Summary**
01 14 00	**Work Restrictions**
01 18 00	**Project Utility Sources**
01 20 00	**PRICE AND PAYMENT PROCEDURES**
01 21 00	**Allowances**
01 22 00	**Unit Prices**
01 23 00	**Alternates**
01 24 00	**Value Analysis**
01 25 00	**Substitution Procedures**
01 26 00	**Contract Modification Procedures**
00 94 00	**Record Modifications**

21 00 00	FIRE SUPPRESSION	
21 01 00	Operation and Maintenance of Fire Suppression	
21 05 00	Common Work Results for Fire Suppression	
21 06 00	Schedules for Fire Suppression	
21 07 00	Fire Suppression Systems Insulation	
21 08 00	Commissioning of Fire Suppression	
21 09 00	Instrumentation and Control for Fire-Suppression Systems	
21 10 00	WATER-BASED FIRE-SUPPRESSION SYSTEMS	
21 11 00	Facility Fire-Suppression Water-Service Piping	
21 12 00	Fire-Suppression Standpipes	
21 13 00	Fire-Suppression Sprinkler Systems	
21 20 00	FIRE-EXTINGUISHING SYSTEMS	
21 21 00	Carbon-Dioxide Fire-Extinguishing Systems	
21 22 00	Clean-Agent Fire-Extinguishing Systems	
21 23 00	Wet-Chemical Fire-Extinguishing Systems	
21 24 00	Dry-Chemical Fire-Extinguishing Systems	
21 30 00	FIRE PUMPS	
21 31 00	Centrifugal Fire Pumps	
21 32 00	Vertical-Turbine Fire Pumps	
21 33 00	Positive-Displacement Fire Pumps	
21 40 00	FIRE-SUPPRESSION WATER STORAGE	
21 41 00	Storage Tanks for Fire-Suppression Water	

Exhibit 17-8

Division 21 fire supression.

23 00 00	HEATING, VENTILATING, AND AIR-CONDITIONING (HVAC)	
23 01 00	Operation and Maintenance of HVAC Systems	
23 05 00	Common Work Results for HVAC	
23 06 00	Schedules for HVAC	
23 07 00	HVAC Insulation	
23 08 00	Commissioning of HVAC	
23 10 00	FACILITY FUEL SYSTEMS	
23 11 00	Facility Fuel Piping	
23 12 00	Facility Fuel Pumps	
23 13 00	Facility Fuel-Storage Tanks	

Exhibit 17-9

Division 23-HVAC.

(Continued)

23 80 00	**DECENTRALIZED HVAC EQUIPMENT**
23 81 00	**Decentralized Unitary HVAC Equipment**
23 82 00	**Convection Heating and Cooling Units**
23 83 00	**Radiant Heating Units**
23 84 00	**Humidity Control Equipment**

Exhibit 17-9
(Continued)

BUDGET PROCESS

Once an estimate has been prepared, it must be presented to the owner and the design team. If the estimate is accepted, then the design process can continue. If the estimate is rejected, then further review of the costs will need to take place by the design team and the CM/GC. Value engineering, as discussed in the next section, may have to be initiated. If the estimate is much greater then the owner's budget, then a change of scope may have to be evaluated. However, the downside of this approach is that inflationary factors may be working against the cost. It behooves the design team and the CM/GC to find a solution to the cost problem immediately.

CSI Number	Item	Value	Cost/S.F.	Bid Value	Difference	Comments
02000	Sitework					
02150	Landscaping					
02360	Pile Foundations					
03300	Concrete					
04200	Brick/Masonry					
05100	Structural Steel					
05500	Misc. Metals					
06100	Rough Carpentry					
06400	Architectural Millwork					
07100	Thermal and Moisture Protection					
07250	Waterproofing					
07500	Roofing/Cladding					
07900	Caulking/Joint Sealer					
08100	Doors/Frames and Hardware					
08300	Entrances and Storefront					
08800	Glass/Glazing					
09250	Gypsum Drywall					
09300	Ceramic Tile					
09500	Acoustic Ceilings					
09680	Carpet					
09900	Painting					
10000	Specialities					
11000	Equipment					
14000	Elevators (Conveying Systems)					
21000	Fire Protection					
22000	Plumbing					
23000	HVAC					
26000	Electrical/Fire Alarm					
	Subtotal					
	General Conditions					
	Subtotal					
	Insurance					
	Subtotal					
	Bond					
	Subtotal					
	Fee					
	TOTAL					

Exhibit 17-10

Cost breakdown based on CSI format.

Exhibit 17-11

HVAC CSI detailed break down.

CSI	Description	Cost	Comments
	Equipment		
233616	VAV Boxes	$ 24,500	
236200	Air Conditioning Units	$ 211,750	
238233	Finned Tube Radiation	$ 21,500	
232129	Condensate Pumps	$ 1,500	
235700	Heat Exchanger	$ 71,500	
232129	Condensate Return Unit	$ 47,000	
233400	Fans	$ 28,000	
230548	Vibration Isolaters	$ 9,000	
	Rigging	$ 6,500	
230593	Balancing	$ 1,500	
	Duct Work and Insulation		
233113	Duct Work	$ 943,250	
230713	Duct Work Insulation	$ 55,000	
233113	Kitchen Exhaust Duct Work	$ 40,000	
233713	Outlets	$ 15,000	
233313	Fire/Smoke Dampers	$ 50,000	
233713	Louvers	$ 60,000	
230800	Commissioning	$ 20,000	
230800	Air Flush Out	$ 15,000	
	Piping and Pipe Insulation		
232100	Piping Work	$ 1,400,000	
220700	Pipe Insulation	$ 110,000	
220800	Commissioning	$ 15,000	
232513	Chemical Cleaning	$ 7,000	
230923	**BMS and Control Wiring**		
230933	Control Equipment	$ 100,000	
230933	Control Wiring	$ 150,000	
230993	Commissioning	$ 9,000	
	Allowances	$ 120,000	
	TOTAL	$ 3,532,000	

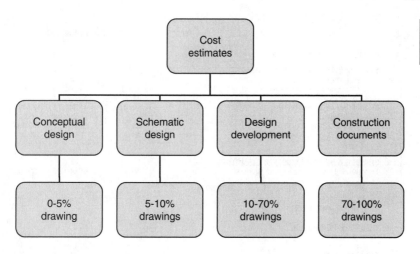

Exhibit 17-12
Cost estimate
phases.

If the estimate is in line with the owner's budget, then the design team proceeds to the next design phase. As indicated in Exhibit 17-12, a revised cost estimate must be prepared after each design phase of the project (schematic design, design development, and construction documents) to make sure the project is on the proper cost track. As more definitive design information is submitted, the cost estimates will be more accurate.

The true test of the validity of the estimate will come once the project is submitted to the subcontractors (see Chapter 16) for bidding. An 80% cost estimate is indicated in Exhibit 17-13. Note that where design of an element has not been clearly defined, an allowance has been indicated. An allowance is an estimate based on the available information prepared by the design team at the time of the estimate.

VALUE ENGINEERING

Value engineering is a method in which the project team reviews the available drawings and specifications and then makes suggestions as to where costs can be reduced without changing the scope of the project (i.e., reducing square footage). The initial intent is to review the finishes and systems for the project. By preceding in this way the project intent does not change and only subtle finish changes and small system changes are introduced.

When major costs need to be trimmed from a project, the only way to achieve this is by reducing the scope of the project. This could entail reducing the square footage of a project. However, this drastic approach may have a major impact on the total viability of the project.

Suggested ways to determine value engineering items are indicated in Exhibit 17-14.

Exhibit 17-13

70 Percent design development estimate.

CSI	DESCRIPTION	QUANTITY	UNIT	UNIT PRICE	AMOUNT ($)	TOTAL
	GENERAL REQUIREMENTS					
01700	**TEMP. SIDEWALK BRIDGE, FENCING, SCAFFOLDING**					
	Local 14 Operating Engineer	65	wks	$ 1,500	$ 97,500	
	Hoisting Allowance	1	Allow	$ 50,000	$ 50,000	
	Sidewalk Bridge & Fence - Rent	10	months	$ 17,500	$ 175,000	
	Sidewalk Bridge Lighting	1	LS	$ 9,000	$ 9,000	
	Scaffolding & Netting - South Side	1	LS	$ 50,000	$ 50,000	
	Visual/Sound Control - South Side	1	Allow	$ 175,000	$ 175,000	
					$ 556,500	
01710	**SHORING & BRACING EXISTING STRUCTURE**					
	Temp shoring and bracing existing structure	1	Allow	$ 50,000	$ 50,000	
	SUBTOTAL					$ 606,500
	SITE CONSTRUCTION					
02112	SELECTIVE DEMOLITION					
	General Demolition	1	Allow	$ 400,000	$ 400,000	
	Probes	1	Allow	$ 20,000	$ 20,000	
	Misc. Demolition	1	Allow	$ 50,000	$ 50,000	
					$ 470,000	
02200	EXCAVATIONS					
	Excavations - Elevator Pits, Service Door, Basement Window, Trenching, Gate Footings	1	Allow	$ 175,000	$ 175,000	
02900	PLANTINGS					
	Landscaping - 4th Floor Terrace	1	Allow	$ 20,000	$ 20,000	
	SUBTOTAL					$ 665,000
	CUTTING AND PATCHING					
02070	CUTTING & PATCHING					
	Allowance for Ext and Int.	1	Allow	$ 250,000	$ 250,000	
	SUBTOTAL					$ 250,000
	CONCRETE					
03300	CONCRETE					
	Foundation & Slab Infill	145	CY	$ 680	$ 98,600	
	Fill on metal deck	1	Allow	$ 11,000	$ 11,000	
	Waterproofing & Insulation	1	Allow	$ 15,000	$ 15,000	
	SUBTOTAL					$ 124,600
	MASONRY					
04050	MASONRY & STUCCO					
	New Masonry at Bulkheads, Interior Walls	4778	SF	$ 45	$ 215,010	
	Stucco on New Masonry	1974	SF	$ 12	$ 23,688	
					$ 238,698	
	SUBTOTAL					$ 477,396
	MASONRY RESTORATION					
04500	MASONRY RESTORATION & CLEANING					
	Exterior - Clean, Repoint, replace stucco, etc.	1	Allow	$ 500,000	$ 500,000	
	SUBTOTAL					$ 500,000
	STEEL					
05100	STRUCTURAL STEEL					
	Structural Steel	24	Tons	$ 9,000	$ 216,000	
	Metal Deck	11	Tons	$ 9,000	$ 99,000	
	Connections - 10% of above	2.4	Tons	$ 9,000	$ 21,600	
					$ 336,600	
05700	MISCELLANEOUS METALS					
	4th Floor Metal & Glass Unit, radiator grills, etc.	1	Allow	$ 125,000	$ 125,000	
	SUBTOTAL					$ 461,600

Exhibit 17-13
(*Continued*)

CSI	DESCRIPTION	QUANTITY	UNIT	UNIT PRICE	AMOUNT ($)	TOTAL
	GENERAL REQUIREMENTS					
	WOOD & PLASTIC					
06400	ARCHITECTURAL WOODWORK					
	Cabinetry - Fabricate & Install	1045	LF	$ 740	$ 773,300	
	Finish Carpentry	1560	LF	$ 14	$ 21,060	
	SUBTOTAL					$ 794,360
	THERMAL & MOISTURE PROTECTION					
07511	ROOFING					
	Roofing	36,400	SF	$ 15	$ 527,800	
	Temporary Roofing	1	Allow	$ 25,000	$ 25,000	
	Roofing Details	1	Allow	$ 50,000	$ 50,000	
					$ 602,800	
07811	SPRAY FIREPROOFING					
	Spray Fireproofing	1	LS	$ 30,000	$ 30,000	
	Patching allowance	1	Allow	$ 10,000	$ 10,000	
					$ 40,000	
07840	FIRESTOPPING					
	Firestops	1	Allow	$ 10,000	$ 10,000	
07900	JOINT SEALERS					
	Misc. Caulking	1	Allow	$ 10,000	$ 10,000	
	SUBTOTAL					$ 662,800
	DOORS & WINDOWS					
08110	HM DOORS FRAMES & HARDWARE					
	HM Double doors and frames	65	Each	$ 850	$ 55,250	
	Hardware	65	EA	$ 500	$ 32,500	
	Allowance for final hardware adjustment, etc.	1	Allow	$ 5,000	$ 5,000	
					$ 92,750	
08310	ACCESS DOORS & PANELS					
	Access panels in finish work for MEP access	1	Allow	$ 12,000	$ 12,000	
08500	REPLACEMENT WINDOWS					
	Wood Replacement Windows	1	Allow	$ 440,000	$ 440,000	
	Allowance for final adjustment, etc.	1	Allow	$ 15,000	$ 15,000	
					$ 455,000	
	SUBTOTAL					$ 559,750
	FINISHES					
09260	CEILINGS					
	Suspended GWB ceilings	21,000	SF	$ 13	$ 293,370	
	Suspended acoustic ceiling	15,400	SF	$ 13	$ 211,750	
					$ 505,120	
09260	INTERIOR GWB PARTITIONS					
	Cellar	5632	SF	$ 13	$ 77,440	
	1st Floor	3644	SF	$ 13	$ 44,550	
	2nd Floor	4017	SF	$ 13	$ 50,213	
	3rd Floor	4849	SF	$ 13	$ 60,613	
	4th Floor	2860	SF	$ 13	$ 35,750	
					$ 269,565	
09260	SHAFT LINERS					
	Cellar	616	SF	$ 15	$ 9,961	
	1st Floor	588	SF	$ 15	$ 9,508	
	2nd Floor	728	SF	$ 15	$ 11,772	
	3rd Floor	728	SF	$ 15	$ 11,772	
					$ 43,012	
09300	CERAMIC & STONE TILE					
	Cellar	7690	SF	$ 18	$ 139,574	
	1st Floor	11225	SF	$ 18	$ 203,734	
	2nd Floor	5775	SF	$ 18	$ 104,816	
	3rd Floor	5450	SF	$ 18	$ 98,918	
	4th Floor	9625	SF	$ 18	$ 174,694	
	Stone Desk & Counter Tops	1	Allow	$ 115,000	$ 115,000	
	Patching allowance - ceramic and stone tile	1	Allow	$ 10,000	$ 10,000	
					$ 846,735	

(*Continued*)

Exhibit 17-13
(*Continued*)

CSI	DESCRIPTION	QUANTITY	UNIT	UNIT PRICE	AMOUNT ($)	TOTAL
	GENERAL REQUIREMENTS					
09681	CARPET & RESILIENT FLOORING					
	Carpet, Resilient Floooring	1	Allow	$ 200,000	$ 200,000	
	Dex-o-Tex flooring in Mech rooms	1	Allow	$ 25,000	$ 25,000	
	Floor leveling allowance	1	Allow	$ 100,000	$ 100,000	
					$ 325,000	
09910	PAINTING AND WALLCOVERING					
	Painting and Wall Covering	1	Allow	$ 215,000	$ 215,000	
	Allowance for Touch-up and Refinishing	1	Allow	$ 25,000	$ 25,000	
	Allowance for Exterior Painting	1	Allow	$ 40,000	$ 40,000	
					$ 280,000	
	SUBTOTAL					$ 2,269,432
	SPECIALTIES & EQUIPMENT					
10150	TOILET PARTITIONS					
	Standard - Powder coated	22	EA	$ 675	$ 14,850	
10425	SIGNAGE					
	Interior Signage - Code related	1	Allow	$ 25,000	$ 25,000	
10505	METAL LOCKERS					
	Lockers - Staff	31	EA	$ 600	$ 18,600	
10523	FIRE EXTENGUISHERS					
	Allow	30	EA	$ 60	$ 1,800	
10651	OPERABLE PARTITIONS					
	Folding Partitions	1	Allow	$ 200,000	$ 200,000	
10520	WINDOW TREATMENTS					
	Electric shades	1	Allow	$ 60,000	$ 60,000	
	Drapery Tracks	1	Allow	$ 14,000	$ 14,000	
					$ 74,000	
10810	TOILET ACCESSORIES (furnish & deliver)					
	Framed Mirrors	1	Allow	$ 8,000	$ 8,000	
	Toilet Accessories	1	Allow	$ 6,500	$ 6,500	
					$ 14,500	
	SUBTOTAL					$ 348,750
	FOOD SERVICE EQUIPMENT					
10410	FOOD SERVICE EQUIPMENT					
	Food Service Equipment	1	Allow	$ 1,800,000	$ 1,800,000	
	SUBTOTAL					$ 1,800,000
	CONVEYING SYSTEMS					
14000	ELEVATORS					
	Two Passenger Elevators	2	Each	$ 267,500	$ 535,000	
	Passenger cab interior	1	Allow	$ 40,000	$ 40,000	
	Refurbish Freight Elevtor	1	Allow	$ 185,000	$ 185,000	
	Freight cab interior	1	Allow	$ 20,000	$ 20,000	
	Maintenance & Repairs during construction	1	Allow	$ 30,000	$ 30,000	
	Protective Mats - 3 elevator cabs	3	Each	$ 500	$ 1,500	
	SUBTOTAL					$ 811,500
	FIRE PROTECTION					
21000	FIRE PROTECTION					
	Riser Piping	1	Allow	$ 46,450	$ 46,450	
	Concealed type sprinkler heads include branch piping	1	Allow	$ 175,900	$ 175,900	
	Upright/pendant sprinkler heads include branch piping	1	Allow	$ 37,000	$ 37,000	
	Sidewall Type springler heads include branch piping	1	Allow	$ 1,000	$ 1,000	
	Dry sprinkler heads include branch piping	1	Allow	$ 8,000	$ 8,000	
	Roof manifold	1	Allow	$ 5,550	$ 5,550	
	Floor Control Assembly Valves	1	Allow	$ 18,800	$ 18,800	
	Furnish and install sprinkler booster pump and jockey pump	1	Allow	$ 56,500	$ 56,500	
	Allowance for additional drafting/field coordination issues	1	Allow	$ 30,000	$ 30,000	
	SUBTOTAL					$ 379,200

CSI	DESCRIPTION	QUANTITY	UNIT	UNIT PRICE	AMOUNT ($)	TOTAL	
	GENERAL REQUIREMENTS						**Exhibit 17-13**
	PLUMBING						(*Continued*)
22000	PLUMBING						
	Underground Piping Work	1	Allow	$ 165,000	$ 165,000		
	Domestic Water System	1	Allow	$ 77,000	$ 77,000		
	Storm & Clear Water Syster	1	Allow	$ 48,000	$ 48,000		
	Sanitary System	1	Allow	$ 85,000	$ 85,000		
	Gas Piping	1	Allow	$ 40,000	$ 40,000		
	Core Drilling	1	Allow	$ 32,000	$ 32,000		
	Equipment: pumps, water heaters, sump pumps, etc.	1	Allow	$ 220,000	$ 220,000		
	Furnish Plumbing Fixtures Package	1	Allow	$ 31,000	$ 31,000		
	Install Plumbing Fixtures Package	1	Allow	$ 15,000	$ 15,000		
	Appliance Hook-ups	1	Allow	$ 20,000	$ 20,000		
	Kitchen Work	1	Allow	$ 48,000	$ 48,000		
	Install Toilet Accessories	1	Allow	$ 2,000	$ 2,000		
	Pipe Insulation	1	Allow	$ 31,000	$ 31,000		
	Temporary Toilets and Water	1	Allow	$ 3,000	$ 3,000		
	Street Service Work	1	Allow	$ 65,000	$ 65,000		
	Standby	1	Allow	$ 24,000	$ 24,000		
	Allowance for additional drafting/field coordination issues	1	Allow	$ 30,000	$ 30,000		
	SUBTOTAL					$ 936,000	
	HVAC						
23000	HVAC SYSTEMS AND DUCTWORK						
	Equipment						
	VAV Boxes	1	Allow	$ 24,500	$ 24,500		
	AHU / Air Condition Units, Cabinet Heaters	1	Allow	$ 211,750	$ 211,750		
	Finned Tube Radiation	1	Allow	$ 21,500	$ 21,500		
	Condensate Pumps	1	Allow	$ 1,500	$ 1,500		
	Pumps / Expansion Tanks / Air Separators / Heat Exchanger	1	Allow	$ 71,500	$ 71,500		
	Condensate Return Unit	1	Allow	$ 47,000	$ 47,000		
	Fans	1	Allow	$ 28,000	$ 28,000		
	VDD's / Starters / Vibration Isolators	1	Allow	$ 9,000	$ 9,000		
	Rigging, set and tie-in Air Handlers, Pumps, Fans, etc.	1	Allow	$ 6,500	$ 6,500		
	Balancing / Service / Startup / Commisioning	1	Allow	$ 1,500	$ 1,500		
					$ 422,750		
	Ductwork & Duct Insulation						
	Ductwork & Duct Insulation	1	Allow	$ 943,250	$ 943,250		
	Ductwork Insulation	1	Allow	$ 55,000	$ 55,000		
	Kitchen Exhaust Ductwork	1	Allow	$ 40,000	$ 40,000		
	Outlets, VAV-CD, WMs, and Grills	1	Allow	$ 15,000	$ 15,000		
	Fire Smoke Dampers / Fire Dampers / Motorized Dampers	1	Allow	$ 50,000	$ 50,000		
	Louvers	1	Allow	$ 60,000	$ 60,000		
	Standby labor for balancing / Service / Startup / Commissioning	1	Allow	$ 20,000	$ 20,000		
	Air Flushout per LEED	1	Allow	$ 15,000	$ 15,000		
					$ 1,198,250		
	Piping & Pipe Insulation						
	Piping Work	1	Allow	$ 1,400,000	$ 1,400,000		
	Pipe Insulation	1	Allow	$ 110,000	$ 110,000		
	Standby labor for balancing / Service / Startup / Commissioning	1	Allow	$ 15,000	$ 15,000		
	Chemical Cleaning / Water Treatment / Misc.	1	Allow	$ 7,000	$ 7,000		
					$ 1,532,000		
	BMS Controls & Control Wiring						
	Control Equipment	1	Allow	$ 100,000	$ 100,000		
	Control Wiring	1	Allow	$ 150,000	$ 150,000		
	Standby labor for balancing / Service / Startup / Commissioning	1	Allow	$ 9,000	$ 9,000		
	Allowance for additional drafting/field coordination issues	1	Allow	$ 120,000	$ 120,000		
					$ 379,000		
	SUBTOTAL					$ 3,532,000	

(*Continued*)

Exhibit 17-13

(*Continued*)

CSI	DESCRIPTION	QUANTITY	UNIT	UNIT PRICE	AMOUNT ($)	TOTAL
	GENERAL REQUIREMENTS					
	ELECTRICAL					
26000	ELECTRICAL					
	Temporary Power and Light - Install & Maintenance	1 Allow		$ 150,000	$ 150,000	
	4000' amp switchboard and feeders, panels	1 Allow		$ 495,000	$ 495,000	
	Branch circuits for light and power	1 Allow		$ 352,000	$ 352,000	
	Electrical Receptacles and Devices	1 Allow		$ 330,000	$ 330,000	
	Fire Alarm System	1 Allow		$ 275,000	$ 275,000	
	Lighting fixtures	1 Allow		$ 465,000	$ 465,000	
	Power to MEP Systems	1 Allow		$ 265,000	$ 265,000	
	Lutron Lighting System - occupancy sensing in offices, conference rooms and classrooms	1 Allow		$ 177,500	$ 177,500	
	Telecommunications - TV/Phone/Data	1 Allow		$ 110,000	$ 110,000	
	Security System - motion sensors, door & window sensore, keypads, etc.	1 Allow		$ 7,500	$ 7,500	
	Power and Controls for Blinds	1 Allow		$ 50,000	$ 50,000	
	Allowance for additional drafting/field coordination issues	1 Allow		$ 30,000	$ 30,000	
	SUBTOTAL					$ 2,707,000
	OVERTIME ALLOWANCE				TOTAL	$ 17,885,888
	Overtime for construction personnel	1 Allow		$ 450,000	$ 450,000	
	SUBTOTAL					$ 450,000
	TOTAL					$ 18,335,888

Exhibit 17-14

Value engineering analysis.

1. Review the various trade breakdowns (HVAC, electrical, structural) and analyze the largest trade items (by percentage of the total cost) to determine where costs can be saved.
2. On complicated projects, it may be helpful to construct a mock-up to see where costs can be saved.
3. Can elements be eliminated that will not materially impact a project (i.e., landscaping in the parking lot)?
4. Make sure that components being considered for value engineering do not create a more costly maintenance problem.
5. The longer the schedule, the higher the costs. Fast tracking should be considered.
6. Material substitutions.
7. Union versus non-union.
8. Lighting levels and type of fixtures.
9. Time of year construction starts.
10. Availability of certain materials.
11. Consideration of pre-purchasing items to reduce inflation costs.
12. Make sure all the mechanical and electrical systems are not over-designed. Look at the energy codes to confirm the engineer's design criteria.
13. Look at hard ceilings versus "soft" ceilings.
14. Review floor material, especially the quality of carpet being specified.
15. Check structural systems (steel versus concrete) and evaluate the total cost differences.
16. Review systems and materials with manufacturers and suppliers to solicit their opinion on ways for proper installation and for potential cost savings.
17. Try not to delay a project when reviewing value engineering items. This will only increase the cost due to inflation.

CHANGE ORDERS

A change order is defined as a cost for a new scope of work that was not indicated on the bid construction documents (drawings, sketches, addendums, bulletins, and specifications). A request for a change order can be initiated by any of the following:

1. The owner
2. The architect or other design consultant
3. Unforeseen conditions (i.e., soil conditions different from the submitted boring logs)
4. Municipal government requirements
5. Missing information as determined by RFIs (request for information)
6. Design discrepancies
7. Latent conditions
8. Potential coordination problems with owner-provided equipment
9. Acceleration to the schedule
10. Damage caused by insurance claims (hurricanes, tornados, earthquakes)
11. Changes to equipment or finishes that were initially specified but no longer manufactured

PROCESSING CHANGE ORDERS

In many instances, the design team prepares drawings or sketches and specifications defining the new scope of work. This information is then reviewed by the CM/GC's team including the PM, estimator, and purchasing department. The CM/GC's in-house estimating department may prepare a preliminary estimate of the change order. Then the change order documents are submitted to the subcontractors who will be performing the work. The subcontractors prepare the change order including their overhead and profit costs (see Exhibit 17-15). The subcontractors' pricing is reviewed by the CM/GC for consistency with the estimator's costs. If the submitted costs are reasonable, then the CM/GC will add his requisite markups, which include:

1. General conditions cost associated with the change
2. Time allocated by the estimating and purchasing departments
3. Overtime that may be required by the superintendent or laborers
4. Any permits that may be required
5. Rental equipment extensions
6. Insurance costs
7. Bond costs
8. General overhead
9. Fee

Exhibit 17-15

Subcontractor's change order costs.

TO: Urban City Contractors

Date:
Job ID:

JOB:
Project: High Rise A

SCOPE:

New duct hanging detail on structural drawing S-2 dated March 24, 2008. All duct to be trapezed with 2" × 2" × ¼" angles, in lieu of hanging from slab per SMACNA standards.

1	4 men × 42 hours @ $130/hr.	$21,840.00
2	material	$2,400.00
3	furnish and install 65' of linear feet (lf) of ductwork @ $9.00/lf	$585.00
4	furnish and install 17' of flow bar @ $165/lf	$2,805.00
5	tax	$276.00
6	overhead (10%)	$2,790.00
7	fee (10%)	$3,070.00
8	insurance (3%)	$1,013.00

PLEASE ISSUE THE NECESSARY P.O. IN THIS AMOUNT------------ $34,779.00

AUTHORIZED BY: _____

NAME & TITLE: _____

DATE: _____

FOR: URBAN CITY CONTRACTORS

SUBMITTED BY:
_____

PROJECT MANAGER
SUBCONTRACTOR HVAC

These markups are added to the change order just as they would if the new work was part of the original bid documents. The total change order (Exhibit 17-16) is attached to the change order request form (Exhibit 17-17) and then submitted to the authorized party. Of utmost importance is to note on the change order request form any time extension that would be required to complete the change order work. The authorized party has to sign the request form prior to starting any work.

Exhibit 17-16

Contractor's change order costs.

CHANGE ORDER NO. 5

Subcontractor HVAC cost	$34,779.00
Urban city contractors cost	
Permit	$ 3,478.00
Subtotal	$38,257.00
General conditions (6%)	$ 2,295.00
Subtotal	$40,552.00
Insurance (2%)	$ 811.00
Subtotal	$41,363.00
Bond (1.5%)	$ 620.00
Subtotal	$41,983.00
Fee (5%)	$ 2,099.00
Total	$44,082.00

CHANGE ORDER

Exhibit 17-17
Contractors
change order
submittal.

PROJECT TITLE _____
PROJECT NO. _____CONTRACT NO._____CONTRACT DATE_____
CONTRACTOR _____

The following changes are hereby made to the Contract documents:

Justification:

CHANGE TO CONTRACT PRICE

Original Contract Price: $ _____

Current contract price, as adjusted by previous change orders: $ _____

The contract Price due to this Change Order will be (increased) (decreased) by $ _____

The new Contract Price due to this Change Order will be: $ _____

CHANGE TO CONTRACT TIME

The Contract Time will be (increased) (decreased) by _____ calendar days.

The date for completion of all work under the contract will be _____

Approvals Required:

To be effective, this order must be approved by the Owner if it changes the scope or objective of the project, or as may otherwise be required under the terms of the Supplementary General Conditions of the Contract.

Requested by _____ Date _____

Recommended by_____ Date _____

Ordered by _____ Date _____

Accepted by _____ Date _____

REJECTION OF CHANGE ORDER COSTS

If the authorized party does not agree with the costs submitted and the CM/GC is required to perform the work per the provisions of the contract, then the required work must proceed. However, in order to avoid any future claims, the cost and any time delay issues must be resolved in an expeditious manner (see Chapter 18). An alternative solution would be to complete the work on a time and material (T&M) basis with a not to exceed (NTX) price established. The PM and the superintendent would have to keep daily logs of the time required time to complete the work along with the material invoices. In addition, the CM/GC would discuss with the subcontractors performing the work the consequences of not doing the work in a timely manner and not producing a quality product. Another suggestion may be (if time is available) to have the owner's design team and the CM/GC review and evaluate alternative methods for performing the work at a lower cost (then what was originally submitted).

CHANGE ORDER WORK AND THE SCHEDULE

Any delay time would have to be shown in detail of how the new work affects the total schedule. This would be submitted with the change order request form. It is not good practice to just submit a change request and ask for an extension of time without supporting documentation. The CM/GC must prove the case to the owner that the requested new work will delay other elements of the project.

With an accelerated schedule request (if timely), all the costs that are required to achieve this objective must be submitted by all affected subcontractors. The CM/GC may have to add additional equipment and staff to expedite the schedule.

TRACKING COSTS DURING CONSTRUCTION

Once the bids have been accepted for the project, the PM now must be concerned with maintaining this cost. Since approximately 80 to 90% of the project costs fall within the purview of the subcontractors, the PM must now exercise a management role for keeping the costs within the approved bid price. This will involve analyzing any additional costs submitted by the subcontractors, such as for:

1. Change orders
2. Out of sequence work
3. Schedule changes
4. Delay claims

A detailed cost summary, as indicated in Exhibit 17-18, must be tracked and updated on a continuous basis.

In addition, overhead costs (see Exhibit 17-19) must be constantly reviewed, along with general conditions costs (see Exhibit 17-20), to make sure that the original costs as prepared for a GPM, lump sum, or CM at risk contract are consistent with the current values. Adjustments may have to be made to keep the general conditions costs on track. This may involve:

1. Reallocation of laborers
2. More efficient use of debris removal
3. Looking for better ways to salvage material

CSI	Trade	Initial Cost	Changes	Total Cost	Payed to Date	1	2	3	4	5	6	7	8	9	10	11	12
											MONTHS						
01700	Temporary construction	$ 556,500		$ 556,500													
01710	Shoring and bracing	$ 50,000		$ 50,000													
02060	Selective demolition	$ 470,000		$ 470,000													
02070	Cutting and patching	$ 250,000		$ 250,000													
02200	Excavation	$ 195,000		$ 195,000													
02250	Concrete	$ 124,600		$ 124,600													
04500	Masonary restoration	$ 500,000		$ 500,000													
04810	Masonry/stucco	$ 477,396		$ 477,396													
05120	Structural steel/metal deck	$ 336,600		$ 336,600													
05700	Miscellaneous metal	$ 125,000		$ 125,000													
06400	Architectural woodwork	$ 794,360		$ 794,360													
07511	Roofing and waterproofing	$ 602,800		$ 602,800													
07811	Spray fireproofing	$ 40,000		$ 40,000													
07840	Fireproofing	$ 10,000		$ 10,000													
07900	Joint sealer	$ 10,000		$ 10,000													
08110	HM doors, frames and hardware	$ 92,750		$ 92,750													
08310	Access doors	$ 12,000		$ 12,000													
08500	Wood windows	$ 455,000		$ 455,000													
09260	GWB assemblies/Act.ic ceilings	$ 817,697		$ 817,697													
09270	Temporary protection	$ -		$ -													
09300	Ceramic and stone tile	$ 846,735		$ 846,735													
09681	Carpet and resilient flooring	$ 325,000		$ 325,000													
09910	Painting and wall covering	$ 280,000		$ 280,000													
10150	Toilet partitions	$ 14,850		$ 14,850													
10425	Temporary signage	$ 25,000		$ 25,000													
10505	Metal lockers	$ 18,600		$ 18,600													
10523	Fire extinguishers	$ 1,800		$ 1,800													
10651	Operable partitions (allowance)	$ 200,000		$ 200,000													
10810	Toilet accessories (allowance)	$ 14,500		$ 14,500													
11410	Food service equipment (allowance)	$ 1,800,000		$ 1,800,000													
12520	Window treatment (allowance)	$ 74,000		$ 74,000													
14000	Elevators	$ 811,500		$ 811,500													
21000	Fire protection	$ 379,200		$ 379,200													
22000	Plumbing	$ 936,000		$ 936,000													
23000	HVAC	$ 3,532,000		$ 3,532,000													
26000	Electrical	$ 2,707,000		$ 2,707,000													
	Over time allowance	$ 450,000		$ 450,000													
	Trade total	$ 18,335,888		$ 18,335,888													
	Construction contigency (5%)	$ 916,794		$ 916,794													
	Subtotal	$ 19,252,682		$ 19,252,682													
	General conditions (8%)	$ 1,540,215		$ 1,540,215													
	Subtotal	$ 20,792,897		$ 20,792,897													
	Insurance (2%)	$ 415,858		$ 415,858													
	Subtotal	$ 21,208,755		$ 21,208,755													
	Overhead and fee (3%)	$ 636,263		$ 636,263													
	TOTAL	$ 21,845,018		$ 21,845,018													

Exhibit 17-18

Cash flow analysis.

Exhibit 17-19

Overhead
checklist.

Project Overhead

_____ [ ] Barricades	_____ [ ] Survey
_____ [ ] Bid bond	_____ [ ] Temporary electrical
_____ [ ] Builder's risk insurance	_____ [ ] Temporary fencing
_____ [ ] Building permit fee	_____ [ ] Temporary heating
_____ [ ] Business license	_____ [ ] Temporary lighting
_____ [ ] Cleaning floor	_____ [ ] Temporary toilets
_____ [ ] Cleaning glass	_____ [ ] Temporary water
_____ [ ] Cleanup	_____ [ ] Transportation of equipment
_____ [ ] Completion bond	_____ [ ] Travel expense
_____ [ ] Debris removal	_____ [ ] Watchman
_____ [ ] Design fee	_____ [ ] Water meter fee
_____ [ ] Equipment floater insurance	_____ [ ] Waxing floors
_____ [ ] Equipment rental	
_____ [ ] Estimating fee	**Company Overhead**
_____ [ ] Expendable tools	_____ [ ] Accounting
_____ [ ] Field supplies	_____ [ ] Advertising
_____ [ ] Job phone	_____ [ ] Automobiles
_____ [ ] Job shanty	_____ [ ] Depreciation
_____ [ ] Job signs	_____ [ ] Donations
_____ [ ] Liability insurance	_____ [ ] Dues and subscriptions
_____ [ ] Local business license	_____ [ ] Entertaining
_____ [ ] Maintenance bond	_____ [ ] Interest
_____ [ ] Patching after subcontractors	_____ [ ] Legal
_____ [ ] Payment bond	_____ [ ] Licenses and fees
_____ [ ] Plan checking fee	_____ [ ] Office insurance
_____ [ ] Plan cost	_____ [ ] Office phone
_____ [ ] Protecting adjoining property	_____ [ ] Office rent
_____ [ ] Protection during construction	_____ [ ] Office salaries
_____ [ ] Removing utilities	_____ [ ] Office utilities
_____ [ ] Repairing damage	_____ [ ] Pensions
_____ [ ] Sales commission	_____ [ ] Postage
_____ [ ] Sales tax	_____ [ ] Profit sharing
_____ [ ] Sewer connection fee	_____ [ ] Repairs
_____ [ ] State contractor's license	_____ [ ] Small tools
_____ [ ] Street closing fee	_____ [ ] Taxes
_____ [ ] Street repair bond	_____ [ ] Uncollectible accounts
_____ [ ] Supervision	
	_____ [ ] **Profit**

Exhibit 17-20

General conditions tracking log.

Description	Monthly Rate	% of Time	Months											
			1	2	3	4	5	6	7	8	9	10	11	12
Account Executive														
Project Manager														
Assistant Project Manager														
General Field Superintendent														
Field Superintendent														
Assistant Superintendent														
Administrative Assistant														
Project Estimator														
Scheduler														
Hoist Engineer														
Crane Engineer														
Operations														
Purchasing														
MEPS														
Field Office Clerk														
Safety Director														
Safety Manager														
Accountant														
Administration (insurance)														
Surveyor														
Legal														
Temporary Probes or Borings														
Labor Regular Time														
Labor Overtime														
Labor Foreman														
Materials Handling Driver														
Containers														
Final Clean Up														
Permits														
Equipment/Misc. Tools														
Travel														
Blueprints														
Field Office Construction														
Field Office Furnature														
Temporary Field Office Utilities														
Computers/Network/Programs														
FAX Machine														
Copier														
Telephone														
Petty Cash														
Office Supplies														
Postage														
Cleaning Supplies														
Messenger														
Fed Ex														
First Aid Kits and Fire Extin.														
Two-Way Radios														
Cell Phones														
Temporary Toilets														
Job Photographs														
Security Service														
Temporary Signage														
Insurance														
Builders All Risk														
Bonds														
Special Union Requirements														
Hoist Rentals														
Crane Rentals														
Other Rentals														

4. Reviewing the costs that are being charged to the subcontractors

5. Reviewing the time management of the project team (superintendent, assistant superintendent, PM, cost estimator, scheduler, accountant, purchaser, etc.). These resources may have to be reallocated.

SUMMARY

- Good cost estimating is the key to a successful project. You also have to make sure that no major surprises will occur when the project starts.
- Several methods are used for cost estimating purposes and a combination of the methods is used to develop the costs. The percentage of construction documents completed will determine which method should be used:
  - Cost per square foot—This method should be used when limited information is known.
  - Take off—This should be used when information is available on the construction documents.
  - Subcontractors—This is the most accurate method for determining costs.
  - Time and material—Used mostly for change order work.
  - Component—Can be used when limited information is available. Also, a very good way to check the current estimate as compared to previously constructed similar projects.
- Whichever method is used, it is very important that factors outside the scope of the project be incorporated. This will include inflation, length of schedule, availability of labor and material, weather conditions, and long lead items.
- When the cost of the project exceeds an approved budget, then value engineering may be required. This would involve a review of the components of the project to determine where costs can be reduced.
- Cost estimates can be prepared using computerized spreadsheet programs, estimating programs, and CAD methods.
- The CSI format should be used for outlining the cost estimate so that all anticipated costs are included.
- Tracking of costs after bids have been approved must be done on a constant basis. This will involve keeping the subcontractors costs within the negotiated prices. This also involves managing the contingency assigned to the project.
- Change orders must be submitted to the owner with all the required back-up material.
- If the change order will require an extension of time, then the CM/GC must submit a detailed schedule indicating how the change affects the total schedule.
- The general conditions and overhead costs must be constantly tracked against the approved budgeted amounts. Adjustments may have to be made to keep the cost in line with approved amounts.

18 Claims and Dispute Resolutions
(You mean I have to keep records in case of a problem?)

WHAT IS A CLAIM?

A claim is a dispute between two or more parties that cannot be resolved by conventional business methods or by negotiations. In a majority of cases, the disputes deal with the amount of money owed to one of the affected parties or schedule modifications.

Some disputes can go on for years before a resolution is obtained. This only means that the financial viability for settling the dispute in a timely manner can be greatly reduced. The adverse affects of prolonged disputes include increased legal fees, reduction in cash flow, and increased time allocated to a problem that could otherwise be used for more productive construction-related situations.

The bottom line is that every means should be utilized to resolve disputes in a timely manner. Better yet, do not have disputes in the first place and especially try to avoid litigation. See Exhibit 18-1.

WHY DO WE HAVE CLAIMS?

Due to the complexity of the construction process and the number of participants involved, there are countless reasons why claims evolve.

1. The design team prepares the construction documents. People make mistakes, so there is always the possibility that items may be missing or unclear details may creep into the construction documents. This will require the construction manager/ general contractor (CM/GC) to submit cost changes for the missing information. The owner's position may be that the CM/GC is responsible for constructing the project with or without "minor" missing information.
2. Requested information may not have been submitted in a timely manner.
3. The construction contract may have been breached.

Exhibit 18-1
Claims and
disputes flowchart.

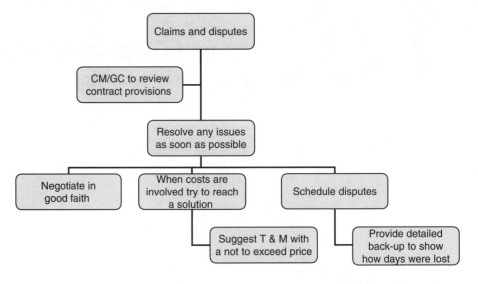

4. The CM/GC may have a different interpretation of the contract than the owner.

5. The owner does not accept cost of change orders.

6. Site conditions differ from what was originally noted.

7. Changes are made to the approved schedule.

8. The schedule is accelerated and with that, the costs increase.

9. Materials are substituted without approval.

10. The municipality requests changes that may involve a change order by the CM/GC. The owner's position may be that the CM/GC is responsible of any requests initiated by the municipality.

11. Accidents occur, which affects the schedule and thus adds additional costs.

12. There is a lack of trades people or a material shortage.

13. There are jurisdictional disputes between union trades.

14. Additional safety regulations may be required after the project starts.

POTENTIAL MAJOR CLAIM AREAS

Some areas of the construction process are more susceptible to claims than others usually due to the ambiguity of these potential problem areas. Following are a few of the areas that should be looked at in more detail when accepting construction work to avoid future claims.

Construction Documents

Construction documents include all drawings, specifications, addendums, sketches, and bulletins that the CM/GC receives in order to perform the contract work. Interpretation of the construction documents is a source of major disputes. As stated in Chapter 8, a thorough review of all of the construction documents should be accomplished as soon as they are received. Any questions must be asked of the design team in a timely manner after the review has been completed. In certain instances, CM/GCs will select lesser material or equipment hoping to make additional money on the vagueness of the construction documents. In these particular cases, let your conscience be your guide. The CM/GCs should be cognizant of the fact that they bought the construction documents as noted. The "dumb" approach by a CM/GC is not recommended and can lead to costly disputes and potential greater costs (for the removal of the product and or equipment). A thorough review early in the process will not uncover all potential problem areas, but it will help in minimizing potential misunderstandings and thus future claims.

Subsurface Conditions

In most cases, the subsurface information (via soil borings or test pits) is given to the CM/GC as part of the construction documents. The problem that arises here is that the actual conditions may vary from what is presented in the documents. This occurs when the CM/GC starts the foundation work and finds a different condition then what was indicated on the drawings. Depending on what the construction contract states, some liability may fall to the CM/GC. Good contract language would have some statement about the condition of the soil found verses what was on the construction documents and any material change would not be the CM/GC's responsibility. Lacking this language, the CM/GC must notify the owner immediately upon discovering a change in the actual soil conditions from those indicated in the construction documents.

It may also be necessary to prepare a cost estimate to indicate what the expense would be to deal with the new subsurface conditions. In addition, a new schedule may have to be prepared showing a possible delay caused by the subsurface condition survey. The new schedule would indicate the additional time required to rectify the problem.

Change Orders

Legitimate change orders that have been initiated by the owner and the design team for scope change or for inconsistency in the drawings are the bane of every CM/GC. As noted in Chapter 17, the process for dealing with change orders is extremely time consuming for the CM/GC. Make sure that all requests for a change by the CM/GC are in writing. Do not accept verbal requests. When submitting the cost for the change order, a description along with any drawings, sketches, and specifications should be provided. A detailed cost breakdown (see Chapter 17 for an example of a cost breakdown) must accompany the documents. It should be noted that all costs (direct and indirect) must be included (general conditions, bonding, fee, insurance, security, trucking). Disputes usually arise over the cost and schedule delays associated with a change request. That is why it is important to give as much detail as possible. As with the site conditions, it is imperative that if the change order creates a delay, it must be shown on the schedule that will be presented with the cost. All change order cost submittals must be signed off by the appropriate owner or approved representative. If total agreement cannot be obtained and the construction contract gives you no recourse except to proceed with the change order work, then try to negotiate at least part of the change order. Also, consider time and material with a "not to exceed" price. Partial approval is better than none at all. The negotiations should continue until the CM/GC and the owner reach an agreement. It should be noted that the CM/GC's bargaining position diminishes as the project comes to a close.

Delays

As stated previously, any changes initiated by the owner or the design team that cause a delay to the project must be noted in writing (with a copy of the revised schedule). The project manager must update the schedule on a constant basis. Events such as severe weather (or weather that is abnormal for the time of year), strikes, and material shortages must be addressed immediately on the schedule, along with a written explanation. If the owner-provided equipment and finishes affect the schedule, the owner has to be notified immediately. The schedule must then be submitted to the owner for approval. If approval is not received from the owner, then you must request reasons for the rejection. On complex jobs where CPM schedules are utilized, any changes should be reflected in the events that are affected. The impact on the critical path should be noted and all parties are to be advised of the change. The logic for any change must be detailed. In certain circumstances it may be beneficial to hire a scheduling consultant who would review and evaluate the logic of the schedule.

If liquated or consequential damages are to be paid for any delay caused by the CM/GC, then all relevant schedule information must be assimilated. This information should indicate the valid reasons for the delay. Some decision may have to be made in regards to evaluating the cost to prove that the delay was not caused by the CM/GC's failures versus the cost of the liquated damages. When evaluating costs for owner-initiated delays, the checklist in Exhibit 18-2 should be reviewed.

Accelerated Schedules

The schedule of a project may have to be accelerated because an owner has requested expediting the project's opening date. If this is the case, then a full cost analysis has to be presented to the owner prior to starting the acceleration process. The cost analysis would include the following:

- Overtime labor costs
- Expediting of material

Exhibit 18-2

Owner-initiated delay checklist.

1. Labor costs with mark-ups
2. Loss of productivity (could be from 5 to 50%)
3. Weather factors (i.e., cold weather protection)
4. Insurance and bond increases
5. Equipment extensions
6. Any union or trade requirements (i.e., Teamster longer on the project)
7. Any interest expenses
8. Material and labor escalations
9. Out-of-sequence work and associated expenses
10. General condition increases
11. Subconstructor's cost
12. Fee

- Management expenses for reviewing and evaluating the schedule and purchasing material
- Subcontractor's cost
- Bond and insurance costs that must be added for the cost increases
- Fees associated with the increased costs
- Adjustments to the general conditions costs
- Costs associated with using the crane and hoists on an overtime basis
- Determination of the number of additional hours the trades have to work in order to make a new accelerated schedule. The CM/GC may find that with limited time left in the original schedule it may not be practical to accelerate the schedule.

An owner may want the CM/GC to accelerate the project because the project is behind schedule. In this particular case, the CM/GC must have all the available information to show that the project is not behind schedule (if this were the case). This would involve the following:

- Walking the project with the schedule and showing the owner how the installed components match up with the schedule.
- If any items are not installed, then a milestone schedule will have to be prepared showing when critical elements will be placed.
- The subcontractor must be part of this milestone schedule.
- If the CM/GC cannot meet the milestone dates, then an accelerated schedule may be required to bring the project back on schedule.
- Any force majeure delays must be discussed with the owner, especially if this type of delay is covered in the contract.

HOW TO MINIMIZE CLAIMS

The CM/GCs must do their utmost in order to minimize a claim being initiated. It should also be noted that in certain instances no matter how much due diligence is taken, a claim may still occur. A thorough review of Chapter 9 should be completed in order to understand some of the implications of clauses stated in the construction contract. In order to minimize claims, the checklist in Exhibit 18-3 should be reviewed.

CLAIM AND DISPUTE RESOLUTION

Prior to starting any project, some formal procedures must be established to resolve a dispute in an expeditious manner. These procedures should be delineated in the construction contract. These procedures benefit the CM/GC as well as the owner. This is achieved by eliminating non-productive time on the part of both parties and eliminating the need for lawyers. The following are descriptions of methods that could be used for resolving disputes.

Exhibit 18-3

Minimize claims checklist.

1. Complete review of all construction documents.
2. Obtain clarifications on any vague items.
3. Constantly review the approved schedule.
4. Obtain approvals from the owner for all schedules.
5. Make sure that all owner-supplied equipment and finishes are incorporated into the schedule.
6. Fully understand any construction document language that requires the CM/GC to adhere to local code requirements (versus what may be in the drawings).
7. Keep all appropriate inspections reports.
8. Have test results in an easy to read format.
9. Take progress photographs (with date stamps).
10. Update RFI logs with the requests and answers (and appropriate dates).
11. Make sure daily logs are kept current and detailed.
12. Have shop drawing logs available. Make sure submittal and approved dates are noted.
13. Keep logs of written instructions to the subcontractors.
14. Have the bid leveling sheets available with the sign-off by the subcontractors.
15. Keep updated correspondence (and emails) to the owner and the developer with responses.
16. Keep track of all weather reports and note when unusual weather occurs.
17. Time sheets for all trades work must be kept current.
18. Details of all change order work and the signed documents by the assigned approval party from the owner and developer should be kept.
19. Keep copies of all field memos and telephone conversations.
20. Have all site surveys approved by the consultants.
21. Have available additional soil reports.
22. Become familiar with the case law of similar types of construction projects.
23. Know the details of the construction contract.
24. Have a safety program in place with a director responsible for its implementation and monitoring.
25. Have a security plan available with details as enumerated in Chapter 11.
26. Make sure all correspondence, reports, logs, and schedules are filed properly for ease of access.
27. Prepare all subcontractors and their associated leveling sheets.
28. Be cognizant of the potential use of a mixture of performance specifications and detailed specifications.
29. Have all signed requisitions and copies of the owner's checks.
30. Establish a team approach with the owner, architect, consultants, and the CM/GC.
31. Avoid any adversarial relationships with the subcontractors, the design team, and especially the owner.
32. Make sure a site walkthrough is accomplished with all the projects' participants (owner, architect, consultants, and CM/GC) prior to start of any work. Any problems should be raised by the CM/GC and followed up with a written summary requesting clarification of the items raised.
33. Assign responsibilities and a definite time table for resolution of any open items.

Partnering

In this method, each party to the construction process selects an individual who would be part of the review team. Thus, the CM/GC, architect, and owner would each select a person to be on the dispute review committee. When a dispute arises, the committee would evaluate the conditions of the dispute. Since all parties are involved in the current construction process, they should have a good understanding of the dispute being presented. The decisions that are reached by this committee are usually not binding. The other factor that has to be considered is that each party has its own agenda, which can lead to a bias that may have an impact on the decision recommended. Partnering is very cost effective and, in most instances, the disputes can be resolved in a timely manner.

Dispute Review Board

The CM/GC and the owner both select an outside construction expert. The two experts then select an outside chairperson. The experts will work together as a review board and will come up with rational dispute resolutions. Depending on the complexity of the project, they may meet on the site once a month to review all outstanding issues. By holding the meetings on site, they are able to access all available information and inspect the site conditions. The owner and the CM/GC usually pay the members of the Dispute Review Board. Depending on how the construction contract is written, the decisions arrived at by the Dispute Review Board are usually non-binding.

The Dispute Review Board is an effective way for all disputes to be resolved. You are dealing with experts in the construction field who have experience with similar types of problems. They also understand the affect that disputes have on all involved parties.

Mediation

The parties select a mediator from the Mediation Board. The mediator does not come up with a decision. The mediator assists the CM/GC and the owner in trying to get the two parties to agree to a solution to the problem. Position papers are usually prepared by each party so that the mediator understands the basic problem being presented. This is a back-and-forth approach where the mediator asks questions to understand each party's position. The mediator then tries to work out some of the problems while negotiating each detail. Mediation is not binding. The Mediation Board receives a fee to perform this work.

Arbitration

The CM/GC and the owner select an arbitrator from the American Arbitration Association (www.adr.org). Arbitrators are usually selected by their expertise in the field of construction. The process works in the same way as mediation, except that the decision reached by the arbitrator is usually binding. In the majority of the arbitration cases, any awards are seldom appealed by the courts. As with the mediator, a fee is paid to the American Arbitration Association.

Mini Trails

In this process, the executives of each party are involved in the negotiations. The lawyers for each of the affected parties prepare briefs and summaries and then present these results to the executives. A neutral party selected by the executives' acts as a mediator between the disputed parties. The rationale here is that the final decision makers are presented with the facts. Hopefully, each executive understands the consequences of not resolving the issues.

Litigation

This should be avoided where practicable. If other methods do not work, then obviously this would be the last resort. It should be understood that litigation has many down sides:

1. High costs for the payment of lawyers and court costs
2. Time required by the CM/GC's personnel to be involved in discovery, depositions, and court procedures
3. Time that money may be tied up
4. Loss of interest on money that is tied up
5. No guarantee of a positive outcome
6. Construction is a complex process and juries and judges may not have the technical expertise to truly evaluate a claim

SUMMARY

- Disputes should be resolved as soon as possible or, better yet, avoid them altogether.
- In order to minimize disputes, documentation, tests, photographs, reports, and logs should be updated continuously.
- All construction documents should be reviewed and evaluated prior to starting any formal work.
- The construction contract should be completely understood as well as high risk factors involved.
- Several methods are available for resolving disputes prior to proceeding to litigation:
  - Partnering
  - Resolution review board
  - Mediation
  - Arbitration
  - Mini trials

19

Design-Build
(An industry buzz word.)

DESIGN-BUILD—A PROJECT DELIVERY SYSTEM

The design-build project delivery system is a method by which the owner contracts with a single entity to provide architectural, engineering, and construction services for a project. The design-build project delivery system was initially slow to catch on, with less than 10% of all projects utilizing the design-build delivery approach prior to 1990. Today it has grown to approximately 30% of all building. The design-build project delivery system is finally gaining acceptance and respect in the industry.

In the design-build project delivery system, the owner must first define the program and requirements for the project. Once the owner retains the design-build entity, then that one entity is responsible for all of the aspects of the design, engineering, and construction. The design-build process was developed to assist owners with one-stop shopping for procuring all of the professional services required to design and build a project, and facilitate a fast track process for the efficient and effective completion of the project. This approach gives the owner one point of contact if there is a problem. In the conventional approach, the owner would first hire an architect and engineers to design the project. Once the design phase was completed, the project would be sent out for competitive bidding for the scope of work defined on the bid documents either to general contractors (GCs) if the project was to be a lump sum bid, or to subcontractors for bidding if the project was being managed by a CM/GC. CM/GC would have to level all bids from the subcontractor. The owner would then have to receive the bids and determine the responsive bidder, which is often the lowest bidder and not necessarily the most responsible and qualified bidder. This separate contracting often leads to conflicts within the overall project team and can result in claims, disputes, and lawsuits.

HISTORY

The design-build process has been used in the United States for quite some time, primarily in the private sector. It is gaining more acceptances in the public sector as well, especially within the power, communications, sports, industrial facilities, and medical segments of the construction industry. Many public projects, under the jurisdiction of government agencies, have been recently procured using the design-build methodology. The first railroads to cross the United States were built utilizing this approach.

The design-build process allows the owner to select from a qualified list of experienced professionals to design and construct the project as a team. Team members have a common vested interest in working together to complete the project on time, within budget, within the project scope requirements, and with good quality. A design-build team can be an individual firm possessing all required professional skills, a consortium of firms with their unique specialized skills, or a joint venture between several firms. Sometimes the project is very large and complex and beyond the capabilities of any one individual firm, resulting in a team or partnering approach.

Design-build has been used for both large and small projects across many different industry sectors. Some contracting firms have taken the lead role in the design-build process by partnering with architectural and engineering firms. The construction manager/general contractor (CM/GC) often takes the lead role in a design-build team in that they have bonding capacity and insurance coverage for the construction process. The CM/GCs must be careful that they do not directly practice architectural or engineering services because their insurance policies typically do not cover them for these services. The architect and engineer should name the CM/GC as a named insured on their professional liability insurance policies, and vice versa, to provide all parties with insurance coverage as a design-build team.

Either way, the team must fulfill all of the project's requirements, based on the program put forth by the client. Over time, these teams can continue to work together on other projects or they may evolve into different teams, depending on the experiences of working together, type of projects, areas of expertise, workload, and market conditions.

Examples of design-build projects are New Foley Square Federal Courthouse in Manhattan, New York; Olympus United States Headquarters in Nyack, New York; and renovation of the landmark Chicago Blackstone Hotel.

Exhibit 19-1 shows the relationship of the design-build team project delivery system and Exhibit 19-2 shows the relationship of a more conventional design, bid, and build project team delivery system.

THE PROJECT PERFORMANCE SPECIFICATION

Definition of the scope of a project by the owner is one of the key aspects to the success of the design-build process. The owner must identify the proper performance specification for the project (program), in order for the perspective design-build team

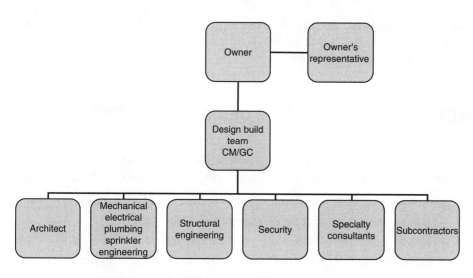

Exhibit 19-1
Design-build
delivery system.

to respond properly to the process and to ensure that the owner's needs are being satisfied, while allowing the design-build team to produce the most efficient and effective approach to the owner's program requirements. The performance specification (program) is much different from the project specifications. The performance specification defines the owner's requirements and program, while the conventional project specifications are more a description of the products, methods, and procedures to be followed by the contractor during construction. The owner's involvement in the design-build process during the early development stages of the project is the one of the essential ingredients to the success of a design-build project. The owner needs to be a reasonably sophisticated and active participant to define the projects requirements and the process. If an owner does not have this experience in house, he/she may wish to retain an experienced professional advisor to assist the owner with all of the aspects

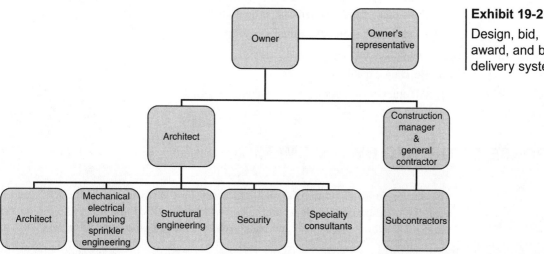

Exhibit 19-2
Design, bid,
award, and build
delivery system.

of the design-build process. With the correct performance specification (program) and a skilled sophiscated owner or advisor, there is no reason why the design-build project approach will not be successful.

THE PROCUREMENT PROCESS FOR THE DESIGN-BUILD ENTITY

The construction manager/general contractor (CM/GC) may wish to offer the owner a design-build approach to their project. The advantages of such a partnering approach are:

1. Involvement in the project from its inception, rather than at the bid phase
2. Ability to be involved with the design process
3. Ability to better control selection of materials, equipment, and systems
4. A unified approach with the architect, engineer, and other consultants
5. Minimize chance for litigation among the design-build team
6. Deliver to the owner a well-designed and well-built project
7. Ability to fast track the design and construction process

Once the owner has decided to utilize the design-build process, developed the project definition and concept, and obtained the required financing, the procurement process is ready to begin. The owner will issue a request for qualification (RFQ) to the perspective design-build entities or teams. The RFQ usually asks for:

1. The organizational structure of the team
2. Resumes of team members
3. List of similar projects and relevant experience
4. Financial information on the companies
5. Current workload
6. Unique approach to the project
7. Principals of the companies and their backgrounds
8. Pending litigation or criminal violations
9. Insurance coverage
10. Bonding capability
11. All other pertinent information about the firms

RESPONSE TO THE RFQ BY THE CM/GC

The design-build teams will be qualified based on their response to the RFQ, and a short list of responsive design-build teams will be identified and invited to bid on the project and submit a technical and managerial approach to the project for the owner to evaluate. Usually the owner will limit the list of design-build bidders to 3 to 5 firms to focus the process and to limit the financial exposure of additional firms responding to the RFQ, which can be rather expensive, especially for a large design-build project. It is important that the CM/GC has knowledge of the project and client, and qualifies on the short list of bidders for the project.

Once the design-build firms have received and reviewed the RFQ, they need to decide if it is in their best interest to respond to the request for proposal (RFP) and if their team has a reasonable chance to be awarded the project. If the design-build team decides to move forward, then they need to define the roles and responsibilities of each team member, determine the methodologies to be used for the process and project, and prepare cost estimates for the professional services and construction work to be performed. The team must identify the best design and construction techniques to ensure the success of the project, for both the owner and the design-build team. When all is said and done, everyone must be a winner in the process.

The owner will issue an RFP to the chosen finalists from the RFQ process. The RFP will include the project performance specification (program) and requirements, form of design-build contract, instructions to the design-build entities, directions for submitting the RFP, and the evaluation and selection process and methodology. The owner usually has the right to award the design-build project to the responsive bidder that provides the best overall value and technical approach to the project, and not necessarily the lowest bidder. It is therefore important that the CM/GCs differentiate themselves from the competitors in a manner that demonstrates the unique qualification, experience, approach to the project, and value that they will add to the process, in addition to a responsive competitive dollar amount of the bid. A prudent owner who recognizes the value that the professional CM/GC can add to the process sometimes awards the project to CM/GCs who are not necessarily the lowest bidders.

RESPONSE TO THE RFP BY THE CM/GC

Once the design-build proposal is received, the owner needs to evaluate the RFP to determine the responsive bidder. The owner will usually assess the technical approach of the team, proposed costs, unique approach to the project, and value added by the team to ultimately determine the best value and responsive bidder to be awarded the project. This is done with the review of the proposals received, interviews with the bidders, addendums issued to clarify questions and project requirements, and a final meeting with the apparent responsive bidder. Some owners use both a quantitative (pricing) and qualitative (schedule, experience, personnel assigned, safety, quality control, past performance, project understanding, value added, and technical approach) approach to reviewing and ranking the bidders to determine the responsive qualified bidder that provides the best value. Sometimes the owner may issue a best and final offer (BAFO) to the design-build teams once they have narrowed down the playing field to two or three responsive bidders to determine the bidder to be awarded the project.

THE DESIGN-BUILD TEAM

The design-build team accepts all responsibilities for the architectural design, engineering, and construction of the overall project, as per the client's performance specifications (program). The team accepts all risks associated with the project. The major challenge facing a design-build team is forming a team that can work together and perform the project requirements for the owner. Each firm brings its technical expertise and professional experience to the project to meet the owner's requirements.

The firms need to have compatibility in their approach to the project and their corporate culture, and they must be willing to work together as partners in the process. Each team member must perform his duties and responsibilities for the overall team, project, and owner for the project to be successful and have a satisfied client, who will recommend the team for the next project. It is important that each team member understands the role that they will play in the process, supports the overall team, and accepts the duties, responsibilities, and risks associated with the process. The design-build team should enter into a teaming agreement prior to submitting qualifications or proposals. The teaming agreement should contain the following:

1. Identity of all parties along with their duties and responsibilities
2. Definition of roles and selection of team leader
3. Interaction with the owner and team members
4. Communications within the team and with the owner
5. Costs for services
6. Making business decisions among the team and with the owner
7. Making technical decisions among the team
8. Actions that build teamwork and trust among the team
9. Avoidance of conflict within the team
10. Resolution of conflict should it arise
11. A homogeneous united entity to provide the design-build services for the project

Teaming agreements are recommended whether the team structure is a single entity, a joint venture, multiple team members, prime and subconsultants, etc.

ADVANTAGES OF THE DESIGN-BUILD PROCESS

The main advantage of the design-build process is that there is one source of responsibility. Projects are typically completed +20% faster, which results in savings of +4% of the overall cost of the project. The major advantages of the design-build project delivery system approach are summarized in Exhibit 19-3.

1. Contract with one entity for the entire design-build process.
2. Avoids the finger-pointing syndrome of who is responsible—the design team or contractor.
3. Fast tracking of the project.
4. Develops an environment for an innovative and creative approach.
5. Encourages teamwork and partnering, with all parties working for the same common interest of a successful project.
6. Encourages an effective and efficient approach to the project.
7. Encourages open, timely, and professional communications within the entire team.
8. Reduces potential litigation among the parties.
9. Can potentially deliver a better overall best value project.
10. Good cost estimating (no questions on the design intent).
11. Good value engineering.

Exhibit 19-3
Advantages of the design-build process.

DISADVANTAGES OF THE DESIGN-BUILD PROCESS

The major disadvantages of the design-build project delivery system approach are summarized in Exhibit 19-4.

WHY CMs/GCs CHOOSE THE DESIGN-BUILD PROCESS

Owners often choose the design-build approach for a project for the considerations outlined in Exhibit 19-5.

1. The normal checks and balances between the design team and construction team are now with one entity.
2. The owner must play a very active role and develop a definitive project program and performance specification.
3. Joint ventures are often required to bring all of the necessary experience and professional talent together to design and build the project. The different firms have to have a mutual respect for each other and a similar mindset and corporate culture to work together in a professional manner in the best interest of the project.
4. Loss of control by the owner of the specific design and construction approach.
5. Depending on which entity is heading up the design-build process, the architect, engineer, contractor, or a developer, the project approach may be biased or compromised by the experience and mindset of the leading entity.
6. Legal aspects between the owner, design-build entity, and within the design-build team have to be defined.

Exhibit 19-4
Disadvantages of the design-build process.

Exhibit 19-5

Why CM/GCs choose the design-build approach.

1. Saves time over a conventional design, bid, and build process.
2. Saves costs over a conventional design, bid, and build process.
3. Creates unique solutions by the project team for the project.
4. Builds on the synergy of the entire architectural, engineering, and CM/GC staff.
5. Promotes teamwork and partnering of the entire team for the project.
6. Personnel working on the project tend to think "out of the box" and innovate.
7. Less potential litigation among the design-build team members.
8. Reduces the risk of awarding the project or a portion thereof to the lowest bidder.
9. Contractual relationships are simplified with one contract with a design-build entity.
10. Potential for higher fees and profits.

SUCCESSFUL DESIGN-BUILD STRATEGIES

Exhibit 19-6 details the successful design-build strategies that the design-build team must address.

Exhibit 19-6

Successful strategies of a design-build team.

1. Assign duties and responsibilities among the team members.
2. Ensure the proper professional resources are retained to perform the work for each unique phase of the project.
3. Build teamwork and partnering for the project.
4. Be innovative and encourage creativity.
5. Have open communications between all parties.
6. Define the team leader.
7. Have in place a dispute resolution mechanism between the team if there are any conflicts, disputes, or problems.
8. Define who will communicate with the owner.
9. Ensure safety for the workers, owner, and the public.
10. Maintain QC procedures.
11. Coordinate all design and construction processes to facilitate fast tracking of the project.
12. Identify project risks, and develop a risk management plan to mitigate them.

SUMMARY

- The design-build project delivery system has become more popular and will continue to grow as a project delivery system.

- The design-build project delivery approach can be a very efficient and effective system, especially for a fast track project.

- The likelihood of success using the design-build project delivery approach is no greater or worse than any other project delivery approach.

- The owner plays a great part in this process, and must know the project's overall needs, requirements, performance specifications, project phasing, budget, schedule, and administrative requirements.

- The design-build team must be able to function as a seamless entity, and have good teamwork, partnering, and communications skills within the team.

- Everyone needs to perform their respective roles, duties, and responsibilities in a timely and professional manner in order for the team and the overall project to be successful (see Chapter 1).

- Everyone should be a winner in the process, and the owner should be happy with the project being completed on schedule, within budget, and with high quality, and meeting the performance specification requirements.

- A true measure of success is if the owner invites the team to bid on the next design-build project.

- The design-build team needs to identify the risks, assess and assign duties and responsibilities, ensure the right professional expertise is available to perform the project, and develop clear procedures to mitigate all risks (see Chapter 2).

20 Requisitions
(We are finally being paid!)

FINANCIAL CONTROL

Financial control is one of the major essential elements of a construction project, in addition to the project schedule, budget, resource control, quality control, safety, and client satisfaction. Financial control is the carrying out of all of the fiscal duties and responsibilities of the project, as required contractually, along with good and prudent business practices. This includes a breakdown of project costs, forecast of project expenditures, progress payments, payments for extra work, processing of change orders, resolution of claims and disputes, waivers of lien, documentation required for payment, reduction of retainer, timely payments, and final payment.

One of the most important requirements for keeping a project healthy is cash flow management. Cash is king and without sufficient cash, it is difficult to meet payrolls and pay union dues, taxes, benefits, suppliers and vendors, equipment rentals and purchases, utility bills, and subcontractors. The timing and amount of payments is a serious matter for the construction manager/general contractor (CM/GC) to pay attention to, because without operating capital, even a highly profitable and successful project can present a problem. The CM/GC must have in place a financial management system to monitor and control project payments, cash flow, disbursements, and control payments to material vendors, subcontractors, and suppliers. In addition, the system needs to ensure that financial standards required contractually, legally, and by established accounting practices are complied with.

When building in the urban environment, some of the major projects are constructed with union labor, which presents further demands on cash flow requirements to pay union dues, benefits, and payroll taxes. These payments must be made to the union and the governing authorities, whether payment of the requisition has been made. Non-union projects do not have this additional cash management challenge.

CM/GCs must ensure that they have sufficient operating capital to fund the day-to-day payments and disbursements for doing business until the owner pays the requisition.

Firms must rely on retained earnings, lines of credit, business loans, private venture capital, factoring, or equity from the public or private markets for operating capital to fund their ongoing operation. It is not sufficient to just have a good income statement and balance sheet; they also need positive cash flow and money in the bank, which is what pays the bills.

If payment is not made for work performed, the CM/GC has the right to place a lien on the property for the value of the work performed and not paid. This is not a measure to be taken lightly, as the owner will be rather upset with the lien. It encumbers the title to the property and makes it difficult for the owner to obtain financing or sell the property. If the owner does not pay a requisition on time, it is recommended that the CM/GC have the accounts receivable department along with the project executive contact the client to find out why the requisition has not been paid, and try to obtain the outstanding payment. Subcontractors, when they have not been paid, have a tendency to slow down the project by not supplying adequate personnel and materials to perform the work, which ultimately can lead to a delay in the overall project. If a lien is placed on the property due to lack of timely payment, it is recommended that the subcontractors, suppliers, and vendors, rather than the CM/GC place the first liens. In this manner, the CM/GC can place pressure on the owner to pay the outstanding requisitions without jeopardizing their relationship. The owner usually has the contractual right to have the CM/GC bond the lien. In this situation, the CM/GC can tell the owner that the subcontractor's lien is beyond his control, and request that the owner make payment of the requisition.

REQUISITIONS

The requisition is the invoice that the CM/GC submits to the owner for payment, usually on a monthly basis, for the work performed on the construction project. The requisition is based on the contract amount, schedule of values for the work, the percentage of work completed for the current period, minus any prior payments received, and minus retainer to be held (which is usually 10%). The contract between the owner and the CM/GC will provide for partial payments of the contract amount based on the progress of the work, on a periodic payment cycle, which is usually monthly. The request for payment or requisition involves a compilation of the cost of the work accomplished since the last payment cycle. Having the architect, engineer, and owner walk the project with the GC to determine the status of the work completed and the amount of money to be paid for the work performed during the current invoicing period usually accomplishes this. This process is called the "pencil copy meeting."

The total value of the work completed can be obtained in different ways, depending on the terms of the contract and the preferences of the owner and CM/GC. Under a lump sum contract, the progress is usually measured by the estimated percentage of completion of each major project component or trade. This is then multiplied by the schedule of values for each item or trade, comprising the total contract value, minus all prior payments to arrive at the payment amount for that period. For a unit price contract, the progress is usually measured by the amount of units of work put in place to which the unit prices apply. One item that the CM/GC must look out for is "front end loading" of the requisition by the subcontractors. This entails invoicing for more money than the value of the work put in place. The process of "front end loading" a project affords the subcontractor access to money and operating capital beyond the value of the work completed. If there is a dispute with the subcontractor, and another subcontractor has to be brought in to complete the work, there may not be sufficient money left over to complete the work. This could also be a potential problem if there is a large punch list item to be resolved and there are insufficient funds withheld to cover the remediation of the problem.

The owner will usually pay for materials stored on- or off-site if the material has been inspected and a certificate of insurance obtained for the material stored off site, protecting all parties' interests. Exhibit 20-1 illustrates a sample pencil copy of requisition. Exhibit 20-2 is a checklist for preparing a requisition.

Small Subcontractors

There are times when a monthly payment cycle is insufficient to meet the cash flow requirements of a project, especially if you are dealing with small businesses. This presents a challenge to the CM/GC to be able to attract small, minority, disadvantaged, local, and woman-owned business enterprises to bid the project and perform the work. Special provisions may need to be made to provide for payment to these smaller start-up firms on a more frequent basis other than monthly, for example, requisitioning twice a month. In addition, owners sometimes delay payments to the CM/GC beyond the contract terms of 30 days, which in turn delays payments to subcontractors, suppliers, and vendors. The owner is then placing the CM/GC in the position of being a "banker" rather than a CM/GC, which is not their business. Many contracts have a "paid when paid" clause, which simply states that the subcontractors will be paid when the owner makes payment to the CM/GC and not before. This contractual provision is not legal in many states, and the subcontractor has the legal right to demand payment from the CM/GC, whether the owner has paid them. If payments go beyond 60 to 120 days, many CM/GCs, subcontractors, suppliers, and vendors find themselves in a cash flow crisis.

Timing of Requisition Payments

Once the requisition is produced, submitted, and approved for payment, the owner usually has 30 days to make payment, as per the terms of the contract. In essence, this means that the work that is requisitioned each month will not be paid for 30 days, or the following month at best. Sometimes payments can be delayed for 60, 90, or 120 days. The CM/GC has to ensure that there are sufficient funds available to fund the construction project for this period. If by chance the owner is not satisfied with a given month's requisition due to an item not be properly documented and invoiced, it is recommended that the CM/GC advise the owner to delete that item from the current requisition and adjust the amount accordingly. This should be done in lieu of returning and resubmitting the requisition and thereby delaying the payment. Another approach is to adjust the item on the next month's requisition, with an appropriate credit, if the owner is cooperative and trusting in the process. The item needing further clarification can then be invoiced on the next month's requisition without adversely affecting the balance of the work to be paid for the current requisition. See also Chapter 17 for processing and payment for change orders.

Month
Application Number
Period From:
Period To:
Project Name:

A	B	C	D	E	F
			WORK COMPLETED		MATERIALS
ITEM NO	DESCRIPTION OF WORK	SCHEDULED VALUE	PERCENT LAST PERIOD	PERCENT THIS PERIOD	CURR STORED (NOT IN D OR E)
001	Demolition	$2,500	90	95	$0
002	Concrete	$12,950	75	90	$0
003	Concrete Allowance	$400	10	30	$0
004	Structural Steel	$62,540	55	70	$5,000
005	Structural Steel Allowance	$2,000	10	30	$0
006	Architectural Millwork	$150,257	5	10	$0
007	Doors/Frames/Hardware	$11,866	5	10	$0
008	Glazing	$39,238	8	12	$7,500
009	Drywall	$181,740	0	0	$0
010	Ceramic Tile	$25,700	0	0	$0
011	Carpet	$17,373	0	0	$0
012	Painting	$13,000	0	0	$0
013	Div 9 Allowance	$7,200	0	0	$0
014	Specialties	$5,907	5	10	$0
015	Sprinkler	$35,367	5	10	$0
016	Plumbing	$26,927	5	12	$0
017	HVAC	$215,500	8	14	$0
018	Div 15 Allowance	$8,300	0	0	$0
019	Electrical	$212,588	10	15	$0
020	Div 16 Allowance	$6,500	4	18	$0
021	Project General Conditions Wages	$118,643	TBD	TBD	$0
022	Project General Conditions Other	$140,061	TBD	TBD	$0
023	Overhead & Profit	$51,862	TBD	TBD	$0
024	Administrative General Conditions	$47,973	TBD	TBD	$0
TOTALS		$1,396,392			$12,500

Exhibit 20-1
Sample pencil copy requisition review.

Item	Yes/No
1. Obtain a schedule of values for each subcontractor.	
2. Hold a pencil copy meeting to determine the percentage complete of each subcontractors work, based on the schedule of values.	
3. Have the pencil copy meeting at least 5 days prior to the requisition closing cycle.	
4. Project work through the end of the requisition cycle.	
5. Identify any special deposits that are required.	
6. Identify any special discounts for early payment.	
7. Identify any materials that may be stored off-site that are to be paid for.	
8. Obtain a certificate of insurance for any materials stored off-site prior to payment.	
9. Submit original copies of all invoices.	
10. Have the subcontractor submit and approve the requisition for their work.	
11. Obtain a partial waiver of lien for all prior payments received.	
12. Submit the requisition to the owner on time.	
13. Respond to any questions about the requisition.	
14. If any adjustments are to be made to the requisition, try to adjust them on the next requisition.	
15. Monitor the payment cycle and due date for payment.	
16. Upon receipt of the requisition monies, disperse the monies to all subcontractors, suppliers, and vendors promptly.	
17. If any payments are to be withheld, notify the owner.	
18. Adjust any non-payments on the next requisition cycle.	
19. Consider reduction in retainer at 50% completion of the project, from 10% retainer to 5%.	
20. Always ensure that subcontractors do not front end load the project for work not completed.	
21. Always ensure that sufficient funds are left over for each subcontractor to finish their work if they were to default.	
22. Obtain a final waiver of lien with the final payment and requisition.	
23. Resolve any claims or disputes prior to the final payment and close-out of the project.	

Exhibit 20-2

Checklist for preparing a requisition.

Deposits

It is important for the GC to identify any special deposits that may be required to secure an order for materials required for a project. Many vendors require a deposit with the initial order for materials for a project. This has become an even bigger issue with the rising cost of construction materials, where certain vendors are willing to give a price based on current market conditions only if they are paid in full for the material.

CM/GCs are often faced with a challenge when a change in the work is encountered and they have to submit a change order for approval (see Chapter 17). The owner, architect, and engineer may not fully approve the change due to a difference in the scope of work, cost, or schedule and logistics impact. Often the owner has the right to direct the contractor to perform the work without a formal approval. This leaves the contractor in a precarious position of having to perform the work without an approved change order, which usually limits their ability to requisition for the progress of the work. In situations like this, it is recommended that the CM/GC try to resolve the issues, get at least a partial approval for the change order, and perform the work on unit prices or on time and material with an upset cost not to exceed. In this manner, progress payments will be able to be made as the work is performed. The CM/GC is not in the banking and financing business, but in the construction business, and should not be expected to have to pay for work that they cannot be reimbursed for. Remember cash is king, and time is money.

Impressed Bank Account

Some owners, especially financial institutions, have utilized an impressed bank account to make payment to the CM/GC. An impressed bank account is a checking account funded by the owner each month to pay the CM/GC the value of the current requisition. The CM/GC is then given a checkbook to draw on the account to make payment to the subcontractors for their work for that month's requisition. The owner/financial institution always has possession and custody of the money and can monitor amounts, timing, and details of payments to all parties. When a CM/GC is paid directly by the owner and deposits the money into the CM/GC's own bank account, they have the use of the float on the money and interest earned. With an impressed account, the owner/financial institution have the benefit or the float and interest earned.

Special Documentation

It is important that the CM/GC fully understand all of the contractual and administrative obligations they have with regard to the documentation needed to be submitted with each requisition. Contracts often call for original invoices, certified payrolls, copies of all signed subcontracts, schedule of values, and insurance premium cost breakouts (this is especially true when the owner is providing an owner controlled insurance policy).

CERTIFICATE OF CAPITAL IMPROVEMENT

A certificate of capital improvement is the document supplied by the owner to the CM/GC to certify that the work being performed is an improvement to the real value of the property, and as such, no sales tax is due on the labor portion of the work. The

CM/GC must have a certificate of capital improvement on file for each project, in order to avoid paying sales tax to the governing authorities for the value of the labor work performed on the project. In addition, some projects can also be eligible to forgo sales tax on the material portion of the work by being given a special tax number by their local economic development authority, which empowers them as the agent for the governing authority and grants them a sales tax total exemption. This application and status must be obtained prior to purchasing any materials for the project, and is difficult to apply retroactively. Exhibit 20-3 is a sample of a certificate of capital improvement. Not every state issues a Certificate of Capital Improvement and thus the CM/GC will have to check the state they are working in to determine if a reduction in sales tax is offered.

ST-8 (1-08, R-12)

State of New Jersey
DIVISION OF TAXATION

SALES TAX

FORM ST-8

CERTIFICATE OF EXEMPT
CAPITAL IMPROVEMENT

To be completed by both owner of real property and contractor, and retained by contractor. Read instructions on back of this certificate. Do not send this form to the Division of Taxation.

A registered New Jersey contractor must collect the tax on the amount charged for labor and services under the contract unless the owner gives him a properly completed Certificate of Exempt Capital Improvement.

MAY BE ISSUED ONLY BY THE OWNER OF THE REAL PROPERTY
MAY NOT BE ISSUED FOR THE PURCHASE OF MATERIALS

_____
(Name of Contractor)

_____
(Address of Contractor)

_____
(Contractor's New Jersey Certificate of Authority Number)

THE FOLLOWING INFORMATION MUST BE FURNISHED:

The nature of the contract is as follows (describe the exempt capital improvement to be made):_____
_____
_____
_____
_____

The address or location where work is to be performed: _____
_____

TOTAL AMOUNT OF CONTRACT $ _____

The undersigned hereby certifies that he is not required to pay sales and use tax with respect to charges for installation of tangible personal property, because the performance of the contract will result in an exempt capital improvement to real property. The undersigned purchaser hereby affirms (under the penalties for perjury and false swearing) that all of the information shown in this Certificate is true.

CONTRACTOR'S CERTIFICATION	PROPERTY OWNER'S SIGNATURE
I certify that all sales and use tax due has been or will be paid by the undersigned on purchases of materials incorporated or consumed in the performance of the contract described herein.	_____ (Name of owner of real property) _____ (Address of owner of real property) By _____
_____ (Signature of Contractor) (Date)	_____ (Signature of owner, partner, (Date) officer of corporation, etc.)

Any person making representations on this certificate which are willfully false may be subject to such penalties as may be provided for by law.

REPRODUCTION OF CERTIFICATE OF EXEMPT CAPITAL IMPROVEMENT FORMS: Private reproduction of both sides of the Exempt Capital Improvement Certificates may be made without the prior permission of the Division of Taxation.

Exhibit 20-3
Certificate of capital improvement.

WAIVERS OF LIEN

With each requisition, the CM/GC must submit a partial waiver of lien to the owner, indicating that the prior payment was received and that all payments have been made to subcontractors, suppliers, and vendors in accordance with the requisition amounts, and that there are no outstanding claims, liens, disputes, and change orders that have not been presented to date. This cycle of presenting lien waivers continues with each requisition cycle until the final and last requisition is presented. In addition, the CM/GC needs to obtain from each subcontractor a partial lien waiver for all prior payments before releasing the current requisition payment. A partial waiver of lien is illustrated in Exhibit 20-4.

A final waiver of lien needs to be submitted by the CM/GC indicating the final contract amount with all approved change orders, which certifies that all prior payments have been made to all subcontractors, suppliers, and vendors in accordance with all prior requisitions, payments, and partial waivers of liens. The final waiver of lien also is an affirmation that

Exhibit 20-4

Partial waiver of lien.

STATE OF _____

COUNTY OF _____

RE: _____

_____

To Whom It May Concern:

For valuable consideration paid to date (\$),_____
does hereby waive, release and relinquish any and all lien, or claim, or right of lien they now have or may have hereafter under the Statutes of the State of relating to Mechanic's Liens on account of labor or materials or both provided in connection with work performed on the above described project located at the premises known and described as:

Further, we hereby represent that all subcontractors and suppliers who have performed construction services or supplied materials for this project on our behalf will be paid in full for their services and/or materials.

In the event that any subcontractor or supplier files a lien against the property, we hereby warrant we will immediately post a bond in satisfaction and proceed to discharge the lien or claim or right of lien for causes of action debts, dues, sums of money, accounts, bonds bills, contracts, controversies, damages, judgments, and demands they may have under the Statutes of the State of against Given under hand and seal this day of

CORPORATE SEAL

NAME_____

SIGNATURE_____

TITLE_____

Exhibit 20-5
Final waiver of
lien.

STATE OF _____

COUNTY OF _____

RE: _____

_____

To Whom It May Concern:

For valuable consideration paid to date ($ ____),_____
does hereby waive, release and relinquish anyand all lien, or claim, or right of lien they
now have or may have hereafter under the Statutes of the State of relating to Mechanic's
Liens on account of laboror materials or both provided in connection with work performed on
the above described project located at the premises known and described as:

Further, we hereby represent that all subcontractors and suppliers who have performed
construction services or supplied materials for this project on our behalf will be paid in
full for their services and/or materials.

In the event that any subcontractor or supplier files a lien against the property, we
hereby warrant we will immediately post a bond in satisfaction and proceed to discharge
the lien or claim or right of lien for causes of action debts, dues, sums of money,
accounts, bonds bills, contracts, controversies, damages, judgments, and demands
they may have under the Statutes of the State of against or the buildings and premises
described above. Given under hand and seal this day of.

CORPORATE SEAL

NAME_____

SIGNATURE_____

TITLE_____

there are no outstanding claims, contractual disputes, open change orders, etc., remaining to be resolved, and that the final payment will constitute payment in full for the work performed on the project in accordance with the terms of the contract. Similarly, the CM/GC must obtain from each subcontractor a final waiver of lien prior to making the final payment. A final waiver of lien is illustrated in Exhibit 20-5.

PROJECT TRADE COST BREAKDOWN

The GC needs to submit to the owner a project cost breakdown by trade for his approval, in accordance with the contract requirements, prior to any submission of the first requisition for payment. This breakdown of the project cost by trade, along with the CM/GC's fees, general conditions, insurance, and bonds is often referred to as a schedule of values. The schedule of values is usually a breakdown of the various trades and their components of the project. Each trade is broken down first, and then the major components within each trade are further broken down into the major components of work, equipment, and systems. In the absence of any special contractual schedule of values breakdown, it is customary to provide the cost breakdown by the work as it appears in the specifications and the Construction Specification Institute (CSI) codes (see Chapter 17 for CSI details) for the project. This process allows the CM/GC, owner, architect, and engineer to agree on the breakdown of the contract overall value, so that when the monthly requisition walkthrough occurs, the only thing that the team has to agree on is the percentage of each line item completed. Be sure to include in the schedule of values a cost for mobilization, set up, and commencement of operations to allow the CM/GC to be paid for these costs at the beginning of the project, rather than prorated with the cost of the balance of the work. Exhibit 20-6 shows a typical project cost breakdown for the HVAC trade.

Exhibit 20-6

Trade cost breakdown–HVAC.

Trade: HVAC Contractor

Item No.	Description of Work	Schedule of Value
1	Mobilization	$ 30,000.00
2	Shop Drawings	$ 40,000.00
3	Insurance and Bond	$ 24,000.00
4	Boiler	$ 53,700.00
5	Expansion Tank	$ 3,500.00
6	Pumps	$ 18,000.00
7	Cooling Tower	$ 53,000.00
8	Heat Exchangers	$ 31,000.00
9	Chillers	$ 150,000.00
10	DX Units	$ 44,000.00
11	Chemical Feed	$ 20,000.00
12	Unit Heaters	$ 5,000.00
13	Lead Lag Controller	$ 1,700.00
14	BMS System	$ 28,000.00
15	Differentail Pressure Control	$ 3,100.00
16	Piping	$ 213,000.00
17	Sheetmetal Sketching	$ 35,000.00
18	Sheetmetal Fabrication	$ 125,000.00
19	Sheetmetal Installation	$ 95,000.00
20	Insulation	$ 98,000.00
21	Air Testing and Balancing	$ 30,000.00
22	Water Testing and Balancing	$ 25,000.00
23	Fan Systems	$ 74,000.00
24	Diffusers and Registers	$ 34,000.00
25	Rigging	$ 28,000.00
26	VF Drives	$ 19,000.00
27	Linear Diffusers	$ 9,800.00
28	Exhaust Fans	$ 23,800.00
29	Smoke Purge System	$ 29,700.00
30	Louvers	$ 3,500.00
31	Chimney Flue Pipe	$ 184,000.00
32	CO/Gas Detection System	$ 8,500.00
	TOTAL	**$ 1,540,300.00**

RETAINER

A retainer is the money held back by the owner with each progress payment until the work is completed. This is to protect the owner if there is a matter that needs to be corrected (see Chapter 21). The contract will spell out the provisions for retainer to be held by the owner with each progress payment. The retainer is held by the owner until the work is accepted by the owner, architect, and engineer, the punch list is completed, all final approvals and certifications are obtained, the owner accepts the project, a letter of substantial completion is issued, and all required releases of lien, certified payrolls, and affidavits are submitted and approved by the owner. Only then does the owner make final payment to the CM/GC, including all retainers that have been withheld during the project.

A retainer amount of 10% is common in most construction contracts. It is important to understand as per the terms of the contract what the retainer clause will apply to, that is, the base contract work, change orders, etc. If possible, the retainer should only apply to the base contract and not change orders. The owner will keep the retainer with each payment. Many construction contracts have had provisions that allow the owner to withhold a 10% retainer for the first half of the project. After that, based on the assessment of the owner, architect, and engineer, the retainer may be reduced to 5% or lower or even eliminated. The owner usually reserves this right until the project has been completed satisfactorily. With profit margins being rather low in a competitive bidding environment, the 10% retainer can represent the GC's profit on the project, as well as actual costs for materials, supplies, labor, benefits, and taxes. Unfortunately, many reputable and successful contractors have been forced out of business when they run out of cash and cannot pay their current obligations.

FINAL PAYMENT

The final payment is usually made by the owner to the GC once the following has occurred: Work is completed, a certificate of substantial completion (see Exhibit 20-6) has been issued by the architect, the project has been accepted by the owner, the punch list is completed and all deficiencies and defects have been corrected, administrative requirements are satisfied, guarantees and warranties are provided, operating and maintenance manuals have been delivered, training has been conducted, spare parts and attic stock have been turned over, and a final inspection has been held. The CM/GC will then prepare a final requisition for payment and close-out of the project. Along with the last and final requisition, a final waiver of lien needs to be submitted by the CM/GC as described previously. The final payment by the owner is usually exchanged for the final waiver of liens from the CM/GC and all of the subcontractors.

CASH FLOW PROJECTIONS

The owner often wants to know what the projected cash flow and payments are to be made on the project. The owner requires this to ensure that adequate funds are available in a timely manner to make prompt payment. It is often said that if you want to keep someone stimulated and challenged, work him or her hard and pay them well and in a timely manner. The CM/GC should also have a vested interest in the projected cash flow and requisition process to anticipate when money will be paid to cover the work performed on the project to date. The CM/GC is often faced with the fiscal challenging of managing cash flow, in that monthly project payments usually exceed the amount of money paid to date due to the timing of the requisition cycle, payment cycle, hold back for retainer, etc. The CM/GC must make up this deficit from working capital, bank line of credit, bank loan, personal loan, or factoring, which can be expensive given the interest rates and cost of money. Chapter 17 has a CSI breakdown.

S CURVE

When you take the monthly individual projected requisition amounts and then sum them for each month, you can develop a baseline "S" curve for the project. The S curve can be used to track the monthly expenditures and progress of the project against the baseline developed with the project schedule and costs. Exhibit 20-7 contains a sample S curve cumulative payment. Exhibit 20-8 contains a sample S curve Earned value analysis.

During the course of the construction of the project, we can monitor the actual money spent and the progress of the work, and note any deviations from the original plan or baseline S curve. Exhibit 20-9 indicates a project that is behind schedule. Exhibit 20-10 indicates a project that is ahead of schedule.

Exhibit 20-7

Sample S curve cumulative monthly payments.

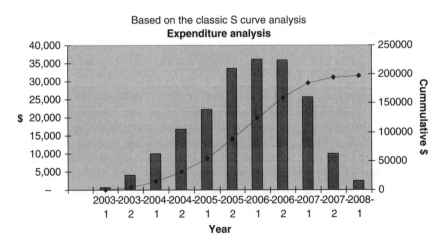

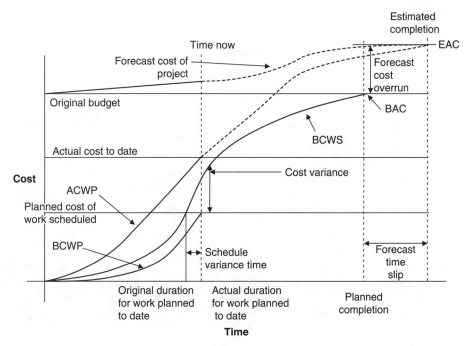

Exhibit 20-8
Sample S curve earned value analysis.

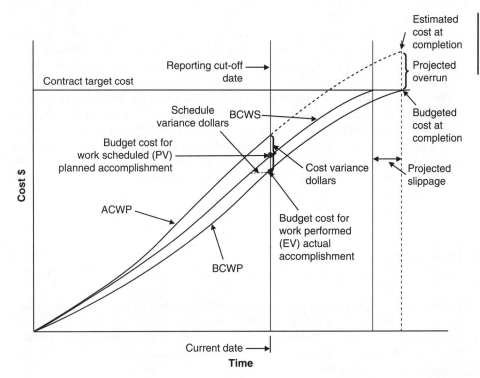

Exhibit 20-9
Sample S curve variance analysis behind schedule.

Exhibit 20-10

Sample S curve variance analysis ahead of schedule.

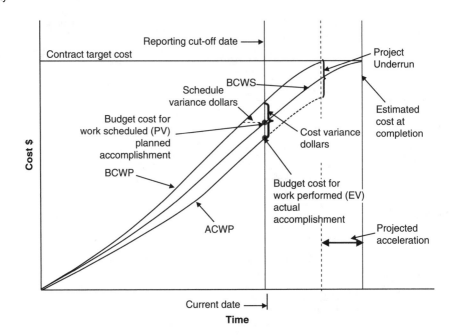

FILING A LIEN

If, during the course of the project, the owner does not make timely payment for work that is being performed for a private sector project, the CM/GC, subcontractors, suppliers, and vendors usually have the right to file a lien for the value of the work. Most contracts have a provision that calls for the CM/GC to bond the lien in the event that a subcontractor files it. The owner does not like to have a lien on the property because it encumbers the title and can cause problems with financing or sale of the property. Work in the public sector usually requires payment bonds. A typical provision in most state lien laws calls for the filing of a lien within 30 days of notice of completion or within 90 days after work was last performed on the project. The laws of each state must be checked because they vary by state. If payment is not received, then an attorney should be consulted.

PROJECT MONTHLY COST REPORT

The owner of the project often requests monthly documentation and reports in addition to the requisition to detail the cost incurred for the project, along with anticipated and pending costs, to develop a projection for the total cost to complete the project. Exhibit 20-11 contains a monthly payment estimate summary. Exhibit 20-12 shows a project monthly cost report.

Exhibit 20-11
Monthly payment estimate summary.

Project Title		Project No.	
Contractor	Address		
Estimate No.	Contract or Spec No.		Date
Period to			

	Amounts		
Description	Previous	This Month	To Date
General Earnings Including Materials on Hand.............			
Deduction __% of Work Done			
Deduction After __% of Work is Done			
Deduction __% of Material on Hand			
Total Deduction.............................			
Net Earning.......................................			

% Time Elapsed	% Work Completed

Contract Completion Data

    Notice to Proceed Received by Contractor .... _____

    * Contract Completion Time .... _____

    Contract Completion Date.. _____

    * As ammended by Change Order No.

Submitted by (Signature & Title)	Administrative Review - Final Payments
Approved by (Signature & Title)	Approved by - Final Payments

Exhibit 20-12

Project monthly cost report.

Project Title_____ Project No._____ Invoice No._____

Owner_____ Contract No. _____ Modif. No. _____ For Period Ending_____

WORK ITEM	WEIGHT VALUE IN PERCENT OF TOTAL	STATUS PREVIOUS REPORT (PERCENT)	STATUS THIS REPORT (PERCENT)	COMPLETED DURING THIS REPORTING PERIOD (PERCENT)	REMARKS
.					
THIS MONTH PROJECT TOTALS	100.00				

SUMMARY

- Cash is king and is the lifeblood of a project.
- If you want to keep subcontractors motivated, work them hard and pay them well and on time.
- Financial control of the project is as important as completing the project on schedule, within budget, with high quality, and with a satisfied client.
- Requisition procedures need to be set up at the beginning of the project to ensure timely payments. Often a schedule of values is developed, which breaks down the contract value by item or trade to facilitate the requisition process.
- Schedule a monthly walkthrough with the owner, architect, and engineer to review the status of the project, and determine the appropriate percentage of completion of each item on the schedule of values for the contract amount. By taking the schedule of values and multiplying it by the percentage of work complete for that month, the new requisition amount can be easily determined.
- Requisitions are normally submitted each month, with a 30-day payment cycle. The contractor needs to have sufficient access to cash to make required payments during the process, from working capital, bank lines of credit, personal loans, factoring, etc.
- At times special requisition arrangements need to be made for start-up companies that are not well capitalized and for special deposits required for orders.

- Waivers of lien need to be submitted with each requisition to document receipt of prior payments and payment of all amounts invoiced for to date, along with the final payment.

- A retainer is often held by the owner to cover items of work not completed in accordance with the contract, punch list, deficiencies, damages, etc.

- Ensure that a line item in the schedule of values is included for project mobilization to allow this cost to be invoiced during the start of the project, rather than prorated with the balance of the work as performed.

- Pay off original invoices, and not Xerox or faxed copies.

21

Project Punch List and Close-out

(What punch list? The job was perfect! More paperwork.)

PROJECT CLOSE-OUT

Close-out of a project is one of the last phases. After the construction manger/general contractor (CM/GC) constructs the project, the punch list and administrative handover of all project documentation in accordance with the contract and contract documents must be complied with. Many contractors neglect this phase of the work and then pay a dear price for it, with both dollars and their reputation. A successful construction project can be adversely affected by improper close-out procedures and poor quality control (QC). Project teams are often reassigned at the completion of the construction of the project, prior to the project close-out and punch list phase. Unfortunately when this happens, the history and continuity of the project is lost, and a separate close-out team is at a disadvantage to perform the close-out task properly. All contract work must be completed satisfactorily, outstanding claims resolved, change orders fully negotiated and processed, extensions of schedule time approved, guaranteed maximum price provisions on the contract properly administered, operations and maintenance manuals submitted and approved, as-built drawings completed, balancing reports submitted, and all other contractual obligations fulfilled. Exhibit 21-1 contains a sample project close-out checklist.

PUNCH LIST

The punch list is the list of defects compiled by the architect, engineer, and owner and then submitted to the CM/GC. It outlines all items that do not conform to the construction documents, are not installed properly, are not operating properly, or have been damaged. A prudent CM/GC will have been paying attention to the QC of the project from Day One with all of the subcontractors, suppliers, and vendors to minimize the punch list at the end of the project. The punch list should be managed throughout the

Exhibit 21-1

Project close-out checklist.

	Date to Owner	Received By

1. As-built drawings
2. O&M manuals
3. Elevator inspections
4. Training
5. Attic stock
6. Guarantees
7. Warranties
8. Spare parts
9. Punch list sign off
10. Balancing reports
11. Controlled inspection reports
12. Removal of temporary facilities
13. Final cleaning
14. Elevator inspections
15. Plumbing inspections
16. Electrical inspections
17. Highway and street inspections
18. Fire department inspections
19. Plumbing inspections
20. Sprinkler inspections
21. Life safety system inspections
22. Construction inspections
23. Other regulatory inspections
24. Keys/keying
25. Architect's certificate of completion
26. Engineer's certificate of completion

project, in order to ensure a manageable list of deficiencies to be dealt with at the end. A procedure used by contractors is to have their superintendents perform a pre-punch list of the project and develop the CM/GC's punch list prior to the completion of the work and the preparation of the final punch list by the architect, engineer, and owner. This pre-punch list enables the CM/GC to eliminate a large percentage of items that would have otherwise appeared on the final punch list, and saves the owner, architect, engineer, and the CM/GC a lot of aggravation, time, and effort. The reduction of the retainer held by the owner, which is usually 10%, is contingent on the contractor meeting the contractual requirements, and having good QC with a minimal punch list. Reduction of the retainer below 10% is also contingent on the completion of the punch list items. It is in the contractor's financial best interest to deal with the punch list items in a timely and proper manner to ensure timely payments. With profit margins on many projects being rather slim in a competitive environment, the retainer often represents the entire profit for the CM/GC and the subcontractors.

The CM/GC should insist on receiving one punch list from the architect, who is usually the leading design consultant and would integrate the punch lists from the architect, engineer, owner, building manager, tenants, and owner representative.

Otherwise, there can be many iterations of the punch list originating from different parties. The contractor also has to watch out that the punch list is a list of deficient items to be corrected, and not a wish list of what the owner, architect, or engineer would like to see in the project but is not called for in the contract documents, such as lighting control systems, building management systems, etc. A sample punch list form is shown in Exhibit 21-2.

Recommendations for preparing and completing the punch list are shown in Exhibit 21-3.

Date: _____

Project Name: _____

Project Number: _____

Subcontractor: _____

Contract Number: _____

Exhibit 21-2
Sample punch list.

Room/ Area	Room/area Item	Spec. Section	Punchlist Item	Accepted by Contractor	Accepted by Owner

Exhibit 21-3

Recommendation for preparing and completing the punch list.

1. Coordinate with the architect, engineer, and owner the development and receipt of one coordinated punch list.
2. Contractor should "pre-punch" the project.
3. Assign subcontractor responsibilities, distribute and maintain the status of the punch list work to be completed and what has been corrected.
4. Monitor contractors' timely and proper completion of the punch list.
5. Prepare final inspection report of all items being completed.
6. Verify final acceptance of the punch list work.
7. Ensure that the space is turned over at the designated time, with the punch list work completed.
8. Punch list work requiring completion after the owner occupancy can be expensive, as it will have to be performed on overtime and will require additional protection.
9. When you have many repetitive items on the punch list, evaluate ways to consolidate them.
10. When constructing apartments, hotel rooms, etc., where there is often a dozen or more items on the typical punch list, evaluate a mechanized system for preparing and managing the punch list. Some systems utilize bar scan codes to identify items of a repetitive nature.
11. Give a subcontractor 72 hours notice that the punch list items need to be completed. If the subcontractor is unresponsive, then put them on notice that you as the CM/GC will arrange to have the punch list completed for their trade work on their behalf and that they will then be back-charged for the associated cost.
12. It is recommended that the CM/GC withhold at least 100% of the value of the punch list as a retainer to ensure that the punch list work is completed. Many CM/GCs will withhold up to 200% of the value of the punch list as insurance that the punch list will be properly completed.

PROJECT CLOSE-OUT MEETING

It is prudent to schedule a close-out meeting with the owners and their representatives. At this meeting, an agenda should be prepared (see Exhibit 21-4 for a sample agenda), and items such as the punch list, schedule for turnover of equipment, acceptance by the owner, training, schedule for turnover of documentation, and format of documentation should be discussed. Minutes of the meeting should be published and the meeting should be videotaped if necessary. Since the reduction of the retainer is often linked to the length and cost of the punch list, it is prudent to manage the completion of the punch list as quickly as possible. Chapter 16 deals with some of the challenges of managing the subcontractors and the completion of their respective punch list items.

1. Names of participants
2. Status of the completion of the work
3. Status of the punch list
4. Owner's acceptance of the work
5. Start up of equipment and training
6. Guarantees and warranties
7. Turnover of attic stock
8. Scheduling of inspections and system testing
9. Completion of all controlled inspections and certifications
10. Balancing of systems and reports
11. Obtaining a temporary/final certificate of occupancy
12. Preparation of as-built drawings
13. Review of outstanding change orders
14. Review of reduction of retainer
15. Final contract amount including all approved change orders
16. Obtaining final waivers of lien
17. Administrative close-out of special programs
18. Final payment

Exhibit 21-4
Sample agenda for a project close-out meeting.

PREPARATION OF AS-BUILT DRAWINGS

As-built drawings are the documents that are prepared at the end of the construction process to reflect the built and installed conditions of the project. The as-built drawings should incorporate the design drawings as well as detailed drawings of the installation of the elements of each system (e.g., mechanical, electrical, plumbing, sprinkler and life safety, etc.) into a coordinated drawing reflecting the as-built conditions for the project. These documents are called for in the contract documents, and must be turned over to the owner in a timely manner for the owner to properly operate and maintain the building after acceptance and occupancy.

Many CM/GCs keep track of the as-built conditions by taking the design drawings, especially the architectural, structural, mechanical, electrical, plumbing, sprinkler, and life safety documents and marking them up with changes as they occur. These documents are often placed on the wall of the CM/GC's field office to allow the field personnel to mark them up and have an ongoing record of the as-built conditions. By doing this, one does not have to try to recapture from memory what changes were actually made. With today's technology, these documents are often developed and delivered as CAD files, thus being more environmentally conscious and making the documentation easily available to all. Each subcontractor should be contractually responsible for preparing their own as-built documents, and submitting them to the CM/GC for review, coordination with the other trades, and submission. Many contracts call for the submission

Exhibit 21-5

Recommendations for preparation of as-built drawings.

1. The preparation of as-built documentation should start from day one on the project with a set of marked-up drawings to reflect the as-built conditions, as the project is being built.

2. Keep a separate set of the contract documents in the CM/GC's construction field office, so that they can be marked up and used as living documents throughout the construction process.

3. Collect all approved shop drawings, catalog cuts, and equipment specification sheets.

4. Collect, compile, organize, and distribute documentation of the as-installed conditions for the equipment on the project and any modifications made to them.

5. Collect all of the coordination drawings prepared by the trades to coordinate the installation of their work and the established rights of way for each trade.

6. Have the subcontractors prepare the as-built documentation for their trade, as per the contract documents, which is often an AutoCAD file.

7. Submit a sample of one of the trades' as-built documentation to the consultants, for review, format, and approval, prior to finalizing all of the trades' as-built documents.

of these documents using AutoCAD systems, to facilitate the archiving, review, and distribution of the information. Exhibit 21-5 contains the checklist for the preparation of as-built documentation.

OBTAINING AND COORDINATING FINAL APPROVALS

The final close-out phase of the project also requires obtaining all required approvals and inspections from all governing agencies, testing laboratories, and licensed architects and engineers. In large urban cities there are often many jurisdictional agencies with which the CM/GC has to deal to obtain final approvals in order to close-out a project. This is discussed in more detail in Chapter 4. Exhibit 21-6 contains recommendations for coordinating and obtaining final approvals for a project.

WAIVERS OF LIEN—PARTIAL AND FINAL

A partial waiver of lien is obtained with each requisition payment to document that the contractor has received the payment and the contractor has paid all of his invoices from subcontractors, suppliers, and vendors, based on the work performed, requisitioned, and paid for to date. During the construction process, partial waivers of lien will be prepared

1. Confirm that the appropriate agencies have performed the required inspections and issued their respective approvals. It is recommended that a checklist of all required inspections and approvals be prepared detailing each inspection, who is responsible for performing it, the date it is scheduled, and its status.
2. Confirm the work is to contractual and governing code standards.
3. Confirm that all life safety, sprinkler, security, and safety features of systems are operating properly in accordance with the contract documents and governing codes to protect the safety of the public at large, construction workers, and the owner's occupants.
4. Coordinate commissioning, which includes the start-up; balancing; water treatment; filling up of the building's mechanical, electrical, sprinkler, and plumbing systems; life safety; vertical transportation; and building management systems.
5. Arrange for the contractually required training of the owners and their staff. Send out notices to all concerned parties of the schedule and location of training, and document the training session. It is good to have operating and maintenance manuals available during the training sessions.
6. Develop a procedure to log in and monitor all warranty and guarantee claims.
7. Maintain contact with the owner throughout the first year, or longer if the warranty period is extended, and respond to all requests for warranty service in a timely manner.
8. Prepare operating and maintenance manuals for all building systems, including cuts of all manufacturer's supplied equipment.
9. Prepare product care instructions for all finished materials installed in the project.
10. Coordinate the architectural, mechanical, plumbing, sprinkler, electrical, and life safety systems into a comprehensive set of as-built documentation.
11. Assemble all product warranties, guarantees, and start dates.
12. Hand over all attic stock, spare parts, keys, etc.
13. Provide any maintenance of equipment that may have been used during the construction period (e.g., fan systems, boilers, chillers, pumps, cooling towers, elevators, etc.) to restore them to their original new condition.

Exhibit 21-6
Recommendations for coordinating and obtaining final approvals.

for each monthly progress payment, and must be received prior to disbursing the current payment due.

When the final requisition is due to be paid, and all retainers released, a final waiver of lien is prepared to document the last final payment in accordance with the contract amount and all approved change orders. The final waiver of lien also acknowledges that the contractor is accepting the last and final payment as payment in full for the work performed, and that there are no outstanding claims. The final waiver of lien must be obtained prior to disbursing the final payment. This is discussed in more detail in Chapter 20. Exhibit 21-7 has a checklist for preparing and assembling the final lien Waiver.

Exhibit 21-7

Recommendations for assembling the final lien waiver.

1. The CM/GC should prepare the partial waiver of lien for each subcontractor to be paid with each requisition cycle. Exchange the signed and executed partial waiver of lien from the subcontractor for the payment for that requisition period. This usually works when the requisition cycle does not exceed 30 days for payment.

2. When payment of the requisition by the owner to the CM/GC exceeds 30 days, the next requisition is usually already being prepared in the monthly requisition cycle. The CM/GC should only sign off for the amount of monies actually received, and not the money requisitioned for in the last requisition.

3. Include with each requisition submitted to the owner the partial waivers of lien for the CM/GC and all subcontractors, as defined in the contract. This will facilitate payment and processing of the requisition.

4. If a requisition must be adjusted by the owner, request that the adjustment be made in the next month's requisition so as not to cause an accounting problem between the requisition, waivers of lien, actual payments, etc.

5. Prior to preparing the final lien waivers, reconcile and finalize each subcontractor's final contract amount including all approved change orders.

6. Set up meetings with the subcontractors to discuss and resolve any open change orders or claims that may still be outstanding and resolve them.

7. Set up a meeting with the owner to discuss and resolve any open issues with regard to claims for monetary compensation to finalize the contract amount.

8. Collect final waivers from contractors before releasing final payment. The CM/GC preparing the final waiver of lien and exchanging an executed signed copy of the final waiver of lien for the final payment usually does this.

9. Review the final payment to ensure that the work being billed has been completed and that final waivers of lien have been prepared for the value of the work being invoiced and paid.

ADMINISTRATIVE CLOSE-OUT PROCEDURES

Many projects today are equally challenging with the construction and administrative processes required to properly complete them. It is not only sufficient to complete the project on time, within budget, and of high quality, but also to properly administer and document the process. This is discussed in more detail in Chapter 14. Exhibit 21-8 is a checklist for the recommended close-out procedures.

TURNOVER AND ACCEPTANCE OF EQUIPMENT

As the project is completed, the CM/GC must turnover the spaces and MEPS equipment to the owner. Sometimes this is done is phases, as the owner wants to occupy and begin utilizing portions of the overall project for their beneficial use. This is a milestone in the project, and usually represents the start of the warranty and guarantee period,

Exhibit 21-8
Project close-out procedures.

1. Obtain a certificate of substantial completion from the owner upon completion of the project

2. Obtain letters of partial acceptance of areas or equipment turned over to the owner for his beneficial use and occupancy, due to project phasing or early occupancy.

3. Determine the start of all turnover dates. Make sure warranty dates are established and communicated with the owner.

4. Determine that a temporary and/or final certificate of occupancy has been issued for the project.

5. Verify that all special administrative programs and requirements have been satisfied with all required documentation; certified payrolls have been submitted, etc.

6. Ensure that the contractor's bonding company (if applicable) has issued a consent of surety.

7. Ensure that the final insurance credits are issued if the project is covered by an owner or contractor controlled insurance program.

8. Obtain evidence that the rental items were transferred as appropriate and related rental payments have stopped.

9. Turnover to the owner any equipment that was purchased specifically for the project by the contractor with the owner's payments for reimbursable general conditions, if required as per the terms of the contract.

10. Ascertain that any petty cash funds or project checking accounts were closed out, if applicable. Ensure all funds were transferred back to the owner as required.

11. Determine the project is "closed" in the accounting system to prevent further charging of costs to it.

12. Prepare and archive a project close-out file and ensure that it contains:

    a. Final signed-off owner's letter of final completion and releases.

    b. Building and other permits, sign offs, and approvals.

    c. Documentation of all inspections.

    d. All general correspondence with the owner, inspectors, etc.

    e. Submittal records and approvals.

    f. Back-up of any computer files relating to the project, clearly labeled.

    g. Final waiver of lien and satisfaction of any liens placed on the project.

    h. Copies of all temporary and final certificates of occupancy.

along with the owner's operation of the MEPS systems. In the urban environment where projects are often very large in scale, this is often a requirement. Exhibit 21-9 is a sample subcontractor completion letter.

Exhibit 21-10 is a sample checklist for the completion and turnover of HVAC equipment. Other MEPS equipment can utilize a similar type of list.

Exhibit 21-9

Sample
subcontractor
completion letter.

Corporate Letterhead

(Date)

To: (CM/GC)

Re: Name, address, and project number for the project

To Whom It May Concern:
This letter is to certify that the (name of subcontracting trade) equipment for the above—mentioned project was substantially completed and turned over for the beneficial use of the owner on (date).

The project is operating as per the plans, specifications, and contract requirements at the time of this letter. During the course of the project we have made periodic visits to the project site. To the best of our knowledge, information, and belief, our work is now substantially complete. All work has been installed as per the contract documents and all controlled inspections, balancing, start up, testing, and commissioning has taken place. All applicable provisions of the building code have also been complied with.

All materials, equipment, and workmanship are covered by a warranty and guarantee for a period of one year. The warranty period will expire on (date).

Thank you for the opportunity to be of service to your firm on this project.

Sincerely,

Principal of Firm
Name of Firm

Exhibit 21-10

Sample checklist
for completion and
turnover of HVAC
equipment.

	Initial if Yes	Date
1. Maintenance manuals		
2. As-built drawings of systems as installed		
3. Balancing reports		
4. Controlled inspection reports		
5. Operating and maintenance manuals		
6. Nameplates on all equipment		
7. All valves tagged with valve tag chart		
8. Attic stock turned over		
9. Filters changed		
10. Warranties and guarantees provided		
11. Training provided		
12. Preventative maintenance program established		
13. Punch list completed		
14. Equipment use permits obtained		
15. List of parts for all equipment		
16. Manufacturer's equipment cuts		
17. Contact numbers for normal maintenance		

Exhibit 21-10
(*Continued*)

	Initial if Yes	Date

18. Emergency contact numbers
19. Maintenance proposal for ongoing maintenance
20. System description and operational sequencing

Acceptance of the system:
By: (print name) _____
Signature: _____
Date: _____

Note: If acceptance is qualified based on any outstanding items, please indicate the items below.

SUMMARY

- The project close-out and punch list phase is an important one and must be managed properly.
- Pre-punching of the project by the contractor is an effective way to manage the punch list and QC program throughout the project, and minimize the time, energy, and aggravation in completing the punch list.
- Plan to have a close-out meeting to identify all of the operational and administrative close-out requirements for the project.
- Contractors often construct a project on time, within budget, and of high quality, and then ruin their reputation by not closing out a project in a timely and proper manner.
- As-built drawings and documents, operation and maintenance manuals, and warranty information should be gathered throughout the project and checked for accuracy, completeness, and adherence to the project specifications.
- Assemble all final waivers of lien.
- Ensure that all appropriate agencies and personnel have performed the required inspections and approvals.
- Ensure that the administrative close-out of all special programs and contractual administrative requirements have been complied with, for example, wrap up insurance policy, EOE, and surety bonds.
- Obtain a job close-out file and ensure that it is archived for future reference.
- When the final requisition is due to be paid, and all retainers are released, a final waiver of lien is prepared to document the last final payment in accordance with the contract amount and all approved change orders.
- The final waiver of lien also acknowledges that the contractor is accepting the last and final payment as payment in full for the work performed, and that there are no outstanding claims.

22 Technology

(They are trying to make my job easier.)

TECHNOLOGY

What is it? Technology, as it relates to construction, is the advancement of materials, tools, equipment, and processing that will enhance the productivity of putting structures together. See the technology flow chart in Exhibit 22-1.

Technology in a broader sense is helping to achieve a better universal environment by having the total construction industry produce materials and systems that will contribute to the new green objective. See Chapter 23 for more information on green construction.

HISTORY

When you look at building construction today, you can see that no dramatic labor changes have occurred since the building of the Great Pyramid at Giza, 4500 years ago. How can this be, considering the fact that today's construction process involves the mechanization of most of the elements that go into constructing a building? However, consider the number of trades people it takes to complete a project, including suppliers and fabricators. The Great Pyramid at Giza was constructed the same way, with many workers performing various tasks.

Yes, we use cranes and hoists to move trades people and material which were not available to the Egyptians. It still takes trades people to operate the equipment, move the material from the loading dock to a particular floor, and then install all the components. It is still a very labor-intensive business.

What the construction industry has done over the years is to find new methods, material, and technology to reduce the time it takes to construct a project and in a small way

Exhibit 22-1

Technology flow chart.

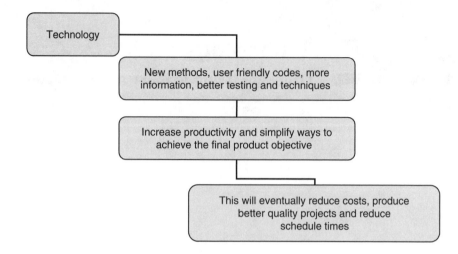

reduce some of the required trades people. We have seen the evolution from using ramps to get materials up and around a project to manually operated cranes to steam operated stationary cranes to mobile cranes to self-climbing cranes. A number of other innovations have helped advance the construction process as noted in Exhibit 22-2.

Thus, the industry has come a long way in the advancement of technology, but we have a long way to go to resolve the constant problems that afflict the construction industry. These problems include the following:

1. Better quality drawings
2. Better quality control
3. Better trained trades people
4. Better safety standards
5. Control of costs
6. Control of schedules
7. Reduction in the number of people it takes to put a project together
8. Reduction in misunderstandings among the construction manager/general contractor (CM/GC), the owner, and the subcontractor
9. Reduction of risks for all parties

The industry has also found ways to reduce some of the problems mentioned above. These include:

1. Use of building information modeling
2. New materials to assist in the construction process
3. Better ways to communicate
4. More accurate ways for surveying by the use of global positioning systems (GPS)
5. Use of cameras that give the CM/GC instant pictures

1. Stone structures to steel and concrete frames
2. Punched windows to curtain walls
3. Natural ventilation to air conditioning
4. Fireplaces to boilers
5. Natural light and gas lamps to electric lamps
6. Thatch roof to single-ply roofing
7. Stairs to elevators/escalators
8. DC electricity to AC electricity
9. Moving concrete by wheel barrows to pump trucks
10. Stone walls to drywall
11. Plaster ceilings to ceiling tile
12. Cast iron to no hub piping
13. Welded pipe to pipe clamp compression fittings
14. 3000 psi concrete to 25,000 psi concrete
15. Levels and plumb bobs to lasers
16. Hand-drawn drawings to CAD
17. Written notes to rotary phones to cell phones
18. Telephone wires to cables
19. Hand-made steel sections to computerized manufacturing
20. Plain concrete to admixtures that can be used for hot and freezing weather-pouring conditions
21. Large steam operated pumps to small electric pumps
22. Exposed electric wires to BX or conduit or copper bus duct
23. Surveying transits and chains to total stations and GPS
24. Adding machines to computerized spreadsheets
25. Paper notes to computers and trios (e-mail, telephone, and two-way radios)
26. Poured in place concrete to precast or pre-stressed concrete
27. Soldier beams with rackers to slurry walls
28. Hand signals for crane operators to two-way radios

Exhibit 22-2

Innovations that have advanced the construction process.

6. Computers that can process more of the contractor's required information
7. Use of the Internet to find relevant information
8. Better-quality testing methods
9. Codes

BUILDING INFORMATION MODELING (BIM)

This method takes CAD one step further and produces drawings in 3D with the ability to review a real time scheduling process. This is a virtual reality step for the construction industry. The CM/GC will now have the ability to review the project from any prospective. As an example, if the CM/GC wants to review a structural detail at a complicated connection, this would be possible with BIM. BIM has many advantages to offer the CM/GC, including the following:

1. Take off of materials would be simplified.
2. Complicated details could be reviewed and analyzed.
3. Coordination of the various trade components can be reviewed for potential "hits."
4. Sequence of putting a project together is enhanced.
5. The fourth dimension (time) can be incorporated to show how fast a project can be put together.
6. Site work elevations between existing conditions and the final elevations can be determined.
7. The best routing for ductwork, pipes, wires, lights, sprinklers, and cables can be reviewed.
8. Logistics of the site with the location of cranes and hoists can be analyzed.
9. Lift schedules for the placement of steel, concrete, and large mechanical and electrical equipment can be determined.
10. The development of schedules and the associated logic will be enhanced.
11. Potential safety problems can be assessed.
12. Can evaluate alternatives in more realistic terms.
13. Can coordinate trades prior to performing the actual work.

Exhibit 22-3 shows a typical BIM building layout, indicating the location of steel, ductwork, lights, and the ceiling within a proposed project. As depicted in this exhibit, the contractor can actually see the elements of a portion of a project without actually constructing that section of the project.

Exhibit 22-3
BIM layout.

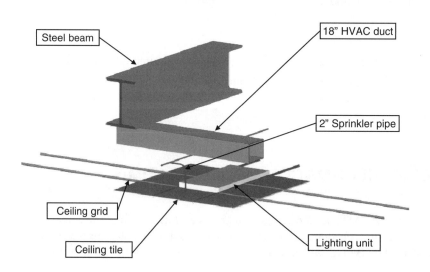

BIM has not been fully integrated with the architectural profession or the contracting community. According to a December 10, 2007 *Engineering News-Record* Special Advertising Section, approximately 48% of architectural firms and 13% of contractors are using BIM. The reluctance to use BIM has to do with costs, lack of industry standards, lack of expertise, and lack of proper training. However, one of the biggest factors is the question of liability and potential exposure to lawsuits. The question is which discipline will take the responsibility for showing these details—the consultants or the CM/GC. However, the article stated more acceptance of BIM would occur within the next year.

The following are some of the software companies offering BIM programs:

1. Autodesk Revit
2. ArchiCAD from Graphicisoft
3. Bently Architecture
4. Digital Project from Gehry Technologies LLC

NEW MATERIALS

New materials are being introduced into the construction industry at a very fast pace. The need for new materials that will help the CM/GC to perform better and to achieve productivity increases is being met. The new buzzword "LEED" (Leadership in Energy and Environmental Design) is also forcing manufacturers to develop new and improved sustainability products. The following are some of the innovations occurring in the construction industry:

1. The use of reinforcing steel that has yield stresses of 125,000 psi. According to an August 27/September 3, 2007 *Engineering News-Record* article, the high strength reinforcing steel makes it easier to pour concrete in columns. The number of reinforcing bars required to support the load was reduced due to the higher strength steel. The higher strength reinforcing steel is more expensive then the standard yield bars of 60,000 psi, but pouring of the concrete into the column forms was simplified.

2 The use of waterless urinals is being touted as a way to not only save water but also to increase plumbing productivity by eliminating water supply pipes, flushometers, and pressure reducing valves.

3. The use of self-compacting concrete (SCC) has made an impact on the way concrete is being poured. According to Concrete Centre, with the use of super plasticizers and a stabilizer the concrete is able to achieve consolidation into the formwork without segregation of the aggregate. The method was developed in Japan during the early 1990s and has now been successfully used in the United States. The use of SCC achieves the following:

   - Vibration crews are eliminated.
   - The over-vibrating of the concrete by the crews with the possibility of aggregate segregation is also eliminated.

- Noise from construction is reduced.
- Pouring concrete is increased.

4. Pumping of concrete has achieved new levels on the Burj Dubai tower project. Pumping to a height of 1971 feet was accomplished in November 2007. As noted by Putzmeister (concrete pump supplier) and Unimix (concrete subcontractor), the concrete consisted of a very low water-to-cement ratio and high strength concrete. Due to the high daytime temperatures, it was necessary to place ice water in the mix and to pump at night.

5. A November 2007 *Civil Engineering Magazine* article stated that fiber reinforced concrete could greatly increase the ability of slab-to-column connections to withstand earthquake forces.

6. The industry has developed a mechanism for eliminating the machine room for traction elevators. In this process, a traction motor is placed on the side of the rail and coated steel belts are used in lieu of cables. Exhibit 22-4 is a diagram of how the Kone traction motor operates without the need for an elevator machine room. This new equipment reduces space in a building and reduces some of the structure that initially was required to support the floor and equipment in an elevator machine room. In addition, the building structure is reduced because the floor no longer has to support the load of the heavy elevator machines.

7. In New Jersey where a new stadium is being constructed for the local football team, the CM/GC is tracking 3200 precast risers by radio frequency identification (RFID), which is embedded in the precast unit. The contractor then can check the status of each piece of riser from fabrication to when it is shipped to the site. In an urban environment, where storage is at a premium, tracking of critical components is a must. The old method of telephoning or E-mailing the supplier to find out the status of the component will now become passé. RFID has the ability to track any product and obtain a real-time status report.

8. The use of more energy efficient products and systems. This includes chillers, cooling towers, lighting, and exterior glass.

9. The industry is finding more-efficient ways to recycle construction material waste.

10. Laser scanners are being used to take site conditions and convert them into real-time 3D drawings, take inventory of materials, and show punch list items.

COMMUNICATIONS

The field of communications has made fantastic strides since the inception of the cell phone. Besides the vocal wireless communications that have been used for a number of years, we now have the ability to:

1. Use E-mail
2. Get spread sheets for budgets, logs, and reports
3. Have pictures sent from a job site and transmit them electronically to any location
4. Take pictures from a camera phone
5. Obtain information on weather conditions immediately

Exhibit 22-4
Machineless
elevator.

The secret behind the KONE MonoSpace® concept is the permanent-magnet, gearless KONE EcoDisc® motor. The revolutionary KONE EcoDisc® weighs less than half of a conventional geared traction machine, has only one moving part, and is roughly twice as efficient. Its efficiency is three times that of a hydraulic power unit and it uses 60% less energy– a savings which can represent half or more of the annual cost of elevator operation.

Maintenance access panel (MAP) is integrated in the landing door frame

KONE MonoSpace® does not need a separate machine room thanks to the KONE EcoDisc® hoisting motor, which is so compact that it can be located in the shaft.

The whole elevator is fitted inside the shaft, simplifying the building interface and saving space for more profitable use. This also speeds up the installation process, since no scaffolding or cranage is needed.

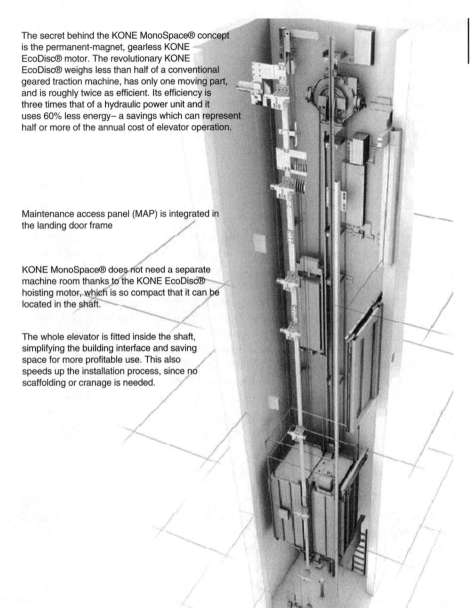

These abilities have made it easier for the project manager (PM) to obtain and give back information on a real-time basis. In previous years, you would have had to wait hours and sometimes days in order to get information and then respond. The new technological revolution has changed all this. Another positive aspect of these innovations is that the PM is only seconds away from obtaining information in which decisions have to be made. Obviously, the down side to this instant communication ability is that the PM can now be available 24/7 for most projects.

GLOBAL POSITIONING SYSTEM (GPS)

GPS has assisted the construction industry in the surveying of buildings. Location and elevations of floors can be checked against conventional surveying methods. The GPS can also confirm property lines and footing locations. GPS can determine the location of concrete trucks and assess the timing of the delivery. GPS can also expedite the delivery of product by assisting drivers in finding an unfamiliar project location, or finding an alternative rout in case of traffic congestion.

CAMERAS

The change from film to digital cameras and videos has made a major difference to the PM. In years past if a problem existed at a site, photos were taken and then time would be lost by the developing and transportation process. With today's digital cameras, a picture of the problem is taken and then it can be documented and transmitted almost instantaneously. This quick response time helps to resolve a problem almost immediately. The quality of digital pictures has improved greatly in the last 2 years. We have gone from 2 pixels to 12 pixels. This capability has greatly improved the image of the picture so a better representation of the field condition can be analyzed. In addition, in some cases the contractor can post video cameras on a pole so that the progress of a project can be monitored remotely in real time. Digital photographs of site conditions distributed to the design team can bring the conditions to their office without the need for a field visit.

COMPUTERS

Up until 5 years ago if you went to a job site, you would not have seen a computer in the field office. This has completely changed. Now every job site has at least one computer in the field office. The computer has drastically changed the way we process information on the job site. Today, information is passed along to the remote project team within seconds. With wireless capability, telephone lines are not required to transmit information. Questions and answers on the project site can now be evaluated in a short period of time. In addition, there is voice recognition software where typing in information will no longer be required. All you have to do is talk into the computer microphone and the message will be placed on the screen for transmitting. The computer assists the PM in many ways:

1. Faster turn around time on requests for information (RFIs)
2. Updates the schedule on a constant basis
3. Keeps all field reports updated
4. Tracks the trades population
5. Can help track weather conditions on a daily basis and provide alerts on potential fronts that may be approaching the site

6. In some locations, can determine the best truck routing by tying into a city's traffic cam

7. Evaluates alternatives based on cost and schedule

8. Tracks long lead times with automatic updating

INTERNET

The Internet has had a dramatic impact on how the construction industry can obtain viable information. New products, new equipment, and new techniques can now be readily available with the click of a button. Some Websites such as Google Earth can assist the PM in assessing the urban area in which they will be working and reviewing potential problem areas. This includes:

1. Traffic patterns

2. Surrounding buildings and their impact on the construction site

3. Best routes for getting material to the site

4. Setting up the logistics plan

If a more sophisticated Website is required, the PM can review Google's Earth Pro website. With this information, the PM can review site data digitally and create 3D drawings.

With all data obtained on the Internet, the CM/GC has to make sure that the information is coming from reliable companies and individuals.

TESTING

New advances in testing procedures have been used to increase quality control in the construction industry. Some of these innovations include the following:

1. Piles—In lieu of load testing the piles, the new method of high strain dynamic load testing is being used. This method uses sensors to analyze the load carrying capacity of the pile. This procedure reduces pile testing costs and time associated with the test.

2. Soil bearing—Pressure meters and Outerberg cell tests are being used to measure *in situ* the bearing capacity of soils and rock. This can save time and in some cases can convince building officials to increase the bearing capacity of the material being investigated (from code-stated values).This could reduce foundation costs.

3. Cameras—Miniature cameras can be sent down small concrete shafts as well as pipes to inspect for any defects.

4. Sensors—Sensors are being placed directly in concrete and on steel to measure the actual stress being imposed on the structural elements when loaded. This is then compared to the theoretical stress that was calculated by the engineers.

5. Stress gauges—These gauges can measure minute movement in buildings that are next to new construction sites. Thus, any vibration that may cause settlement, cracking, or failure is noted immediately.

6. Hand-held laser tapes—These tapes can accurately measure distances without the aid of a target and can be used by one inspector.

7. Infrared sensing—This sensing device can determine hot spots in electrical equipment and wires. It is also used to determine if a building façade is losing heat and the location of wet insulation on roofs.

8. Optical scanners—These devices have the ability to evaluate concrete structures and determine the location and condition of reinforcing steel.

CODES

The new International Building Code (IBC) allows the industry to use new materials by having them tested by a recognized laboratory. This will encourage the construction industry to find new materials and methods that are less costly and easier to use. In addition, as stated in Chapter 23, the new materials that are being produced will also be environmentally friendly.

OTHER RESOURCES

1. *Engineering News-Record* (ENR), the weekly magazine for the construction industry, offers many articles on technological advances in the construction industry. The Website for ENR is www.enr.com

2. Purdue University has a Website that keeps the construction industry updated on new technological advances. The Website is www.ect-purdue.org

SUMMARY

- The construction industry is still based largely on using excessive amounts of labor.

- Technology has helped to speed the construction process and to reduce some labor.

- Great strides in technology are still needed to reduce:
  - Poor quality drawings
  - Quality control problems
  - Poorly trained labor
  - Accidents and deaths
  - Costs

- Schedules
- Amount of labor
- Conflicts between owner and contractor
- Risks
- Over the past several years, technological advances have helped the construction process. These have included:
  - Use of BIM
  - Use of new and labor saving materials
  - More advanced communications systems
  - Use of GPS
  - Digital cameras
  - Computers with wireless capability
  - The Internet

23 Green Construction
(Saving future generations.)

GREEN BUILDING

The intention of green environmentally friendly buildings is to provide environment, economy, health and community benefits by constructing buildings that reduce energy consumption, improve air and water quality, reduce solid waste, conserve our natural resources, have the potential for reducing operating costs, protect our ecosystems and biodiversity, improve employee productivity and satisfaction in the workplace and home, enhance the comfort and health of building occupants, optimize life cycle economic performance, be socially responsible, minimize the demands on the infrastructure of the world and our country, and make a positive contribution to the overall quality of our lives. "Going green" in building projects has become especially important in the urban environment, where large projects and large corporations want to focus on sustainability and environmental responsiveness during the construction process. More and more owners as well as private and public construction programs are realizing the benefits of being recognized for their environmentally friendly approach to building or renovating structures. In addition, the government is starting to pass new rules, regulations, and legislation that encourage green building along with potential tax benefits. This applies to new commercial and residential buildings, schools, interiors, retail, healthcare, renovations, and core and shell construction. The buildings that we construct have a very significant impact on our natural environment, economy, productivity, health, and quality of our lives as well as the overall well-being of our planet. Construction also transforms the land in the urban environment with an impact on the overall ecology. In the United States, buildings currently account for:

- +65% of total electrical consumption
- +35% of overall energy use
- +30% of all greenhouse gas emissions

- +30% of all supplies and raw materials used
- +30% of solid waste annually (+135 million tons annually)
- +10% of total potable water consumption

Making buildings more energy efficient will certainly have a positive effect on the environment and help to slow global warming.

HISTORY

The green movement started in the 1970s with the oil embargo and gasoline crisis. The cost of fossil fuels rose dramatically. Buildings systems starting taking a different approach by introducing variable air volume (VAV) systems, which could throttle the air based on the heat or cooling load. In addition, lighting systems were revamped to change the type of light fixtures to be more energy efficient, and to reduce the total electrical lighting load from approximately 5 watts per square foot to 1.5 watts per square foot. The benefits for building green are obvious. Global warming has become a major concern. The burning of fossil fuels is certainly not helping the situation with the carbon emissions that we are producing and placing into the Earth's atmosphere. In addition, the cost of petroleum is skyrocketing and having an impact on the economy throughout the world.

Although there are many benefits of building green, based on construction industry research the overall cost to construct a project to conform to Green LEED (Leadership in Energy and Environmental Design) standards is approximately 10% to 15% higher than a conventional project. In addition, given that many Green LEED projects are new, there is not a lot of history on the durability and maintenance of these building systems.

Owners, developers, legislators, environmentalists, and community organizations are asking construction managers/general contractors (CMs/GCs) to construct green buildings using environmentally friendly materials to minimize the "carbon footprint" that we are leaving in the construction process. The term "carbon footprint" means the amount of fossil fuels that we burn to produce a project, directly or indirectly. This means that people have to build smarter, more efficiently, and more effectively, and look for materials that are environmentally friendly and readily available locally to minimize the impact on the environment. Many construction firms have developed an in-house group of professionals to assist the owner and design team during the preconstruction and construction phases of a project in qualifying for LEED certification from the United States Green Building Council (USGBC). Many owners want to build a green building for economic benefits where feasible, especially with the rising cost of oil and gas, but also for the social benefits of making their employees, community groups, special interest groups, shareholders, clients, regulatory authorities, etc., feel better about the buildings they are building and enhancing the firm's social awareness of saving the environment.

Achieving Green LEED construction credits requires extensive documentation, along with knowledge of the building processes. The CM/GC must coordinate and manage the contractors during the construction process, if a project is to be successfully Green LEED certified.

BENEFITS OF BUILDING GREEN

There are many benefits of building green. Exhibit 23-1 is a summary of the benefits of building green.

While many of these issues directly benefit the owner and community at large, the CM/GC can also benefit from Green LEED construction. Exhibit 23-2 is a summary of the benefits to the CM/GC for building green.

1. Reduced material usage
2. Savings in construction waste
3. Tax credits and other incentives
4. Reduced operating costs
5. Reduced infrastructure costs
6. Ability to obtain faster construction permits and project approval by complying with recommendations and programs with the governing authorities
7. Lower water consumption and reduce water pollution
8. Reduced toxic emissions
9. Reduced air pollution, emissions, and ozone depletion
10. Potential for lower energy costs
11. Improved durability of equipment and minimized repairs
12. Reduced cost of reconfiguring an office space by installing raised flooring and open landscape layouts, which provide flexibility for future modifications
13. Reduced solid waste generation
14. Potential increase in sales and leasing rates
15. Employees more satisfied in the workplace, more productive, and lower turnover
16. Enhanced comfort and health of employees
17. Potential increase in property values
18. Complying with regulations and being ahead of the environmental curve
19. Reduced lawsuits
20. Support of organizations with socially and environmentally responsible policies
21. New business opportunities by having a specialized line of business in Green LEED environmentally friendly construction
22. Participate in programs that have a positive impact on the environment

Exhibit 23-1
Benefits of building green to the environment.

Exhibit 23-2

Benefits to the CM/GC for building green.

1. Creates new business opportunities
2. Provides a strategic advantage over your competitors
3. Reduces construction material usage
4. Reduces construction material waste
5. Reduces energy usage
6. Potentially lowers overall construction costs
7. Minimizes the manufacturing of new construction materials and their shipping by recycling and reusing construction materials
8. Facilitates construction permits and approvals
9. Generates Community good will
10. Participation in programs with a positive impact on the environment
11. Receives recognition by industry organizations for achievement
12. Receives marketing exposure for new projects and clients

THE LEED RATING SYSTEM AND STANDARDS

The USGBC has developed the LEED certification program, which is aimed at making a positive impact on the environment and peoples' overall health and well-being. It also strives to reduce operating costs, increase a building's marketability, increase the productivity of the occupants, and create an overall sustainable community. LEED has created checklists for various types of construction and renovation projects.

The LEED rating systems are available at their website www.leedbuilding.org in compliance with the USBGC recommended policies and procedures. These rating systems are voluntary, and based on the established engineering, energy, environment, best practices, and principles that have been developed by a consensus of professionals in the design and construction industry. To obtain LEED certification, the project team must register online at the LEED Website, which will guide you through the application and certification process. The website also has technical support, credit interpretations, and training programs available. It is important to note that the project must satisfactorily document all of the requirements and the minimum number of points to qualify in the area of site sustainability, water efficiency, energy use, materials and resources, indoor quality environment, and innovation in design.

There is the potential for 69 total points in the LEED program. A project can qualify for certification as silver, gold, or the highest certification, platinum. The project checklist for Green LEED is available at the LEED website. Additional information is also available at the USGBC Website www.usgbc.org.

SUSTAINABLE CONSTRUCTION IN THE URBAN ENVIRONMENT

Many of the large cities throughout the United States and world have encouraged the building of Green LEED sustainable construction in their jurisdiction. Los Angeles recently passed ordinances that require all new developments in excess of 50,000 square feet to meet LEED requirements, which is estimated to cover over 150 projects annually. The city is promising expedited review and approvals of these projects. It is also creating a sustainability team for green projects. The Battery Park City Authority in lower Manhattan, New York, has a similar mandate for their more recently constructed buildings. Other examples of large Green LEED projects are: Hilton San Diego Convention Center, San Diego, California; Merch Serono, Antidote, Geneva, Switzerland; Burg, Dubai; Taipei 101, Moscow Triumph Palace, Moscow, Russia; Palm Jumeirah, Dubai; and the planned Palm Deiva, Dubai. In Abi Dubai in the Persian Gulf, a new city is being built that will be the first to be carbon neutral.

At a recent conference, the mayors of major cities within the United States indicated that they are pushing for legislation to make their cities sustainable. A resolution was passed that calls for all buildings to be carbon neutral by 2030. Many of the mayors called for the federal government to pass energy and green building legislation, with economic aid and incentives to fund these initiatives and programs. Chicago has over 3 million square feet of green roofs, and four platinum certified buildings, the largest number in the nation. New York City has its first platinum LEED new commercial office building being constructed, the Bank of America Tower at Bryant Park. The city of Austin, Texas, has passed legislation that all new single-family homes constructed after 2015 will have to be "zero energy capable." This means that the home will have to be self-sufficient and produce as much energy as it consumes. Even smaller cities, such as Grand Rapids, Michigan, have passed legislation that all new government buildings must be constructed to pass LEED certification. In addition, they have put in place incentives for owners and developers to build green.

In the urban environment there are many unique challenges one must face, especially the CM/GC, in order to build green, environmentally friendly and sustainable buildings. First, there are the various government agencies that have jurisdiction over the construction work, which can include various authorities, agencies, and local, state, and federal regulations. They often represent layers of applications and approvals, which take time and money to deal with. Some municipalities have passed or have pending legislation that requires buildings to comply with Green LEED certified projects. Some municipalities are considering legislation and new codes, which will not approve a building permit for a new building if it is not Green LEED certified. Often professional expeditors are needed to facilitate the review, approval, and building permitting process. Then there is the issue of congestion in large cities, with restrictions on site access, lane closings, sidewalk closings, moratorium periods, storage of materials around the site, constrained site logistics, neighbors, community boards, etc., which present challenges. The location of materials and manufacturing facilities within a 50-mile radius of the site often limits which materials and supplies can be utilized to minimize trucking and carbon emissions. Many construction materials are made close to the source of the raw materials and labor to manufacture them. Unfortunately, there are a limited number of local manufacturers and suppliers near large urban population centers. There is also a fair amount of documentation that the CM/GC and subcontractors have to submit in order to comply with Green LEED programs. Many CM/GCs are not familiar with the process, and need training and qualified personnel to assist them in compliance.

ENVIRONMENTALLY RESPONSIBLE CONSTRUCTION

There are many items to consider for environmentally responsible construction in the urban environment. Exhibit 23-3 is a checklist for constructing environmentally friendly buildings.

Exhibit 23-3

Checklist of items to consider for environmentally responsible construction.

1. Evaluate with the architect, engineer, and owner the renovation of an existing building as an approach to sustainable construction.
2. Minimize the use of trucking, automobiles, and fossil fuels for equipment used in the construction process.
3. Provide responsible on-site construction management to absorb water runoff.
4. Utilize materials that avoid ozone depletion.
5. Utilize materials and products that are durable.
6. Evaluate the use of low embodied energy materials that are not energy intensive to manufacture.
7. Where possible purchase local building products.
8. Where possible use recycled materials.
9. Provide responsible on-site construction management to recycle construction materials.
10. Evaluate subcontractors who can recycle construction waste products such as concrete, cement block, copper, aluminum, steel, glass, and wood products.
11. Where possible utilize wood suppliers that use lumber from well-managed forests or engineered wood products.
12. Minimize the use of pressure-treated lumber, and evaluate alternative materials such as plastic lumber.
13. Minimize the use of materials that will give off gas with pollutants, such as formaldehyde and volatile organics compounds (VOCs).
14. Minimize the use of packaging protection when shipping and handling materials.
15. Utilize water-efficient equipment during the construction process, such as water conserving toilets and faucets.
16. Utilize waterless vacuum urinals, which do not use any water, thus saving on the size of water and sanitary piping.
17. Construct roof gardens (green roofs), which help to insulate the roof and building, along with absorbing and retaining water.

18. Utilize smart glass to provide solar shading and reduce the solar gain from the sun.
19. Utilize high efficiency lighting such as compact fluorescent and light emitting diodes (LEDs) for temporary light and power during the construction process.
20. Utilize energy-efficient equipment, motors, fans, heaters, air conditioners, etc., during the construction process.
21. Protect trees and topsoil while performing work at the site.
22. Avoid the use of pesticides and other chemicals that can find their way into the environment.
23. Minimize the job site waste generated during the construction process by recycling materials, and carefully sort the construction materials for optimal utilization.
24. Where possible make your construction business operations as environmentally responsible and friendly.
25. Provide education to your clients, architects, engineers, employees, subcontractors, suppliers, vendors, and the general public to minimize environmental impacts and provide a green sustainable environment.
26. Use concrete that contains fly ash in lieu of cement.

Exhibit 23-3
(Continued)

SUMMARY

- The approach to building green buildings is to provide benefits to the environment, economy, people's health, and the community.
- The buildings that we build have a very significant impact on our natural environment, economy, productivity, health, and quality of our lives.
- In the United States, buildings account for +65% of electricity consumption, +35% of overall energy use, and +30% of all greenhouse gas emissions.
- While the owner and community at large are the recipients of many of the benefits of the Green LEED program the CM/GC can also benefit from Green LEED construction with new business opportunities, having a strategic advantage over their competitors, reduction in construction material usage and waste, recycling of construction materials, facilitation of construction permits and approvals, community good will, and participation in programs with a positive impact on the environment.
- The USGBC has developed a LEED certification program, which is aimed at making a positive impact on the environment and people's overall health and well-being. It also looks to reduce operating costs, increase a building's marketability, increase the productivity of the occupants, and create an overall sustainable community.
- There is the potential in the LEED program to have the construction project be certified silver, gold, or platinum.
- Many of the large cities throughout the United States and world have encouraged the building of Green LEED sustainable construction in their jurisdiction.

- In a conference of mayors of large cities in the United States, a resolution was passed which calls for all buildings to be carbon neutral by 2030.
- In the urban environment, there are many unique challenges one must face, especially the CM/GC, in order to construct green and environmentally sustainable buildings, for example, several jurisdictional agencies to deal with, congestion of large cities, restrictions on site access and logistics, materials and manufacturing facilities within a 50-mile radius of the site, the amount of administration and documentation required, and the availability of trained and qualified personnel.

Epilogue

Constructing or renovating a building in a large urban environment is a unique challenge, and must be properly planned and professionally managed. This is especially due to the limited site sizes, congested streets and sidewalks, large populations of people living in the area, car and truck traffic, government agencies, administrative rules and regulations, underground utilities, underground transit systems, buses, lane and sidewalk closing restrictions, special interest groups, and the availability of qualified personnel and firms who can perform the work properly in this challenging environment.

THE FUTURE OF CONSTRUCTION PROJECTS

Throughout the United States and the world, construction is booming with large skyscrapers, housing developments, retail centers, medical facilities, educational facilities, infrastructure projects, the Olympic site in Beijing, and an entirely new city in Dubai (sections of which are "carbon neutral"), being built. This has lead to a major strain on the resources required to properly plan and construct these projects. There has been a shortage of qualified labor at the managerial and construction level throughout the industry. The shortage of personnel has been further impacted by the retirement of the Baby Boomer Generation. The industry has developed training programs and recruitment programs to encourage workers to enter the industry. Years ago, there were probably five people to train one new person coming into the industry. Today, it is the reverse, one person to train five new people. In addition, more non-union workers have entered the construction marketplace. The use of non-union labor presents opportunities for lower costs, without bureaucratic union work rules, but also comes with the challenges of safety and quality control, especially where there is a language barrier with the workers. In many cases, non-union personnel are immigrants to the United States who have come to the United States seeking employment and a better life for themselves and their families. They usually do not have good English language skills, and rely on their boss or foreman to communicate for them. They also come from countries and cultures where construction safety and quality control may not be a high priority, other than to build it fast. Often things get lost in the translation. Many states have right to work laws, such as Florida and Texas, where non-union personnel are performing the majority of the work.

We have also experienced a major shortage of raw materials used in the construction process such as structural steel, concrete, plywood, framing lumber, insulation, copper, glass, aluminum, and cement products. Many of these products have been shipped abroad to foreign countries such as China, India, and Dubai. These locations are also undergoing a massive construction program along with their own industrial revolutions, bringing them into the 21st century. This places a further demand on petrochemical resources, and results in higher prices for all goods and services throughout the world.

In addition to the shortage of raw materials, there is also a shortage of the manufacturing facilities to manufacture the finished construction products. Curtain wall manufacturers are in short supply, with only a handful of suppliers left. The concrete industry has a shortage of concrete mixing plants. The steel industry in the United States has been severely curtailed due to high labor rates, environmental considerations, unions, and outdated technology in the manufacturing process. Many of these products are manufactured abroad due to cheaper labor, no unions, little environmental regulations, and modern technology that other countries are using to manufacture construction materials. In a global economy, the United States has lost its competitive position in manufacturing.

The next 10 years promise to be challenging and demanding times for the construction industry to build the quantity and complexity of projects that are being planned. The projects are becoming even more challenging as available land to build on is becoming scarce. This results in construction taking place in a congested urban environment. The construction industry is slowly embracing technology to assist in building more efficiently and effectively. Only by attracting, retaining, and educating the construction personnel needed in the industry to manage and construct these massive projects, will we as an industry succeed in our mission to build, improve, and modernize the United States and the world.

OUR CONSTRUCTION PHILOSOPHY— LAMBECK AND ESCHEMULLER ISMS

Over the years, the authors have thought of simple ways to enhance the project managers' understanding of construction in the urban environment. We refer to these thoughts—our construction philosophy—as Lambeck and Eschemuller isms, which are as follows:

- Do not trust anyone until you can confirm his or her honesty and integrity.
- Avoid litigation whenever possible.
- Everything should be in writing and documented. We all have bad memories.
- Principals of the construction company should visit the owner and the construction site.

- Negotiate in good faith.
- Have a checklist and action plan that will prepare you for unexpected events.
- Evaluate the risks of constructing the project.
- Treat subcontractors with respect and fairness.
- Always present the facts, not conjecture.
- Always solicit opinions from other people and professionals to assist in problem solving. Two heads are better than one.
- Always focus on the problem and not the symptoms.
- When communicating in writing, keep it simple and use the KISS (Keep it Simple, Stupid) method.
- Study and learn the construction documents, both the plans and specifications.
- Always visit the site to determine the existing and as-built conditions.
- Always ask questions. No question is too foolish.
- Read technical publications to learn more about new technology being used in the industry.
- Never be pushed into making a decision until you understand the ramifications.
- Get all approvals in writing with approved signatures and dates.
- When you make a mistake, admit it and move on.
- Learn from your mistakes.
- Have all your facts together prior to advising as to which way to proceed.
- Know when to back down (you may have won the "battle" but lost the "war").
- Always prepare an agenda for every meeting and try to keep meetings as short and focused as possible.
- Request information by a definitive date, not just ASAP.
- Keep records of all important phone calls.
- Never assume anything, especially as it relates to dimensions on drawings.
- Where possible, use available information. Try not to "reinvent the wheel."
- Always review your work prior to submitting it to an owner.
- Learn something new about the construction industry every day.
- Make sure you understand the whole person you are dealing with—both the technical, rational side and the emotional side—and be sensitive to the person's needs and situation.
- Do not put off until tomorrow what you can do today, as you may need the time tomorrow to deal with unforeseen matters that arise.
- Expect the unexpected, and be prepared with contingency plans to deal with it.
- Use the best talent, resources, systems, and approach to build in an urban environment.
- Train your personnel in both the technical and managerial aspects of the construction process.
- Ensure that all team members within the construction management/general contractor firm understand their roles, duties, and responsibilities.

- Define who the team leader is and give him or her the authority, along with the responsibility, to get the job done.
- Identify all of the project stakeholders and try to address all of their concerns and special interests.
- Establish lines of communication with all of the stakeholders.
- Try to develop a "win–win" strategy when dealing with team members and project stakeholders.
- Make sure that safety is priority number one.
- A clean project is a safe project.
- Establish a good quality control program on the project.
- Identify project risks and develop a risk management plan to mitigate them.
- Plan for the orderly growth and succession of the management of your firm.
- Make sure you enjoy your job.

Now, let's go build an urban project!

References

Alan, Edward, *Fundamentals of Building Construction Materials and Methods*, 3d ed., John Wiley & Sons, New York.

American Subcontractors Association, "Model Ethics to Help Subcontractors Uphold 'Letter and Spirit' of Sarbines-Oxley Act," news release, September 22, 2006.

"Award of Excellence: Clyde N. Baker Jr.," *Engineering News-Record*, pp. 36–43, April 7/14, 2008.

Cardno, Catherine A., "Fiber-Reinforced Concrete Withstands Strong Earthquake," *Civil Engineering Magazine*, pp. 30–31, November 2007.

"Construction Engineering Site Serves as Information Portal," *Civil Engineering Magazine*, p. 37, December 2007.

Construction Financial Management Association, "CFO's Tool Box: Contractor's Check-up," p. 55, January/February 2006.

Construction Financial Management Association, "CFO's Tool Box: Surety 101: How Sureties Evaluate Contractors," p. 51, January/February 2006.

Construction Financial Management Association, "Tax and Legislation: Lien Laws and Out of State Construction Projects," pp. 59–63, November/December 2006.

"Cranes: Stalled Federal Rules Prompt State Action," *Engineering News-Record,* pp. 12–13, March 10, 2008.

CSI—Construction Specification Institute, 2004 format.

"Design-Build Continues to Grow Despite Wariness and Price Concerns," *Engineering News-Record*, pp. 38–39, June 12, 2006.

The DR Construction Group, "Concrete Advice—What Interests a Surety Company Most," pp. 1–2, February 2008.

EPA Website: www.epa.gov (general information).

Eschemuller, John, New York University Schack Institute of Real Estate, CM Capstone Project's documentation prepared by graduate students, 2003–2008.

GREEN Website: www.BuildingGreen.com (general information).

GreenSource Blogs: Green Buildings 2007 (general information).

GreenSource Website: www.greensource.construction.com (general information).

Haplin, David W., and Woodhead, Ronald W., *Construction Management*, 2d ed., John Wiley & Sons, New York.

"High Strength Rebar Called Revolutionary," *Engineering News-Record*, pp. 10–11, August 27/September 3, 2007.

"Hourly Union Pay Scales, September, 2007," *Engineering News-Record*, pp. 27–38, September 17, 2007.

Kerzner, Harold, *Project Management: A Systems Approach to Planning, Scheduling and Controlling*, 8th ed., John Wiley & Sons, New York.

Kordahi, Ray, "Underpinning Strategies for Buildings with Deep Foundations," masters thesis, Massachusetts Institute of Technology, June 2004.

"The Law: Design/Build Exposure: Implied Duties," *Civil Engineering Magazine*, p. 104, November 2005.

"The Law: Design/Builder Mistakenly Relies on Owner's Preliminary Design," *Civil Engineering Magazine*, p. 96, May 2006.

LEED Website: www.leedbuilding.org (general information).

Levitt, Raymond E., and Samelson, Nancy M., *Construction Safety Management*, 2d ed., John Wiley & Sons, New York.

Levy, Sidney, *Project Management in Construction*, 4th ed., McGraw-Hill, New York.

"Making a Case for Green Buildings," *Environmental Building News Magazine*, April 2005.

McGraw-Hill, new blog: To Live and Build in LA, April 26, 2008.

McHayes, Douglas, Sparks, Ian R., and Van Cranenbroeck, Joel, "Core Wall Survey Control Systems for High-Rise Building," October 2006.

Meredith, Jack R., and Mandel, Samuel J., *Project Management: A Managerial Approach*, 5th ed., John Wiley & Sons, New York.

"Missing Connection to Wall May Have Weakened Garage," *Engineering News-Record*, pp. 14–15, November 17, 2003.

"Modeling Supply Chains," *Engineering News-Record*, pp. 24–27, April 28, 2008.

New Jersey Department of Taxation Website: www.state.nj.us (certificate of capital improvement).

"New Planning Method Can Deliver Better Design-Build Projects Faster," *Design Build Magazine*, pp. 33–36, November/December 2005.

New York Building Congress, "New York City Construction Outlook—Construction Forecast 2007–2009," October 2007.

New York City Building Code, 1968 edition.

New York University Schack Institute of Real Estate, Risk Management Breakfast Notes, April 22, 2008.

Occupational Safety and Health Administration, U.S. Department of Labor, "Log of Work-Related Injuries and Illnesses," "Summary of Work-Related Injuries and Illnesses," and "Injury and Illness Incident Report."

Occupational Safety and Health Administration, U.S. Department of Labor, OSHA Quick Cards.

Occupational Safety and Health Administration, U.S. Department of Labor, "Section 1926 of the OSHA Regulations."

"Open Shop Wage Rates for Journeymen," *Engineering News-Record*, p. 35, September 17, 2007.

Ouchi, Masahiro, Nakamura, Sade-aki, Osterberg, Thomas, and Hallberg, Sven, "Applications of Self-Compacting Concrete in Japan, Europe and the United States," July 2006.

Palmer, William J., Coombs, William E., and Smith, Mark A., *Financial Accounting and Financial Management*, McGraw-Hill, New York.

"Penalties in Collapse Are Silent on Trigger," *Engineering News-Record,* pp. 10–11, May 10, 2004.

Putzmeister, "Putzmeister Pump Breaks World Record at Burj Dubai," February 2008 press release.

"Reasons Why Owners Choose Design-Build," *Engineering News-Record*, p. 41, May 2006.

Regional Scaffolding Hoist and Scaffolding Diagrams, May 2008.

Shapiro, Larry, Construction Project Logistical Plans, April 2008.

Simmons, Leslie, and Olin, Harold, *Construction*, 7th ed., John Wiley & Sons, New York.

Smith, Craig B., *How the Great Pyramid Was Built*, Smithsonian Books.

Smith, James Allan, Currie, Overton A., and Hancock, E. Reginald, *Common Sense Construction Law*, 2d ed., John Wiley & Sons, New York.

"Speedy and Efficient Design-Build Lite Is Easing Owners' Worries," *Engineering News-Record*, pp. 26–27, November 13, 2006.

Subcontractors Trade Association, "Best Practices for New York's Construction Industry," *New York Construction,* special advertising supplement, December 2007.

"Surety Market Report 2007: Rising to the Occasion," *Engineering News-Record*, p. S3, June 18, 2006.

"Technology for Construction," *Engineering News-Record,* special advertising section, pp. T1–T19, December 10, 2007.

U.S. Bureau of Labor Statistics, U.S. Department of Labor, "Distribution of Fatalities by Selected Occupations in the Private Construction Industry, 2005–06," 2008.

U.S. Bureau of Labor Statistics, U.S. Department of Labor, "Number and Rate of Fatal Occupational Injuries by Industry Sector, 2006," 2008.

USGBC Website: www.usgbc.org (general information).

"When It's Green to Build," *Environmental Building News Magazine*, September 2007.

"Why Contractors Fail," *Engineering News-Record*, pp. 26–27, June 5, 2006.

"Workers Compensation Rates," *Engineering News-Record*, pp. 32–33, September 17, 2007.

Index

Note: Page numbers referencing figures are italicized and followed by an "*f*"; page numbers referencing tables are italicized and followed by a "*t.*"

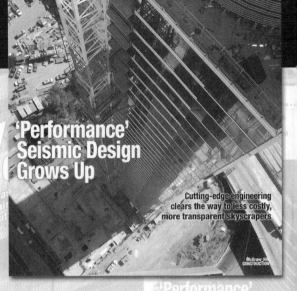